Machine Learning and Knowledge Discovery for Engineering Systems Health Management

Chapman & Hall/CRC
Data Mining and Knowledge Discovery Series

SERIES EDITOR

Vipin Kumar

University of Minnesota
Department of Computer Science and Engineering
Minneapolis, Minnesota, U.S.A

AIMS AND SCOPE

This series aims to capture new developments and applications in data mining and knowledge discovery, while summarizing the computational tools and techniques useful in data analysis. This series encourages the integration of mathematical, statistical, and computational methods and techniques through the publication of a broad range of textbooks, reference works, and handbooks. The inclusion of concrete examples and applications is highly encouraged. The scope of the series includes, but is not limited to, titles in the areas of data mining and knowledge discovery methods and applications, modeling, algorithms, theory and foundations, data and knowledge visualization, data mining systems and tools, and privacy and security issues.

PUBLISHED TITLES

UNDERSTANDING COMPLEX DATASETS: DATA MINING WITH MATRIX DECOMPOSITIONS
David Skillicorn

COMPUTATIONAL METHODS OF FEATURE SELECTION
Huan Liu and Hiroshi Motoda

CONSTRAINED CLUSTERING: ADVANCES IN ALGORITHMS, THEORY, AND APPLICATIONS
Sugato Basu, Ian Davidson, and Kiri L. Wagstaff

KNOWLEDGE DISCOVERY FOR COUNTERTERRORISM AND LAW ENFORCEMENT
David Skillicorn

MULTIMEDIA DATA MINING: A SYSTEMATIC INTRODUCTION TO CONCEPTS AND THEORY
Zhongfei Zhang and Ruofei Zhang

NEXT GENERATION OF DATA MINING
Hillol Kargupta, Jiawei Han, Philip S. Yu, Rajeev Motwani, and Vipin Kumar

DATA MINING FOR DESIGN AND MARKETING
Yukio Ohsawa and Katsutoshi Yada

THE TOP TEN ALGORITHMS IN DATA MINING
Xindong Wu and Vipin Kumar

GEOGRAPHIC DATA MINING AND KNOWLEDGE DISCOVERY, SECOND EDITION
Harvey J. Miller and Jiawei Han

TEXT MINING: CLASSIFICATION, CLUSTERING, AND APPLICATIONS
Ashok N. Srivastava and Mehran Sahami

BIOLOGICAL DATA MINING
Jake Y. Chen and Stefano Lonardi

INFORMATION DISCOVERY ON ELECTRONIC HEALTH RECORDS
Vagelis Hristidis

TEMPORAL DATA MINING
Theophano Mitsa

RELATIONAL DATA CLUSTERING: MODELS, ALGORITHMS, AND APPLICATIONS
Bo Long, Zhongfei Zhang, and Philip S. Yu

KNOWLEDGE DISCOVERY FROM DATA STREAMS
João Gama

STATISTICAL DATA MINING USING SAS APPLICATIONS, SECOND EDITION
George Fernandez

INTRODUCTION TO PRIVACY-PRESERVING DATA PUBLISHING: CONCEPTS AND TECHNIQUES
Benjamin C. M. Fung, Ke Wang, Ada Wai-Chee Fu, and Philip S. Yu

HANDBOOK OF EDUCATIONAL DATA MINING
Cristóbal Romero, Sebastian Ventura, Mykola Pechenizkiy, and Ryan S.J.d. Baker

DATA MINING WITH R: LEARNING WITH CASE STUDIES
Luís Torgo

MINING SOFTWARE SPECIFICATIONS: METHODOLOGIES AND APPLICATIONS
David Lo, Siau-Cheng Khoo, Jiawei Han, and Chao Liu

DATA CLUSTERING IN C++: AN OBJECT-ORIENTED APPROACH
Guojun Gan

MUSIC DATA MINING
Tao Li, Mitsunori Ogihara, and George Tzanetakis

MACHINE LEARNING AND KNOWLEDGE DISCOVERY FOR ENGINEERING SYSTEMS HEALTH MANAGEMENT
Ashok N. Srivastava and Jiawei Han

Machine Learning and Knowledge Discovery for Engineering Systems Health Management

Edited by
Ashok N. Srivastava
Jiawei Han

CRC Press
Taylor & Francis Group
Boca Raton London New York

CRC Press is an imprint of the
Taylor & Francis Group, an **informa** business
A CHAPMAN & HALL BOOK

First published 2012 by Chapman & Hall

Published 2019 CRC Press
Taylor & Francis Group
6000 Broken Sound Parkway NW, Suite 300
Boca Raton, FL 33487-2742

ISBN 13: 978-1-4398-4178-5 (hbk)

Library of Congress Cataloging-in-Publication Data

Machine learning and knowledge discovery for engineering systems health management / editors, Ashok N. Srivastava and Jiawei Han.
p. cm. -- (Chapman & Hall/CRC data mining and knowledge discovery series)
Includes bibliographical references and index.
ISBN 978-1-4398-4178-5 (hardback)
1. System failures (Engineering)--Prevention--Data processing. 2. Machine learning. I. Srivastava, Ashok N. (Ashok Narain), 1969- II. Han, Jiawei.

TA169.5.M33 2011
620'.00452--dc23 2011027204

Visit the Taylor & Francis Web site at
http://www.taylorandfrancis.com

and the CRC Press Web site at
http://www.crcpress.com

*A.N.S. dedicates this book to his late father,
Dr. Jagdish N. Srivastava, whose guidance, broad
vision, and fundamental insights helped shape him.*

Table of Contents

List of Figures

List of Tables

Machine Learning and Knowledge Discovery for Engineering Systems Health Management

Ashok N. Srivastava and Jiawei Han

As the complexity of engineered systems grows, it is becoming more important than ever to develop tools and technologies that can manage the complex interactions within and between these systems to maintain a high degree of reliability. These systems are often instrumented with sophisticated sensors that are used to monitor and in some cases control the system. This book explores the development of state-of-the-art tools and techniques that can be used to automatically detect, diagnose, and in some cases, predict the effects of adverse events in an engineered system on its ultimate performance. This gives rise to the field systems health management, in which methods are developed with the express purpose of monitoring the condition or "state of health" of a complex system, diagnosing faults, and estimating the remaining useful life of the system. If an anomaly is detected, a diagnosis or discovery of the root cause of the anomaly is usually necessary to enact an appropriate mitigation strategy. In some cases, it may become essential to assess, again in an automated fashion, the impact of the anomaly on overall system performance. This relies on performing prognostics, or an estimation of the remaining useful life of the system.

For the purposes of this discussion, there are fundamentally two approaches that can be taken to address systems health management. The first approach has been taken and explored in the engineering community for many decades and that is to build high-fidelity, physics-based models of the system under study. These physics-based models are generally built on sets of coupled, potentially nonlinear partial differential equations that describe the evolution of the system through time and space given an initial condition and a set of forcing functions. The models are based on first principles and thus, by definition, obey the laws of physics from the outset. For example, in an engine simulation the laws of thermodynamics, Newton's laws, and other fundamental physical understandings are implicitly and explicitly expressed. This degree of fidelity to physical laws gives these methods enduring power and predictive capability.

At the other extreme, one can take a purely statistical or data-driven approach, in which one takes the data obtained from sensors on a real system (or the output of simulated sensor readings from an engineering simulation) and builds a statistical model that relates the inputs to the system with the measured outputs. This model could be based on a potentially complex, nonlinear mapping of inputs to outputs that is estimated through the minimization of a loss function. This machine-learning–based approach is extremely powerful because it uses the observed data to make an internal model of the system under study. In many cases, this model does not lend itself to physical interpretability, but it can have predictive power to estimate the evolution of the system in both time and space.

Certainly, the view presented above is simplified: many machine learning systems encode knowledge of the system under study, and similarly, many physics-based systems can adapt to data as it is observed. Most Bayesian approaches to machine learning fundamentally encode a statistical model that describes a system based on information prior to the arrival of data and then adapts the model as new data is observed. The degree to which they encode physical laws varies based on the nature of the model and its flexibility. However, the bridge between physics-based approaches and data-driven approaches needs to be explored further in the context of systems health management. This book attempts to show fundamental algorithms and their operation in a wide variety of domains that can further help bridge this divide.

Many engineering systems differ fundamentally from other complex systems in that they obey physical principles. We note, however, that not all engineered systems must obey physical laws. For example, a complex software system may have no relation to physical laws and thus may not require the same type of modeling as a hardware-based system. If a software system is simulating a physical system, those physical laws must be encoded in the software with sufficient fidelity for a realistic simulation.

Another aspect of engineering systems is that there are often maintenance or other types of reports written about them that can describe off-nominal behavior and the events that lead up to such behavior. These documents can be in the form of maintenance reports or usage reports that are written by different authors. For example, in the case of aviation and space systems, there is often extensive documentation that describes fault and failure modes that are written when anomalies are detected, diagnosed, and then when a mitigation strategy is employed. These reports must be analyzed not only by humans but processed by machines to detect emergent safety or related issues.

The subject matter of this book therefore spans many dimensions. Section I begins with five chapters describing data-driven methods for anomaly detection, diagnosis, and prognostics for analyzing massive data streams and associated performance metrics. It also describes the analysis of text reports using novel approaches in machine learning to help detect and discriminate between failure modes. Section II covers physics-based methods for diagnostics and prognostics, in which these methods adapt to observed data. The chapters in this section cover physics-based, data-driven, and hybrid approaches to studying damage propagation and prognostics with applications in composite materials and solid rocket motors. Section III discusses application domains such as distributed data centers, aircraft engines, and embedded real-time software systems. Given the interdisciplinary nature of the field, there is considerable overlap in these sections as reflected in the chapters in each section.

- Section I: Data-Driven Methods for Systems Health Management. This section covers data-driven methods for anomaly detection, diagnosis, and prediction and methods to analyze massive data streams arising from complex systems with associated performance metrics, including
    - Mining Data Streams: Systems and Algorithms: Aggarwal and Turaga

    - A Tutorial on Bayesian Networks for Systems Health Management: Choi, Zheng, Darwiche, and Mengshoel
    - Anomaly Detection in a Fleet of Systems: Oza and Das
    - Discriminative Topic Models: Shan, Agovic, and Banerjee
    - Prognostic Performance Metrics: Goebel, Saxena, Saha, and Celaya

- Section II: Physics-Based Methods for Systems Health Management. This section discusses physics-based diagnostic and prognostic methods that incorporate and adapt to observed data, including
    - Gaussian Process Damage Prognosis under Random and Flight Profile Fatigue Loading: Chattopadhyay and Mohanty
    - Bayesian Analysis for Fatigue Damage Prognostics and Remaining Useful Life Prediction: Guan and Liu
    - Physics-Based Methods of Failure Analysis and Diagnostics in Human Space Flight: Smelyanskiy, Luchinsky, Hafiychuk, Osipov, Kulikov, and Patterson-Hine
    - Model-Based Tools and Techniques for Real-Time System and Software Health Management: Abdelwahed, Dubey, Karsai, and Mahadevan

- Section III: Applications. Application domains such as distributed computer data centers, aircraft engines, and embedded real-time software systems are discussed in this section, including
    - Real-Time Identification of Performance Problems in Large Distributed Systems: Goldszmidt, Woodard, and Bodik
    - A Combined Model-Based and Data-Driven Prognostic Approach for Aircraft System Life Management: Orchard, Vachtsevanos, and Goebel.
    - Hybrid Models for Engine Health Management: Volponi and Rajamani
    - Extracting Critical Information from Free Text Data for Systems Health Management: Kao, Poteet, and Augustine

SECTION I: DATA-DRIVEN METHODS FOR SYSTEMS HEALTH MANAGEMENT

Chapter 1: Aggarwal and Turaga begin this portion of the book by giving a comprehensive treatment on the subject of monitoring, managing, and extracting real-time information from massive data streams. This is central to the theme of the book since many complex engineering systems are instrumented with high-frequency sensors producing enormous amounts of data. The authors discuss the architecture of a distributed stream mining system and also review algorithms for classic data mining and machine learning tasks such as classification, clustering, pattern extraction, and time series analysis. The authors also describe methods to deal with nonstationary systems that are prevalent in application domains.

Chapter 2: The second chapter of this book gives a tutorial on Bayesian networks, which comprise the backbone of many probabilistic reasoning systems for diagnosing faults in complex systems. The chapter discusses this modeling technique utilizing a set of examples designed to show the use of Bayesian networks for modeling and learning in the context of systems health management. The authors describe how these networks naturally capture both uncertainties due to both measurement and modeling error and discuss reasoning and learning in complex systems. Choi et al. conclude this chapter with a discussion of a complex real-world application to electrical power systems.

Chapter 3: A key feature of many, but not all, engineering systems is that there may be numerous copies of engineering systems operating in different environments. In this chapter, Oza and Das consider the problem of mining data from fleets of systems (such as aircraft) where data from continuous sensors, discrete switches, and text reports must be combined to assess the overall health of the system. The chapter features an in-depth discussion of previous attempts to mine data from fleet-wide sources and then discusses novel statistical and kernel-based algorithms for anomaly detection of the data types that arise from fleets. The chapter concludes with a discussion of the impact of these methods on safety critical systems.

Chapter 4: The majority of this book primarily concerns itself with analysis and interpretation of numeric data coming from hardware or software systems. This chapter differs in that it addresses the analysis of text documents. Although text mining is a vast area of research by itself, we included two chapters on the topic area because they address key problems that have applicability both to systems health management and machine learning. Shan et al. discuss discriminative topic models, which are a family of models that simultaneously address the problem of grouping documents with similar topics together in an automated fashion and then classifying these documents into different categories. These models are shown to be competitive with existing methods with the additional benefit of extracting interesting topics on real-world data from the Aviation Safety Reporting System.

Chapter 5: For many machine learning algorithms, the performance metric is specified and the learning system then optimizes the performance metric based on the observed data. This chapter discusses performance metrics for prognostics, where one estimates the remaining useful life of a component in an engineering system. The chapter gives an overview of the prognostics subdiscipline of systems health management and then discusses the development of performance metrics that are tailored to the needs of the prognostics community. Issues such as end-user requirements, nonstationary, and different time scales are also discussed. The chapter features a set of guidelines on choosing different prognostic metrics based on both user requirements and the characteristics of the observed data.

SECTION II: PHYSICS-BASED METHODS FOR SYSTEMS HEALTH MANAGEMENT

Chapter 6: In contrast to the topics in Section I, the chapters in Section II feature discussions of methods that are physics-based but also incorporate a data-driven element. Chattopadhyay and Mohanty discuss a novel application of Gaussian process regression to the problem of estimating fatigue damage propagation using these methods. They also give a theoretical understanding of the physics underlying fatigue damage prognosis. The chapter gives a clear introductory treatment of Gaussian processes and then shows the performance of these methods on data from simulations as well as real-world experiments. A key issue that the chapter addresses is the fact that these materials are subjected to different loading conditions in many aeronautical applications. The authors demonstrate how these methods can accommodate loadings that are observed in flight as well as random loading.

Chapter 7: Guan and Liu provide a comprehensive treatment to the problem of fatigue damage prognosis using probabilistic methods to appropriately model uncertainties. A probabilistic prognosis framework is given, which shows how prior probability distributions, simulation methods, and response measures can be combined to help estimate remaining useful life of structures. They show a hierarchical Bayesian method for modeling numerous sources of uncertainty including uncertainties due to modeling, variability, and measurement error. The authors show how a Bayesian approach can be used to model prior information and can be updated as new data is observed. This approach can lead to computational burdens that are also addressed through the development of a decoupled Markov chain Monte Carlo sampler. The authors demonstrate their technique on experimental data.

Chapter 8: The dynamics of an engineering system guide its evolution and can thus play a central role in diagnosing an observed anomaly. In Chapter 8, Smelyanskiy et al. show a novel method to perform dynamical inference with stochastic nonlinear models. The analytical approach is based on path integrals, and a method is shown to infer model parameters in a dynamical noise environment. The research results are presented on the familiar Lorenz chaotic system and then extended to a discussion of an application of these methods to fault diagnosis of solid rocket motors. The authors discuss further research for using dynamical inference algorithms to predict fault dynamics in the solid rocket motors.

Chapter 9: Although many examples given in this book are related to physical engineering systems, we also treat the issues of detecting faults in software systems. Software systems differ in numerous ways from hardware systems. A key distinction in the context of this book is that most software systems do not obey physical laws and thus require a new approach to modeling. This chapter discusses an approach known as time failure propagation graphs, which is traditionally applied to hardware systems but can be adapted to address software systems. The formal structure of the TFPG is discussed and a reasoning algorithm to perform diagnosis on software systems is presented. The authors also address the issue of prognostics of impending faults.

SECTION III: APPLICATIONS

Chapter 10: The applications section of the book focuses on the use of machine learning and physics-based approaches to analyze complex systems. The first chapter in this section discusses an approach to real-time identification of performance problems in large distributed data centers. These complex systems are becoming commonplace in their implementation at major Internet companies and have extremely high reliability requirements. These requirements pose major technical challenges, one of which is to rapidly determine whether an observed problem has been seen before or whether this is a new problem. The authors describe this so-called crisis identification problem and provide a method to extract the signature of these problems. They then discuss a full Bayesian approach to clustering these signatures. The results of applying these methods on a real-world data center are also discussed.

Chapter 11: Orchard et al. discuss the particle filter algorithm and its application to estimating the probability density function of a state vector in a system with known dynamics. This probability density function is updated as new data is observed. The technique is described in the context of prognostics. A key issue that arises in prognostics is modeling the effects of uncertainty due to model misspecification, measurement noise, and other sources of uncertainty in the estimate of remaining useful life. The authors present the algorithms and discuss the results of their methods on the problem of fatigue crack growth.

Chapter 12: Volponi and Rajamani describe hybrid models for analyzing aircraft engines to detect, diagnose, and predict the future trends of system faults. The book takes an approach of combining physics-based and data-driven techniques to derive benefit from both approaches. The combined approach employs physics-based models to learn operating set points rapidly and then data-driven techniques to adapt to the change in operating point. The chapter features a description of a typical onboard engine model architecture and then a contrasting hybrid architecture and accentuates the differences between these two approaches. A key contribution of this chapter is the inclusion of a discussion on the verification and validation of hybrid models in the engine community and the requirements that are given on such models from major regulators.

Chapter 13: A key application area of text mining in the context of systems health management is in understanding the linkage between maintenance reports and free narratives written by the users of engineering systems. This chapter discusses an approach to linking information from maintenance reports regarding aircraft and narratives from the Aviation Safety Reporting System. The chapter discusses an approach that combines natural language processing and text mining techniques to address this goal.

This book is the culmination of the hard work of numerous individuals including, first and foremost, the chapter authors. Their dedication and willingness to share their insights is invaluable. We also thank the reviewers for key insights and contributions and Randi Cohen at Taylor & Francis for her outstanding help. A. N. Srivastava wishes to thank the NASA Aviation Safety Program System-Wide Safety and Assurance Technology Project for supporting this work.

The MathWorks, Inc.
3 Apple Hill Drive
Natick, MA 01760-2098 USA
Tel: (508) 647-7000
Fax: (508) 647-7001
E-mail: info@mathworks.com
Web: http://www.mathworks.com

Editors

Ashok N. Srivastava is the project manager for the System-Wide Safety and Assurance Technologies Project at NASA. He is formerly the principal investigator for the Integrated Vehicle Health Management research project at NASA. His current research focuses on the development of data mining algorithms for anomaly detection in massive data streams, kernel methods in machine learning, and text mining algorithms. Dr. Srivastava is also the leader of the Intelligent Data Understanding group at NASA Ames Research Center. The group performs research and development of advanced machine learning and data mining algorithms in support of NASA missions. He performs data mining research in a number of areas in aviation safety and application domains such as earth sciences to study global climate processes and astrophysics to help characterize the large-scale structure of the universe.

Jiawei Han is a professor of computer science at the University of Illinois. He has been researching data mining, information network analysis, and database systems, with more than 500 publications. He is the founding editor-in-chief of *ACM Transactions on Knowledge Discovery from Data* (*TKDD*) and on the editorial boards of several other journals. Jiawei has received IBM Faculty Awards, HP Innovation Awards, ACM SIGKDD Innovation Award (2004), IEEE Computer Society Technical Achievement Award (2005), and IEEE W. Wallace McDowell Award (2009). He is a fellow of ACM and IEEE. He is currently the director of Information Network Academic Research Center (INARC) supported by the Network Science–Collaborative Technology Alliance (NS-CTA) program of the U.S. Army Research Lab. His book *Data Mining: Concepts and Techniques* (Morgan Kaufmann) has been used worldwide as a textbook.

Contributors

Sherif Abdelwahed
Mississippi State University
Mississippi State, Mississippi

Charu C. Aggarwal
IBM T. J. Watson Research Center
Yorktown Heights, New York

Amrudin Agovic
University of Minnesota
Twin Cities
Minneapolis, Minnesota

David Augustine
Boeing Research and Technology
Seattle, Washington

Arindam Banerjee
University of Minnesota
Twin Cities
Minneapolis, Minnesota

Peter Bodik
University of California, Berkeley
Berkeley, California

Jose Celaya
SGT, Inc.
Moffett Field, California

Aditi Chattopadhyay
Arizona State University
Tempe, Arizona

Arthur Choi
Computer Science Department
University of California,
Los Angeles
Los Angeles, California

Adnan Darwiche
Computer Science Department
University of California,
Los Angeles
Los Angeles, California

Santanu Das
NASA Ames Research Center
Moffett Field, California

Abhishek Dubey
Vanderbilt University
Nashville, Tennessee

Kai Goebel
NASA Ames Research Center
Moffett Field, California

Moises Goldszmidt
Microsoft Research
Mountain View, California

Xuefei Guan
Clarkson University
Potsdam, New York

Vasyl V. Hafiychuk
NASA Ames Research Center
SGT Inc.
Moffett Field, California

Anne Kao
Boeing Research and Technology
Seattle, Washington

Gabor Karsai
Vanderbilt University
Nashville, Tennessee

Igor Kulikov
California Institute of Technology
Pasadena, California

Yongming Liu
Clarkson University
Potsdam, New York

Dmitry G. Luchinsky
NASA Ames Research Center
Mission Critical Technologies Inc.
Moffett Field, California

Nagabhushan Mahadevan
Vanderbilt University
Nashville, Tennessee

Ole J. Mengshoel
Carnegie Mellon University
Moffett Field, California

Subhasish Mohanty
Argonne National Laboratory
Argonne, Illinois

Marcos Orchard
Universidad de Chile
Santiago, Chile

Viatcheslav V. Osipov
NASA Ames Research Center
Mission Critical Technologies Inc.
Moffett Field, California

Nikunj Oza
NASA Ames Research Center
Moffett Field, California

Ann Patterson-Hine
NASA Ames Research Center
Moffett Field, California

Stephen Poteet
Boeing Research and Technology
Seattle, Washington

Ravi Rajamani
Pratt & Whitney
Hartford, Connecticut

Bhaskar Saha
MCT, Inc.
Moffett Field, California

Sankalita Saha
MCT, Inc.
Moffett Field, California

Abhinav Saxena
SGT, Inc.
Moffett Field, California

Hanhuai Shan
University of Minnesota
Twin Cities
Minneapolis, Minnesota

Vadim N. Smelyanskiy
NASA Ames Research Center
Moffett Field, California

Deepak S. Turaga
IBM T. J. Watson Research Center
Yorktown Heights, New York

George Vachtsevanos
Georgia Institute of Technology
Atlanta, Georgia

Allan J. Volponi
Pratt & Whitney
Hartford, Connecticut

Dawn Woodard
Cornell University
Ithaca, New York

Lu Zheng
Carnegie Mellon University
Moffett Field, California

I

Data-Driven Methods for Systems Health Management

CHAPTER 1

Mining Data Streams: Systems and Algorithms

Charu C. Aggarwal

Deepak S. Turaga

CONTENTS

1.1 INTRODUCTION

Recent advances in hardware and software technology have enabled the world to get more instrumented and interconnected. In several domains, there are large deployed infrastructures of distributed, heterogeneous sensors generating continuous streams of digital data. There is a tremendous need to design, develop, deploy, and manage applications that process, analyze, and extract intelligence from these high-volume data streams. A set of sample applications are illustrated in Figure 1.1.

As is evident from Figure 1.1, these applications are drawn from very diverse areas such as systems health monitoring for large-scale data centers; stock ticker analysis for market making in financial services; process control for manufacturing; analysis of sensors streams in natural/physical sciences; multimodal surveillance for law enforcement, fraud detection, and prevention in different settings; physiological sensor monitoring for health care; and call data and record processing for telecommunications. Although the nature of the sensors, the complexity of the analysis, the constraints, and the scale of the problems vary across these different applications, they all demonstrate the need for real-time stream mining. There are several other examples where large volumes of streaming data need to be mined for interesting and relevant information to support a wide variety of applications.

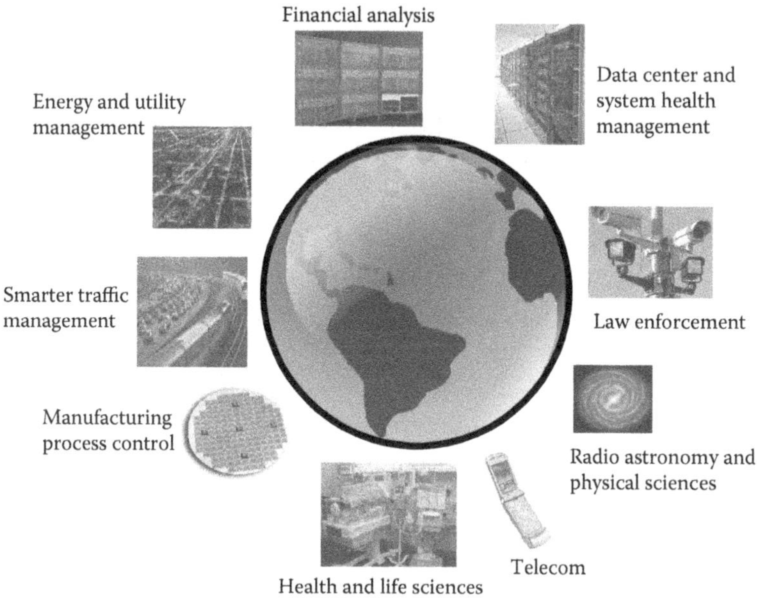

FIGURE 1.1 Example stream processing applications in different domains.

There is increasing dependence within the academic, business, and scientific communities on large IT systems to handle the complexity of the underlying processes that are critical to their functioning. System health monitoring and management are prerequisites for providing a reliable infrastructure for these IT operations. This is complicated by the fact that these IT systems may be extremely large—many current data centers contain several thousands of compute and storage nodes, underlying network and power systems, and environmental control mechanisms. These large IT systems are instrumented with multiple sensors that monitor the energy usage, environmental settings (temperature, humidity), compute and storage system use (CPU, memory, network, disk, virtual machines), and in many cases even application level performance. Information security is also a major concern, and these IT systems need to be actively monitored to protect against different kinds of malware, attacks, and network intrusions, as well as to ensure the integrity and privacy of data and applications.

With the recent interest in hosted compute environments (cloud computing) the infrastructure and applications may need to be changed dynamically based on the different requirements of the large number of users of the system. Hence, all monitoring needs to be continuous to optimize the functioning of the IT system and enable early detection (and prediction) of fault/failure conditions to support real-time requirements on incident management. This results in massive volumes of streaming data collected from these systems in several heterogeneous formats that include structured data streams (sensor readings, CPU measurements, memory, disk, storage use), semistructured data streams (error and information logs generated by the applications), and unstructured data streams (surveillance video, user information). These different data streams need to be ingested, analyzed, and correlated with each other, in combination with other static forms of information, such as network topology, system capacity, and others, to support the reliability requirements. This necessitates the use of large-scale stream mining for system health monitoring in these settings. Work in this space of real-time stream mining for system health monitoring has focused both on research [1–5] as well as commercial efforts [6,7].

To build these large-scale stream mining applications, several data management and analysis challenges need to be overcome. Large volumes of streaming data need to be ingested from heterogeneous and distributed sources, often with varying levels of connectivity. The data itself may consist of various formats, including structured, unstructured, and nonrelational content. There are often strong performance requirements in terms of latency, error resilience, or scaling. The analytical requirements may be arbitrarily complex, and any analysis needs to capture the temporal properties of the data. Additionally, mining algorithms need to be designed to account for the inability to store and process all the data and be adaptive to time-varying resource availability and load requirements.

Distributed stream mining systems [8–15] have been recently developed to tackle several stream data management issues, including providing distributed processing and handling the high data volumes and computational requirements. These systems provide application designers the infrastructure, tooling, and computing paradigms to build large-scale, adaptive mining algorithms. In addition to these systems, there has also been significant research into developing efficient stream mining algorithms for data reduction, classification, clustering, pattern mining, and other such mining tasks.

In this chapter, we provide an overview of the key challenges in stream mining and describe stream mining systems, and algorithms designed to overcome these. The rest of this chapter is organized as follows. In Section 1.2, we discuss the generic challenges that stream mining poses to a variety of data management and data mining problems. In Section 1.3, we provide an introduction to stream mining and processing systems that tackle different types of stream data management and application development issues. Following this, we focus on different algorithms for stream data reduction (Section 1.4) and stream mining (Section 1.5). In Section 1.6, we discuss approaches for optimizing and adapting stream mining applications under time-varying resource availability. We conclude in Section 1.7 with a discussion on the open problems and research directions in this field.

1.2 STREAM PROCESSING AND MINING CHALLENGES

The key stream processing and mining challenges include stream data management, relational processing on streams, stream indexing, and stream mining. We describe these in greater detail.

1.2.1 Stream Data Management

There are several data management challenges. First, data sources tend to be distributed across several different locations with varying levels of connectivity, which may vary over time. Second, the data streams are typically heterogeneous in format, content, rates, information, and noise levels, making processing and analysis difficult. Data streams may also consist of unstructured data types such as audio, video, and text that cannot easily be handled using traditional data management infrastructures. Last but not least, any stream mining solution needs to utilize distributed processing infrastructures, adapt to dynamic data rates, characteristics, and resource availability while scaling to meet performance objectives such as latency and throughput. For instance, in mining of data from distributed sensor networks, it is often desirable to perform in-network processing of data stream to overcome limitations of limited processing and memory [16,17].

1.2.2 Relational Data Processing on Streams

Different relational operations, such as *join*, *aggregate*, *sort*, need to be performed on the data to make it amenable for mining. For instance, stream join is a fundamental operation that allows combining information from different streams. This is especially useful in many applications such as sensor networks in which the streams arriving from different sources may need to be related with one another. Input tuples arrive continuously, and result tuples need to be produced continuously as well. We cannot assume that the input data is already stored or indexed or that the input rate can be controlled by the query plan. Standard join algorithms that use blocking operations, e.g., sorting, no longer work in this setting. Conventional methods for cost estimation and query optimization are also inappropriate because they assume finite input. Moreover, the long-running nature of stream queries calls for adaptive processing strategies that can react to changes and fluctuations in data and stream characteristics. The "stateful" nature of stream joins adds another dimension to the challenge. In general, to compute the complete result of a stream join, we need to retain all past arrivals as part of the processing state, because a new tuple may join with an arbitrarily old tuple arrived in the past. This problem is exacerbated by unbounded input streams, limited processing resources, and high performance requirements, as it is impossible in the long run to keep all past history in fast memory. A survey on different join processing techniques in data streams may be found in [18].

1.2.3 Stream Indexing

The problem of indexing data streams attempts to create an indexed representation, so that it is possible to efficiently answer different kinds of queries such as aggregation queries or trend-based queries. This is especially important in the data stream case because of the huge volume of the underlying data. We note that most traditional indexes require multiple passes to create. We do not have this luxury in the data stream model because of one-pass constraints. A discussion of several models for data stream computation may be found in [18].

1.2.4 Stream Mining Algorithms

The types of analysis that need to be performed on streaming data can be arbitrarily complex and may include several sophisticated learning techniques. In most cases, there is an inherent temporal component to the stream mining process. This is because the data evolves over time and exhibits *temporal locality*. Therefore, a straightforward adaptation of existing data mining algorithms may not be an effective solution to the task of stream mining. Additionally, the high volume of streaming data often makes it impossible to fully store and process all the data from disk. It is thus no longer possible to process the data efficiently using multiple passes.

Rather, one can process a data item at most once. This leads to constraints on the implementation of the underlying algorithms. Therefore, stream mining algorithms typically need to be designed so that the algorithms work with one pass on the data. Several stream mining algorithms have been designed for data reduction, synopsis construction, classification, clustering, frequent pattern mining, and data cubing, and we will discuss these in the following sections.

In the rest of this chapter we focus on the stream data management and mining issues, and describe different systems and algorithms designed to tackle these.

1.3 STREAM PROCESSING SYSTEMS: ARCHITECTURAL ISSUES

Stream processing systems are typically designed in the form of distributed systems that are designed to tackle the complex data management and processing issues associated with high-volume data streams. The distributed nature of stream processing systems is quite common because both the distributed data collection and the immense processing requirements of different kind of applications may necessitate a distributed architecture. Therefore, a general purpose stream-processing system is often likely to be constructed in the form of a distributed architecture. These systems leverage computational resources from a set of distributed processing nodes and provide the framework to deploy and run different stream mining applications on various resource topologies. The specific components of a stream processing system include:

- *Distributed runtime*, which supports the management and monitoring of distributed computational resources, workloads, data transport, and application cooperation issues and fault tolerance issues.
- *Application development environment*, which includes new programming models such as declarative languages, application composition languages, and application development models. This also includes tooling for visualization and application understanding, as well as support for debugging.
- *Optimization tools*, which include different compile and run-time optimizations and schedulers. These allow intelligent fusion, parallelization, placement, load balancing, and adaptation of applications.

As a specific example of a stream processing system, we consider System S [10], which has been developed by IBM, and is now officially available as an IBM product. It is one of the few *commercially available* stream products available and is therefore a natural choice for exposition of the issues that arise in the context of designing practical stream processing systems. It has an associated declarative engine that provides the necessary operators for stream processing in a user-friendly way. The idea of such an engine is to abstract out the application to the user, while making the lower-level distributed computations transparent to a user. This is referred to as the associated application composition language Stream Processing Application Declarative Engine (SPADE) [19,20]. The System S architecture is shown in Figure 1.2. It consists of an operator abstraction layer, a resource and graph manager, and a data fabric that may be deployed across a set of compute nodes, with potentially different processor architectures.

On the System S platform, complex applications are decomposed as a network of operators performing feature extraction, classification, aggregation, and correlation and deployed as a processing pipeline that analyzes streaming data from a set of raw data sources to extract valuable information in real time. A flow graph consists of a set of processing elements (PEs) connected by streams, where each stream has a fixed schema and carries a series of tuples. Stream tuples are typed, employing application-defined type systems, which are built by combining system-supported basic types. The PEs are containers that host operators implementing data stream analytics and are distributed on compute nodes. Computing nodes are organized as a shared-nothing cluster of workstations or as the

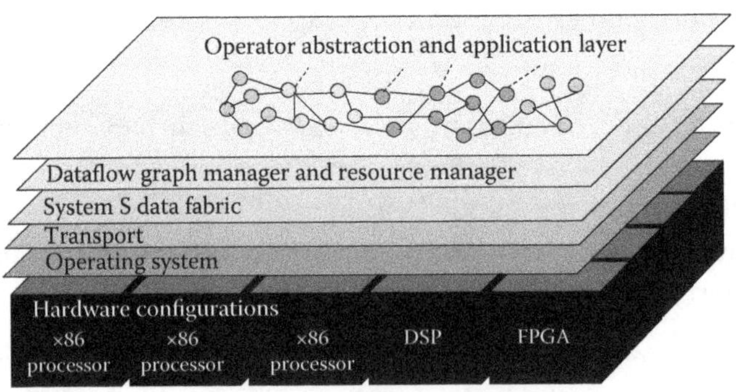

FIGURE 1.2 Stream processing system architecture.

execution nodes in a large supercomputer. PEs communicate with each other via their input and output ports, connected by streams. The PE ports as well as PEs can be explicitly connected using hard-coded links or through implicit links that rely on properties of streams, which can be subscribed to. The latter type of connections is *dynamic*. The subscriptions are evaluated dynamically, as applications are submitted or removed from the system, and the resulting connections are made upon the availability of a stream that matches a subscription expression. This mechanism allows System S to support incremental application development and deployment. This is also key to providing general purpose functionality for a wide variety of applications. Besides these fundamental functionalities, System S provides several other services, such as fault tolerance [21], scheduling, placement optimization [22], distributed job management, storage services, and security.

The System S application development engine SPADE includes several key functionalities. It provides a stream-centric, operator-level programming model. The operator logic can optionally be implemented in a lower-level language, like C++, whereas the SPADE language is used to compose these operators into *logical* data flow graphs. It also includes a compiler that optimizes logical dataflow graphs comprising multiple operators into *physical* ones, comprising a collection of PEs by *fusing* several operators to "fit" in the available compute nodes. It consists of a type generic streaming operator model that captures fundamental concepts associated with streaming applications, such as *windows* on input streams, aggregation functions on windows, output attribute assignments, parameters that are configurable via expressions, and *punctuations* (i.e., stream markers denoting logically related tuples) in streams.

System S also includes a *stream relational* toolkit of operators implementing relational algebra operations in the streaming context. This toolkit includes the following operators: *join* (for correlating two streams based on windows), *aggregate* (for aggregating tuples based on grouping attributes and a window definition), *functor* (to perform selection, projections, and simple tuple transformations), and *sort* (to order tuples within a window). Along with this, there is a broad range of edge adapters, which are used to ingest data from external sources and publish results to external consumers, such as network sockets, databases, file systems, as well as proprietary middleware platforms. Finally, there is support for extending the language with new type-generic, configurable, and reusable operators.

However, despite the benefits of this new computing paradigm, it is nontrivial to design and develop applications that efficiently make use of the underlying system resources while providing the necessary analytical capabilities for users. Application developers need to design intelligent one-pass mining algorithms, then perform intelligent application decomposition, i.e., careful mapping of application requirements onto processing flow graphs of operators, distribute and deploy the application across the processing infrastructure, and finally tune and adapt performance over the lifetime of the application. Hence, in the rest of this chapter, we focus on the design of different stream processing and mining algorithms and their associated challenges.

1.4 STREAM DATA REDUCTION

A key first step in stream mining application involves reduction of the data stream volumes to make any analysis feasible. The large volume of data streams poses unique space and time constraints on the computation process. Many query processing and mining algorithms require efficient execution, which can be difficult to achieve with a high-volume data stream. Furthermore, since it is impossible to fit the entire data stream within the available space, the space efficiency of the approach is a major concern. In many cases, it may be acceptable to generate *approximate solutions* for many problems by summarizing the data in a time- and space-efficient way. In recent years, a number of *synopsis structures* have been developed, which can be used in conjunction with a variety of mining and query processing techniques [23].

The challenges that arise in the context of synopsis construction of data streams are as follows:

1.4.1 Broad Applicability

The synopsis structure is typically used as an intermediate representation, which is then leveraged for a variety of data mining and data management problems. Therefore, the synopsis structure should be constructed in such a way that it has applicability across a wide range of problems.

1.4.2 One-Pass Constraint

As in all data stream algorithms, the one-pass constraint is critical to synopsis construction algorithms. We would like to design all synopsis construction algorithms in one pass, and this is not the case for most traditional methods. In fact, even simple methods such as sampling need to be redesigned to handle the one-pass constraint.

1.4.3 Time and Space Efficiency

Since data streams have a very large volume, it is essential to create the synopsis in a time- and space-efficient way. In this sense, some of the probabilistic techniques such as sketches are extremely effective for counting-based applications, since they require constant space for provable probabilistic accuracy. In other words, the time and space efficiency depend only upon the accuracy of the approach rather than the length of the data stream.

1.4.4 Data Stream Evolution

Since the stream evolves over time, a synopsis structure that is constructed from the overall behavior of the data stream is not quite as effective as one that uses recent history. Consequently, it is often more effective to create synopsis structures that either work with sliding windows or use some decay-based approach to weight the data stream points.

Key techniques for stream data reduction include (but are not limited to) *sampling*, *sketches*, *quantization*, *transform-based summarization*, *summary statistic computation*, and *dimensionality reduction*. Besides these data reduction techniques, it is also necessary to design *resource-adaptive* analytic algorithms that adjust the amount of processing to the available computational cycles. Examples can be found in the signal-processing domain, where the analytics may have several *operating points*, corresponding to different resource-accuracy tradeoffs. Variations and spikes in load can then be dealt with by employing appropriate data reduction and dynamic analytic adaptation techniques.

In this section, we will discuss the broad classes of techniques that are used for synopsis construction in data streams. Each of these techniques has their own advantages in different scenarios, and we will take care to provide an overview of the different array of methods that are used for synopsis construction in data streams.

1.4.5 Sampling

Sampling methods are widely used for traditional applications and are extremely popular because of their broad applicability across a wide array of tasks in data streams. A further advantage of sampling methods is that, unlike many other synopsis construction methods, they maintain their inter-attribute correlations across samples of the data. It is also often possible to use probabilistic inequalities to bound the effectiveness of a variety of applications with sampling methods.

However, a key problem in extending sampling methods to the data stream scenario is that one does not know the total number of data points to be sampled in advance. Rather, one must maintain the sample in a dynamic way over the entire course of the computation. A method called reservoir sampling was first proposed in [24], which maintains such a sample dynamically. This technique was originally proposed in the context of one-pass access of data from magnetic storage devices. However, the techniques also naturally extend to the data stream scenario.

Let us consider the case, where we wish to obtain an unbiased sample of size n from the data stream. To initialize the approach, we simply add the first n points from the stream to the reservoir. Subsequently, when the $(t + 1)$th point is received, it is added to the reservoir with probability $n/(t + 1)$. When the data point is added to the reservoir, it replaces a random point from the reservoir. It can be shown that this simple approach maintains the uniform sampling distribution from the data stream. We note that the uniform sampling approach may not be very effective in cases where the data stream evolves significantly. In such cases, one may either choose to generate the stream sample over a sliding window or use a decay-based approach to bias the sample. An approach for sliding window computation over data streams is discussed in [25].

A second approach [26] uses biased decay functions to construct synopsis from data streams. It has been shown in [26] that the problem is extremely difficult for arbitrary decay functions. In such cases, there is no known solution to the problem. However, it is possible to design very simple algorithms for some important classes of decay functions. One of these classes of decay functions is the *exponential decay function*. The exponential decay function is extremely important because of its *memoryless property*, which guarantees that the future treatment of a data point is independent of the past data points that have arrived. An interesting result is that by making simple implementation modifications to the algorithm of [24] in terms of modifying the probabilities of insertion and deletion, it is possible to construct a robust algorithm for this problem. It has been shown in [26] that the approach is quite effective in practice, especially when there is significant evolution of the underlying data stream.

While sampling has several advantages in terms of simplicity and preservation of multidimensional correlations, it loses its effectiveness in the presence of data sparsity. For example, a query that contains very few data points is unlikely to be accurate with the use of a sampling approach. However, this is a general problem with most techniques that are effective at counting frequent elements but are not quite as effective at counting rare or distinct elements in the data stream.

1.4.6 Sketches

Sketches use some properties of random sampling to perform counting tasks in data streams. Sketches are most useful when the *domain size* of a data stream is very large. In such cases, the number of possible distinct elements becomes very large, and it is no longer possible to track them in space-constrained scenarios. There are two broad classes of sketches: *projection* and *hash based*. We will discuss each of them in turn.

Projection-based sketches are constructed on the broad idea of random projection [27]. The most well-known projection-based sketch is the AMS sketch [28,29], which we will discuss below. It has been shown in [27] that by randomly sampling subspaces from multidimensional data, it is possible to compute ε-accurate projections of the data with high probability. This broad idea can easily be extended to the massive domain case by viewing each distinct item as a dimension, and the counts on these items as the corresponding values. The main problem is that the vector for performing the projection cannot be maintained explicitly since the length of such a vector would be of the same size as the number of distinct elements. In fact, since the sketch-based method is most relevant in the distinct element scenario, such an approach defeats the purpose of keeping a synopsis structure in the first place.

Let us assume that the random projection is performed using k sketch vectors, and r_i^j represents the jth vector for the ith item in the domain being tracked. To achieve the goal of efficient synopsis construction, we store the random vectors implicitly in the form of a seed, and this can be used to dynamically generate the vector. The main idea discussed in [30] is that it is possible to generate random vectors with a seed of size $O(\log(N))$, provided that one is willing to work with the restriction that $r_i^j \in \{-1,+1\}$ should be four-wise independent. The sketch is computed by adding r_i^j to the jth component of the sketch for the ith item. In the event that the incoming item has frequency f, we add the value $f \cdot r_i^j$. Let us assume that there are a total of k sketch components, which are denoted by $(s_1 \ldots s_k)$.

Some key properties of the pseudorandom number generator approach and the sketch representation are as follows:

- A given component r_i^j can be generated in polylogarithmic time from the seed. The time for generating the seed is polylogarithmic in the domain size of the underlying data.
- A variety of approximate aggregate functions on the original data can be computed using the sketches.

Some example of functions that can be computed from the sketch components are as follows:

- Dot product of two streams: If $(s_1 \ldots s_k)$ be the sketches from one stream and $(t_1 \ldots t_k)$ be the sketches from the other stream, then $s_j \cdot t_j$ is a random variable which is the expected value of the dot product.
- Second moment: If $(s_1 \ldots s_k)$ be the sketch components for a data stream, it can be shown that the expected value of s_j^2 is the second moment. Furthermore, using Chernoff bounds, it can be shown that by selecting the median of $O(\log(1/\delta))$ averages of $O(1/\varepsilon^2)$ copies of $s_j \cdot t_j$, it is possible to guarantee the accuracy of the approximation to within $1 \pm \varepsilon$ with probability at least $1 - \delta$.
- Frequent items: The frequency of the ith item in the data stream is computed by multiplying the sketch component s_j by r_i^j. However, this estimation is accurate only for the case of frequent items, since the error estimation is proportional to the overall frequency of the items in the data stream.

More details of computations that one can perform with the AMS sketch are discussed in [28,29].

A second kind of sketch that is used for counting is the *count-min* sketch [31]. The count-min sketch is based upon the concept of hashing and uses $k = \ln(1/\delta)$ pairwise-independent hash functions, which hash onto integers in the range $(0 \ldots e/\varepsilon)$. For each incoming item, the k hash functions are applied and the frequency count is incremented by 1. In the event that the incoming item has frequency f, then the corresponding frequency count is incremented by f. Note that by hashing an item into the k cells, we are ensuring that we maintain an overestimate on the corresponding frequency. It can be shown that the minimum of these cells provides the ε-accurate estimate to the frequency with probability at least 1 – δ. It has been shown in [31] that the method can also be naturally extended to other problems such as finding the dot product or the second-order moments. The count-min sketch is typically more effective for problems such as frequency estimation of individual items than the projection-based AMS sketch. However, the AMS sketch is more effective for problems such as second-moment estimation.

1.4.7 Quantization

This set of techniques are widely used in data compression and involve reducing the fidelity of representing individual data samples using either scalar, vector, uniform, or nonuniform techniques. There has been significant research into the design of optimal quantizers for different types of data and objective functions. Well-known techniques such as the Lloyd's algorithm for scalar quantization and the Linde-Buzo-Gray (LBG) algorithm* for vector quantization have been used in many applications. However, these presume knowledge of the data characteristics or samples during quantizer design—which are not always available in a streaming scenario.

In practice, several simple suboptimal techniques have been used for stream data quantization. Bagnall et al. [32] propose a *binary clipping* method for time series data, where the streaming data is windowed and data samples are converted into 0 and 1 based on whether they lie above or below the mean value baseline in that window. This simple clipping based representation has interesting theoretical underpinnings and has been applied for speeding up the execution of different stream mining algorithms. Similar work has also been proposed in [33], which proposes threshold based representations for querying and indexing time series data. Megalooikonomou et al. [34] present a piecewise vector quantized approximation for time series data, which preserves with high accuracy the shape of the original sequences.

Recently, there has been some effort of applying moment preserving quantization (MPQ) to time series data streams. MPQ has been used in the image processing literature [35] for retaining the texture properties of an image while providing good compression. The key idea behind MPQ may be described as follows. Consider a set of t 1-D samples x_j $(0 \leq j \leq t-1)$

* LBG has strong analogies with the *k*-means algorithm for clustering.

with sample mean μ and sample variance σ^2. Consider a 1-bit quantizer designed as follows:

$$\hat{x}_j = \begin{cases} \mu + \sigma\sqrt{\dfrac{t - t_g}{t_g}} & x_j \geq \Delta \\ \mu - \sigma\sqrt{\dfrac{t_g}{t - t_g}} & x_j < \Delta \end{cases} \tag{1.1}$$

where t_g is the number of samples that have magnitude greater than or equal to Δ, any chosen threshold. This is shown in Figure 1.3.

It is straightforward to show that the resulting quantizer guarantees the preservation of the first two moments [35] of this set of samples—independent of the choice of the threshold Δ. Through smart selection of Δ, we can also use this 1-bit quantizer to retain the first 3 moments of the signal. In general, using a B bit quantizer, we can retain the first $2B + 1$ moments of the signal. Now since the moments also control the shape of the signal, this is also referred to as shape-preserving quantization.

The moment preserving properties of this quantization scheme allow the preservation of different results of stream mining applied to the quantized data. For instance, it has been shown that the result of k-means clustering is preserved by this scalar quantization scheme [36]. This allows significant benefits in data reduction (with compression ratios of over 20:1 for common time series) while allowing effective stream mining. These quantization techniques may also be used to anonymize the data to support privacy preserving mining. Additional discussion of this is included in [36].

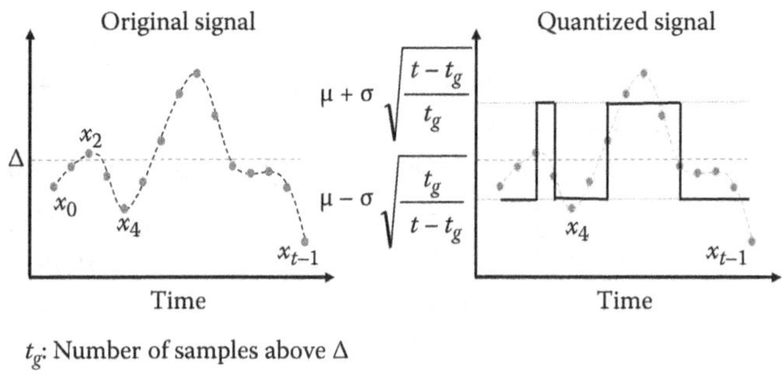

FIGURE 1.3 Moment preserving quantization.

1.4.8 Transform Domain Summarization

Another widely known synopsis representation in data stream computation uses different data transformations to capture time-frequency properties of the signal. Commonly used data transformations include the Fourier transform, the discrete cosine transform (DCT), and multiresolution transformations such as the wavelet transform representation. The Fourier transform and the DCT compute frequency characteristics of the signal using different basis functions (complex exponentials, and cosine functions respectively). In general, the DCT has been shown to have excellent energy compaction properties, i.e., the ability to capture the energy of most natural signals into a small set of frequency coefficients, and has been used extensively in audio (MP3), image (e.g., JPEG), and video (e.g., MPEG 1, 2, 4, H.264) compression.

The interest in multiresolution decompositions that capture the time-frequency scale properties of data stream signals has led to the extensive use of the wavelet transform on time series data. One of the most widely used representations is the Haar wavelet.

This technique is particularly simple to implement and is widely used in the literature for hierarchical decomposition and summarization. The basic idea in the wavelet transform is to create a decomposition of the data characteristics into a set of time- and frequency-scaled basis functions. Higher-order coefficients of the decomposition illustrate the broad trends in the data, whereas the more localized trends are captured by the lower-order coefficients.

We assume for ease in description that the length q of the series is a power of 2. This is without loss of generality because it is always possible to decompose a series into segments, each of which has a length that is a power of two. The Haar wavelet decomposition defines 2^{k-1} coefficients of order k. Each of these 2^{k-1} coefficients corresponds to a contiguous portion of the time series of length $q/2^{k-1}$. The ith of these 2^{k-1} coefficients corresponds to the segment in the series starting from position $(i - 1) \cdot q/2^{k-1} + 1$ to position $i * q/2^{k-1}$. Let us denote this coefficient by ψ_k^i and the corresponding time series segment by S_k^i. At the same time, let us define the average value of the first half of the S_k^i by a_k^i and the second half by b_k^i. Then, the value of ψ_k^i is given by $(a_k^i - b_k^i)/2$. More formally, if Φ_k^i denotes the average value of S_k^i, then the value of ψ_k^i can be defined recursively as follows:

$$\psi_k^i = (\Phi_{k+1}^{2\cdot i-1} - \Phi_{k+1}^{2\cdot i})/2 \tag{1.2}$$

The set of Haar coefficients* is defined by the ψ_k^i coefficients of order 1 to $\log_2(q)$. In addition, the global average Φ_1^1 is required for the purpose of perfect reconstruction. We note that the coefficients of different order provide an understanding of the major trends in the data at a particular level of granularity. For example, the coefficient ψ_k^i is half the quantity by which the first half of the segment S_k^i is larger than the second half of the same segment. Since larger values of k correspond to geometrically reducing segment sizes, one can obtain an understanding of the basic trends at different levels of granularity. We note that this definition of the Haar wavelet makes it very easy to compute by a sequence of averaging and differencing operations, and allows incremental window-based implementations of the transform.

In general, the entire decomposition may be represented as a tree of depth ($\log_2 q$), which captures the hierarchical representation of the entire series. This is also referred to as the *error tree*. The nodes in the tree contain the values of the wavelet coefficients, except for a special *super-root* node, which contains the series average. This super-root node is not necessary if we are only considering the relative values in the series, or the series values have been normalized so that the average is already zero. The original series can be reconstructed by adding or subtracting the values in the nodes along the path leading to that value. As a matter of convention, the coefficient in a node should be added, if we use the left branch below it to reach to the series values. Otherwise, it should be subtracted. This natural decomposition means that an entire contiguous range along the series can be reconstructed using only the portion of the error-tree that is relevant to it. Furthermore, we only need to retain those coefficients whose values are significantly large and, therefore, affect the values of the underlying series. In general, we would like to minimize the reconstruction error by retaining only a fixed number of coefficients, as defined by the space constraints. This property of the wavelet transform has allowed its use in compression algorithms, such as those used in JPEG-2000 image compression.

While wavelet decomposition is easy to perform for multidimensional data sets, it is much more challenging for the case of data streams. This is because data streams impose a one-pass constraint on the wave-

* To have a normalized set of basis functions, the weight of $\sqrt{2}$ is often applied to each coefficient. This ensures energy preservation.

let construction process. A variety of one-pass algorithms for wavelet construction are discussed in [23].

1.4.9 Summary Statistic Computation

This involves computing different summary statistics from one or more data samples that may be used for quantitative data analysis and visual exploration [37]. Examples of summary statistics include statistical means (e.g., arithmetic, geometric, harmonic), moments (e.g., mean, variance, skew, kurtosis, higher-order moments), densities (e.g., histogram), coefficient of variation, counts, distinct counts, quartiles, top *k* values, contingency tables among others [18] that capture the characteristics of a stream.

Significant research interest has centered on the construction of histograms from streaming data. The technique of histogram construction is closely related to that of wavelets. In histograms, the data is binned into a number of intervals along an attribute. For any given query, the counts from the bins can be utilized for query resolution. A simple representation of the histogram method would simply partition the data into equidepth or equiwidth intervals. The main inaccuracy with the use of histograms is that the distribution of data points within a bucket is not retained and is therefore assumed to be uniform. This causes inaccuracy because of extrapolation at the query boundaries. A natural choice is to use an equal number of counts in each bucket. This minimizes the error variation across different buckets. However, in the case of data streams, the boundaries to be used for equidepth histogram construction are not known a priori. We further note that the design of equidepth buckets is exactly the problem of quantile estimation, since the equidepth partitions define the quantiles in the data. Another choice of histogram construction is that of minimizing the variance of frequency of different values in the bucket. This ensures that the uniform distribution assumption is approximately held, when extrapolating the frequencies of the buckets at the two ends of a query. Such histograms are referred to as V-optimal histograms. Algorithms for V-optimal histogram construction are proposed in [38,39]. A more detailed discussion of several algorithms for histogram construction may be found in [18].

1.4.10 Dimensionality Reduction and Forecasting in Data Streams

Because of the inherent temporal nature of data streams, the problems of dimensionality reduction and forecasting are particularly important. When there are a large number of simultaneous data stream, we can use the correlations between different data streams to make effective predictions

[40,41] on the future behavior of the data stream. In particular, the well-known MUSCLES method [41] is useful in applying regression analysis to data streams. The regression analysis is helpful in predicting the future behavior of the data stream. A related technique is the SPIRIT algorithm, which explores the relationship between dimensionality reduction and forecasting in data streams. The primary idea is that a compact number of hidden variables can be used to comprehensively describe the data stream. This compact representation can also be used for effective forecasting of the data streams. Other dimensionality reduction techniques include Incremental PCA [42], and a discussion of several of these techniques (including SPIRIT) is provided in [18].

1.5 STREAM MINING ALGORITHMS

In this section, we will discuss the key stream mining problems and will discuss the challenges associated with each problem. We will also provide a broad overview of the different directions of research for these problems.

1.5.1 Data Stream Clustering

Clustering is a widely studied problem in the data mining literature. However, it is more difficult to adapt arbitrary clustering algorithms to data streams because of one-pass constraints on the data set. An interesting adaptation of the k-means algorithm has been discussed in [43], which uses a partitioning based approach on the entire data set. This approach uses an adaptation of a k-means technique to create clusters over the entire data stream. However, in practical applications, it is often desirable to be able to examine clusters over user-specified time horizons. For example, an analyst may desire to examine the behavior of the clusters in the data stream over the past one week, the past one month, or the past year. In such cases, it is desirable to store *intermediate cluster statistics*, so that it is possible to leverage these to examine the behavior of the underlying data.

One such technique is microclustering [44], in which we use cluster feature vectors [45] to perform stream clustering. The cluster feature vectors keep track of the first- and second-order moments of the underlying data to perform the clustering. These features satisfy the following critical properties that are relevant to the stream clustering process:

- Additivity property: The statistics such as the first- or second-order moments can be maintained as a simple addition of statistics over data points. This is critical in being able to maintain the statistics efficiently over a fast data stream. Furthermore, additivity also implies subtractivity; thus, it is possible to obtain the statistics over a particular time horizon, by subtracting out the statistics at the beginning of the horizon from the statistics at the end of the horizon.

- Computational convenience: The first- and second-order statistics can be used to compute a vast array of cluster parameters such as the cluster centroid and radius. This is useful to be able to compute important cluster characteristics in real time.

It has been shown in [44] that the microcluster technique is much more effective and versatile than the k-means–based stream technique discussed in [43]. This broad technique has also been extended to a variety of other kinds of data. Some examples of such data are as follows:

- High-dimensional data: The stream clustering method can also be extended to the concept of projected clustering [46]. A technique for high dimensional projected clustering of data streams is discussed in [47]. In this case, the same microcluster statistics are used for maintaining the characteristics of the clusters, except that we also maintain additional information that keeps track of the projected dimensions in each cluster. The projected dimensions can be used in conjunction with the cluster statistics to compute the projected distances that are required for intermediate computations. Another innovation proposed in [47] is the use of decay-based approach for clustering. The idea in the decay-based approach is relevant for the case of evolving data stream model and is applicable not just to the high dimensional case, but any of the above variants of the microcluster model. In this approach, the weight of a data point is defined as $2^{-\lambda \cdot t}$, where t is the current time instant. Thus, each data point has a half-life of $1/\lambda$, which is the time in which the weight of the data point reduces by a factor of 2. We note that the decay-based approach poses a challenge because the microcluster statistics are affected at each clock tick, even if no points arrive from the data stream. To deal with this problem, a *lazy approach* is applied to decay-based updates, in which we update the decay behavior for a microcluster only if a data point is added to it. The idea is that as long as we keep track of the last time t_s at which the microcluster was updated, we only need to multiply the microcluster statistics by $2^{-\lambda(t_c-t_s)}$, where t_c

is the current time instant. After multiplying the decay statistics by this factor, it is possible to add the microcluster statistics of the current data point. This approach can be used since the statistics of each microcluster decay by the same factor in each track, and it is therefore possible to implicitly keep track of the decayed values, as long as a data point is not added to the microcluster. In the latter case, the statistics need to be updated explicitly, while other counts can still be maintained implicitly.

- Uncertain data: In many cases, such as in sensor networks, the underlying data may be noisy and uncertain. In such cases, it may be desirable to incorporate the uncertainty into the clustering process. To do so, the microcluster statistics are appended with the information about the underlying uncertainty in the data. This information can be used to make more robust clustering computations. The advantages of using the uncertainty into the clustering process are illustrated in [48].
- Text and categorical data: A closely related problem is that of text and categorical data. The main difference with the quantitative domain is the nature of the statistics that are stored for clustering purposes. In this case, we maintain the counts of the frequencies of the discrete attributes in each cluster. Furthermore, we also maintain the inter-attribute correlation counts that may be required in a variety of applications. In [49], an efficient algorithm has been proposed for clustering text and categorical data streams. This algorithm also allows for a decay-based approach as in [47].

1.5.2 Data Stream Classification

The problem of classification is perhaps one of the most widely studied in the context of data stream mining. The problem of classification is made more difficult by the evolution of the underlying data stream. Therefore, effective algorithms need to be designed to take temporal locality into account. The concept of stream evolution is sometimes referred to as *concept drift* in the stream classification literature. Some of these algorithms are designed to be purely one-pass adaptations of conventional classification algorithms [50], whereas others (such as the methods in [51,52]) are more effective in accounting for the evolution of the underlying data stream. The broad methods that are studied for classification in the data stream scenario are as follows:

1.5.2.1 VFDT Method

The VFDT (very fast decision trees) method has been adapted to create decision trees that are similar to those constructed by a conventional learner with the use of sampling based approximations. The VFDT method splits a tree using the current best attribute, taking into consideration the fact that the number of examples used is sufficient to preserve the Hoeffding bound in a way that the output is similar to that of a conventional learner. The key question during the construction of the decision tree is the choice of attributes to be used for splits. Approximate ties are broken using a user-specified threshold of acceptable error measure for the output. It can be shown that for any small value of δ, a particular choice of the split variable is the correct choice with probability at least $1 - \delta$ if a sufficient number of stream records have been processed. This number has been shown in [50] to increase at a relatively modest rate of $\ln(1/\delta)$. This bound can then be extended to the entire decision tree, so as to quantify the probability that the same decision tree as a conventional learner is created. The VFDT method has also been extended to the case of evolving data streams. This framework is referred to as CVFDT [52], and it runs VFDT over fixed sliding windows to always have the most updated classifier. Jin and Agrawal [53] have extended the VFDT algorithm to process numerical attributes and reduce the sample size that is calculated using the Hoeffding bound. Since this approach reduces the sample size, it improves efficiency and space requirements for a given level of accuracy.

1.5.2.2 On-Demand Classification

While most stream classification methods are focused on a training stream, the on-demand method is focused on the case when both the training and the testing stream evolves over time. In the on-demand classification method [48], we create class-specific microclusters from the underlying data. For an incoming record in the test stream, the class label of the closest microcluster is used to determine the class label of the test instance. To handle the problem of stream evolution, the microclusters from the specific time-horizon are used for the classification process. A key issue in this method is the choice of horizon that should be used to obtain the best classification accuracy. To determine the best horizon, a portion of the training stream is separated out and the accuracy is tested over this portion with different horizons. The optimal horizon is then used to classify the test instance.

1.5.2.3 Ensemble-Based Classification

This technique [54] uses an ensemble of classification methods such as C4.5, RIPPER, and naive Bayes to increase the accuracy of the predicted output. The broad idea is that a data stream may evolve over time, and a different classifier may work best for a given time period. Therefore, the use of an ensemble method provides robustness in the concept-drifting case.

1.5.3 Frequent Pattern Mining

The problem of frequent pattern mining was first introduced in [55] and was extensively analyzed for the conventional case of disk resident data sets. In the case of data streams, one may wish to find the frequent itemsets either over a sliding window or the entire data stream [56,57]. In the case of data streams, the problem of frequent pattern mining can be studied under several models.

1.5.3.1 Entire Data Stream Model

In this model, the frequent patterns need to be mined over the entire data stream. Thus, the main difference from a conventional pattern mining algorithm is that the frequent patterns need to be mined in one pass over the entire data stream. Most frequent pattern mining algorithms require multiple passes to estimate the frequency of patterns of different sizes in the data. A natural method for frequent pattern counting is to use sketch-based algorithms to determine frequent patterns. Sketches are often used to determine heavy-hitters in data streams and, therefore, an extension of the methodology to the problem of finding frequent patterns is natural. Along this line, Manku and Motwani [58] proposed the first one-pass algorithm called *lossy counting* to find all frequent itemsets over a data stream. The algorithm allows false-positives, but not false-negatives. Thus, for a given support level s, the algorithm is guaranteed to contain all frequent itemsets whose support is greater than $s - \varepsilon$. Another interesting approach in [59] determines all the frequent patterns whose support is greater than s with probability at least $1 - \delta$, which the value of δ is as small as desired, as long as one is willing to add space and time complexity proportional to $\ln(1/\delta)$. Thus, this model does not allow false-positives but may miss some of the frequent patterns. The main advantage of such a technique is that it is possible to provide a more concise set of frequent patterns at the expense of losing some of the patterns with some probability, which is quite low for practical purposes.

1.5.3.2 Sliding Window Model

In many cases, the data stream may evolve over time, as a result of which it is desirable to determine all the frequent patterns over a particular sliding window. A method for determining the frequent patterns over a sliding window is discussed in [60]. The main assumption of this approach is that the number of frequent patterns is not very large, and therefore, it is possible to hold the transactions in each sliding window in main memory. The main focus of this approach is to determine closed frequent itemsets over the data stream. A new mining algorithm called MOMENT is proposed, and the main idea is based on the fact that the boundary between closed frequent itemsets and frequent itemsets moves very slowly. A closed enumeration tree is developed to keep track of the boundary between closed frequent itemsets and the rest of the itemsets. Another method that is able to mine frequent itemsets over arbitrary time granularities is referred to as FPSTREAM [61]. This method is essentially an adaptation of the FP-tree method to data streams.

1.5.3.3 Damped Window Model

We note that pure sliding windows are not the only way by which the evolution of data streams can be taken into account during the mining process. A second way is to introduce a *decay factor* into the computation. Specifically, the weight of each transaction is multiplied by a factor of $f < 1$ when a new transaction arrives. The overall effect of such an approach is to create an exponential decay function on the arrivals in the data stream. Such a model is quite effective for evolving data stream, since recent transactions are counted more significantly during the mining process. An algorithm proposed in [62] maintains a lattice for recording the potentially frequent itemsets and their counts. While the counts of each lattice may change upon the arrival of each transaction, a key observation is that it is sufficient to update the counts in a lazy way. Specifically, the decay factor is applied only to those itemsets whose counts are affected by the current transaction. However, the decay factor will have to be applied in a modified way by taking into account the last time that the itemset was touched by an update. In other words, if t_c be the current transaction index and the last time the count for the itemset was updated was at transaction index $t_s < t_c$, then we need to multiply the current counts of that itemset by $f^{t_s - t_c}$ before incrementing the count of this modified value. This approach works because the counts of each itemset reduce by the same decay factor in each iteration, as long as a transaction count is not added to it. We note that such a lazy approach is also applicable to other mining problems, where statistics are represented as the sum of decaying values. For example, in [47], a similar lazy approach is used to maintain decay-based microcluster statistics for a high-dimensional projected stream clustering algorithm.

1.5.4 Change Detection in Data Streams

As discussed earlier, the patterns in a data stream may evolve over time. In many cases, it is desirable to track and analyze the nature of these changes over time. In [63–66], a number of methods have been discussed for change detection of data streams. In addition, data stream evolution can also affect the behavior of the underlying data mining algorithms since the results can become stale over time. The broad algorithms for change diagnosis in data streams are as follows:

1.5.4.1 Velocity Density Estimation

In velocity density estimation [63], we compute the rate of change of data density of different points in the data stream over time. Depending upon the direction of density rate of change, one may identify regions of *dissolution*, *coagulation*, and *shift*. Spatial profiles can also be constructed to determine the directions of shift in the underlying data. In addition, it is possible to use the velocity density concept to identify those combinations of dimensions that have a high level of evolution. Another technique for change quantification is discussed in [64], which uses methods for probability difference quantification to identify the changes in the underlying data. In [65], a method is discussed to determine statistical changes in the underlying data. Clustering [44] can be used to determine significant evolution in the underlying data. In [44], microclustering is used to determine significant clusters that have evolved in the underlying data.

A separate line of work is the determination of significant changes in the results of data mining algorithms because of evolution. For example in [44], it has been shown how to determine significant evolving clusters in the underlying data. In [51], a similar technique has been used to keep a refreshed classification model in the presence of evolving data. In this respect, microclustering provides an effective technique, since it provides a way to store intermediate statistics of the underlying data in the form of clusters. In [51], a microcluster-based nearest-neighbor classifier is used to classify evolving data streams. The key idea is to construct class-specific microclusters over a variety of time horizons, and then utilize the time horizon with the greatest accuracy to perform the classification process. The issue of stream evolution has been extended to many other problems such as synopsis construction and reservoir sampling [26]. We will provide a detailed discussion of some of the methods such as reservoir sampling slightly later.

1.5.4.2 Stream Cube Analysis of Multidimensional Streams

Much of stream data resides at a multidimensional space and at rather low level of abstraction, whereas most analysts are interested in relatively high-level dynamic changes in some combination of dimensions. To discover high-level dynamic and evolving characteristics, one may need to perform multilevel, multidimensional online analytical processing (OLAP) of stream data. Such necessity calls for the investigation of new architectures that may facilitate online analytical processing of multidimensional stream data [67,68]. A stream cube architecture was proposed in [67,68] that effectively performs online partial aggregation of multidimensional stream data, captures the essential dynamic and evolving characteristics of data streams, and facilitates fast OLAP on stream data. Stream cube architecture facilitates online analytical processing of stream data. While this is needed for online stream mining and query processing, the process of model creation requires effective management of the incoming data.

1.5.4.3 Distributed Mining of Data Streams

In many instances, streams are generated at multiple distributed computing nodes. An example of such a case would be sensor networks in which the streams are generated at different sensor nodes. Analyzing and monitoring data in such environments requires data mining technology that requires optimization of a variety of criteria such as communication costs across different nodes, as well as computational, memory, or storage requirements at each node. There are several management and mining challenges in such cases. When the streams are collected with the use of sensors, one must take into account the limited storage, computational power, and battery life of sensor nodes. Furthermore, since the network may contain a very large number of sensor nodes, the effective aggregation of the streams becomes a considerable challenge. Furthermore, distributed streams also pose several challenges to mining problems, since one must integrate the results of the mining algorithms across different nodes. A detailed discussion of several distributed mining algorithms is provided in [18].

1.6 OPTIMIZING STREAM MINING APPLICATIONS

Joint application system adaptation is required to support the performance requirements of stream processing applications under huge data volumes, limited compute resources, and variability in data and resource characteristics. The first step in coping with high data volumes is to perform basic low-cost filtering as well as load balancing, spreading the work across multiple operators, potentially distributed over multiple compute nodes. This progressive filtering approach takes advantage of pipelined parallelism, where later stages in the pipeline can perform more expensive tasks, thanks to the rate reduction performed by the earlier stages. This also allows incremental information extraction, whereby to keep up with high data rates, operators close to the sources parse just enough information to allow early discard of irrelevant data using *loadshedding strategies* (Section 1.4) [69,70]. Later stages then may extract, and expose more information as necessary on the lower-volume data stream. In other words, a *rough*, but quick, algorithm is applied to preclassify the incoming data and route it to a more in-depth downstream analysis.

On the other hand, a load balancing strategy coupled with data and task parallel approaches may be used for effective processing and data reduction without data discard [71]. In these cases, the data is routed to different chains that either (i) perform the same processing on different tuples at reduced rates (data parallelism) or (ii) perform different processing on same or different tuples in parallel (task parallelism). Indeed, in many cases data streams are inherently parallelizable. For instance, stock-options data streams can be trivially parallelized since multiple logical streams are multiplexed into a single physical data source, e.g., streams of trading transactions relating to IBM in a stock market feed. It is important to note that parallelization may not be feasible in all stages of the processing [71,72], especially if correlations need to be drawn across data streams as is the case in many high-frequency or algorithmic trading applications or when data ordering must be preserved.

There has been research on *operator placement* [73,74] and *stream query optimization* [75] to improve the performance of distributed stream processing applications. These approaches solve the problem of operator placement across a distributed infrastructure to facilitate operator reuse, and optimize appropriate query specific metrics such as delay and throughput. There has also been recent interest developing optimized operator topology construction, scheduling, and adaptation strategies for large-scale stream mining. Specifically for queries posed as a set of conjunctive filtering, classification, and mining operations, as is common in search and knowledge extraction applications (e.g., find data that has attribute A and attribute B but not attribute C) [76]. These approaches focus on maximizing the quality of the mining results while optimizing for performance metrics such as delay and throughput. This work builds on research in large-scale stream mining system configuration [77,78], and work on ordering conjunctive filtering operations on streaming data, as part of the broader pipeline-ordering problem [79,80] and the set-cover problem [81]. These approaches design low-complexity algorithms to order a set of filters applied to streaming data, to minimize the end-to-end processing time. There has also been some recent work on enabling fault-tolerant placement of stream mining operators [82]. More details on this topic of stream mining optimization, and recent results may be obtained from [83].

1.7 CONCLUSIONS AND RESEARCH DIRECTIONS

There is a growing interest in building stream processing and mining applications to support the requirements of systems health monitoring of large complex IT infrastructure. This is also supported by the advances in hardware and software sensor technology, distributed compute paradigms, and efficient stream mining algorithms.

In this chapter we provide an overview of the underlying complexities unique to stream mining, and describe the state-of-the-art in stream mining systems, algorithms, and optimization strategies. While significant progress has been made, there are several open research challenges in each of these different domains that need to be overcome for the successful deployment of large-scale stream processing applications.

There are several systems challenges in terms of improving the performance (throughput, latency), scaling behavior, state management, and fault tolerance of stream processing systems to support larger and more complex applications, with long-running requirements. Correspondingly, there is also need to extend the programming models and associated integrated development and visualization environments to support these requirements. Research into formalizing the fundamental principles of stream processing application design and development is also key to enabling different developers to build these applications.

There are also significant advances required in the development of novel stream mining and optimization algorithms to handle the large volumes, temporal locality, and dynamic characteristics of different data streams. There has only been limited work on dealing with the time-varying nature of the data streams and capturing the temporally local properties of the data. Additionally, there are multiple challenges associated with handling noisy, missing, uncertain, or imperfect data. These include techniques for data cleaning that may involve either imputing or modeling the underlying uncertain data in real time, techniques that allow the use of this imperfect data in mining algorithms, while minimizing degradation of any derived results. Some recent research addresses the issue of clustering uncertain data streams [47], however, this work is still preliminary. There are also several unique challenges associated with privacy preserving mining of data streams that have not been sufficiently examined. It is often a challenge to perform privacy transformations of continuously arriving data, since newly arriving data may compromise the integrity of earlier data.

Finally, there are several domains ranging from cybersecurity to financial services where there is growing need to develop "stream exploration applications." These applications require automated exploration of the huge volumes of streaming data to detect novel or surprising events or trends and provide human analysts with supporting evidence gathered from various diverse sources. For instance, this could involve an automated system to monitor network traffic and identify potentially new threats to the security. These applications require the development of several new online learning and stream mining algorithms (including active learning, learning in adversarial settings), combined with strategies for automated composition, planning, management, scheduling, and optimization of these analytics across different compute platforms (e.g., Stream Processing, Hadoop, Cloud).

REFERENCES

[1] http://kdd.ics.uci.edu/databases/kddcup99/kddcup99.html.

[2] P. Patnaik, M. Marwah, R. Sharma, N. Ramakrishnan. Sustainable operation and management of data center chillers using temporal data mining. In *Proceedings of the 15th ACM SIGKDD International Conference on Knowledge Discovery and Data Mining,* 2009.

[3] S. Kandula, S. Sengupta, A. Greenberg, P. Patel, R. Chaiken. The nature of data center Traffic: Measurements and analysis. In *Proceedings of the 9th ACM SIGCOMM Conference on Internet Measurement Conference,* 2009.

[4] K. Yoshida. Sampling-based stream mining for network risk management. *Lecture Notes in Computer Science*, 4834/2007:374–386, 2007.

[5] M. Masud, J. Gao, L. Khan, J. Han, B. Thuraisingham. Peer to peer botnet detection for cyber-security: A data mining approach. In *Proceedings of the 4th Annual Workshop on Cyber Security and Information Intelligence Research: Developing Strategies to Meet the Cyber Security and Information Intelligence Challenges Ahead*, 2008.

[6] http://ganglia.sourceforge.net/.

[7] http://www-01.ibm.com/software/tivoli/products/prov-mgr/.

[8] D. Abadi, Y. Ahmad, et al. The design of the borealis stream processing engine. In *Proceedings of the Innovative Data Systems Research Conference (CIDR)*, 2005.

[9] A. Arasu, B. Babcock, et al. STREAM: The stanford stream data manager. *IEEE Data Engineering Bulletin*, 2003.

[10] N. Jain, L. Amini, et al. Design, implementation, and evaluation of the linear road benchmark on the stream processing core. In *Proceedings of the ACM International Conference on Management of Data (SIGMOD)*, 2006.

[11] H. Balakrishnan, M. Balazinska, et al. Retrospective on Aurora. *Very Large Databases Journal, VLDBJ, Special Issue on Data Stream Processing*, 2004.

[12] S. Chandrasekaran, O. Cooper, et al. TelegraphCQ: Continuous dataflow processing for an uncertain world. In *Proceedings of the Innovative Data Systems Research Conference (CIDR)*, 2003.

[13] V. Kumar, Z. Cai, et al. iFlow: Resource-aware overlays for composing and managing distributed information flows. In *Proceedings of the European Conference in Computer System (EuroSys)*, 2006.

[14] W. Thies, M. Karczmarek, S. Amarasinghe. StreamIt: A language for streaming applications. In *Proceedings of the International Conference on Compiler Construction (CC)*, 2002.

[15] L. Girod, Y. Mei, et al. XStream: A signal-oriented data stream management system. In *Proceedings of the IEEE International Conference on Data Engineering (ICDE)*, 2008.

[16] G. Cormode, M. Garofalakis. Sketching streams through the net: Distributed approximate query tracking. In *VLDB Conference*, 2005.

[17] G. Kollios, J. Byers, J. Considine, M. Hadjielefttheriou, F. Li. Robust aggregation in sensor networks. *IEEE Data Engineering Bulletin*, 2005.

[18] C. Aggarwal. *Data Streams: Models and Algorithms*. Springer, 2007.

[19] B. Gedik, H. Andrade, et al. SPADE: The System S declarative stream processing engine. In *Proceedings of the ACM International Conference on Management of Data (SIGMOD)*, 2008.

[20] M. Hirzel, H. Andrade, et al. Spade—language specification. *IBM Research Tech Report RC24760*, 2009.

[21] G. Jacques-Silva, B. Gedik, H. Andrade, K.-L. Wu. Language level checkpointing support for stream processing applications. In *Procedings of the 2009 International Conference on Dependable Systems and Networks (DSN)*, 2009.

[22] B. Gedik, H. Andrade, K.-L. Wu. A code generation approach to optimizing high-performance distributed data stream processing. In *Proceedings of the Conference on Information and Knowledge Management (CIKM)*, 2009.

[23] M. Garofalakis, J. Gehrke, R. Rastogi. Querying and mining data streams: You Only Get One Look (A Tutorial). In *SIGMOD Conference*, 2002.

[24] J.S. Vitter. Random sampling with a reservoir. *ACM Transactions on Mathematical Software*, 11(1): 37–57, 1985.

[25] B. Babcock, M. Datar, R. Motwani. Sampling from a moving window over streaming data. In *SIAM Symposium on Discrete Algorithms (SODA)*, 2002.

[26] C. Aggarwal. On biased reservoir sampling in the presence of stream evolution. In *VLDB Conference*, 2006.

[27] W. Johnson, J. Lindenstrauss. Extensions of lipshitz mapping onto hilbert space. *Contemporary Mathematics*, 26:189–206, 1984.

[28] N. Alon, P. Gibbons, Y. Matias, M. Szegedy. Tracking joins and self-joins in limited storage. In *ACM PODS Conference*, 1999.

[29] N. Alon, Y. Matias, M. Szegedy. The space complexity of approximating frequency moments. In *The Space Complexity of Approximating Frequency Moments*, pp. 20–29, 1996.

[30] P. Indyk. Stable distributions, pseudorandom generators, embeddings and data stream computation. *IEEE FOCS*, 2000.

[31] G. Cormode, S. Muthukrishnan. An improved data stream summary: The count-min sketch and its applications. *LATIN*, pp. 29–38, 2004.

[32] A. Bagnall, C. Ratanamahatana, et al. A bit level representation for time series data mining with shape based similarity. *Data Mining and Knowlegde Discovery 13:1*, 2006.

[33] J. Aßfalg, H.-P. Kriegel, et al. T-time: Threshold-based data mining on time series. In *Proceedings of IEEE ICDE*, 2008.

[34] V. Megalooikonomou, G. Li, Q. Wang. A dimensionality reduction technique for efficient similarity analysis of time series databases. In *Proceedings of CIKM*, 2004.

[35] E. Delp, M. Saenz, P. Salama. Block Truncation Coding (BTC). *The Handbook of Image and Video Processing*. Academic Press, 2000.

[36] D.S. Turaga, M. Vlachos, O. Verscheure. On K-means cluster preservation using quantization schemes. In *IEEE International Conference on Data Mining (ICDM)*, 2009.

[37] M. Datar, A. Gionis, P. Indyk, R. Motwani. Maintaining stream statistics over sliding windows. *SIAM Journal on Computing*, 31(6):1794–1813, 2002.

[38] Y. Ioannidis, V. Poosala. Balancing histogram optimality and practicality for query set size estimation. In *ACM SIGMOD Conference*, 1995.

[39] H. Jagadish, N. Koudas, S. Muthukrishnan, V. Poosala, K. Sevcik, T. Suel. Optimal histograms with quality guarantees. In *VLDB Conference*, 1998.

[40] Y. Sakurai, S. Papadimitriou, C. Faloutsos. BRAID: Stream mining through group lag correlations. *ACM SIGMOD Conference*, 2005.

[41] B.-K. Yi, N.D. Sidiropoulos, T. Johnson, H.V. Jagadish, C. Faloutsos, A. Biliris. Online data mining for co-evolving time sequences. *ICDE Conference*, 2000.

[42] M. Artac, M. Jogan, A. Leonardis. Incremental PCA for on-line visual learning and recognition. In *Proceedings of the 16th International Conference on Pattern Recognition (ICPR)*, 2002.

[43] S. Guha, N. Mishra, R. Motwani, L. O'Callaghan. Clustering data streams. In *IEEE FOCS Conference*, 2000.
[44] C. Aggarwal, J. Han, J. Wang, P. Yu. A framework for clustering evolving data streams. In *VLDB Conference*, 2003.
[45] T. Zhang, R. Ramakrishnan, M. Livny. BIRCH: An efficient data clustering method for very large databases. In *ACM SIGMOD Conference*, 1996.
[46] C. Aggarwal, C. Procopiuc, J. Wolf, P. Yu, J.-S. Park. Fast algorithms for projected clustering. In *ACM SIGMOD Conference*, 1999.
[47] C. Aggarwal, J. Han, J. Wang, P. Yu. A framework for high dimensional projected clustering of data streams. In *VLDB Conference*, 2004.
[48] C. Aggarwal and P. Yu. A framework for clustering uncertain data streams. In *ICDE Conference*, 2008.
[49] C. Aggarwal, P. Yu. A framework for clustering massive text and categorical data streams. *SIAM Data Mining Conference*, 2006.
[50] P. Domingos, G. Hulten. Mining high-speed data streams. In *Proceedings of the ACM KDD Conference*, 2000.
[51] C. Aggarwal, J. Han, J. Wang, P. Yu. On-demand classification of data streams. In *ACM KDD Conference*, 2004.
[52] G. Hulten, L. Spencer, P. Domingos. Mining time changing data streams. In *ACM KDD Conference*, 2001.
[53] R. Jin, G. Agrawal. Efficient decision tree construction on streaming data. In *ACM KDD Conference*, 2003.
[54] H. Wang, W. Fan, P. Yu, J. Han. Mining concept-drifting data streams using ensemble classifiers. *ACM KDD Conference*, 2003.
[55] R. Agrawal, T. Imielinski, A. Swami. Mining association rules between sets of items in large databases. In *ACM SIGMOD Conference*, 1993.
[56] C. Giannella, J. Han, J. Pei, X. Yan, P. Yu. Mining frequent patterns in data streams at multiple time granularities. In *Proceedings of the NSF Workshop on Next Generation Data Mining*, 2002.
[57] R. Jin, G. Agrawal. An algorithm for in-core frequent itemset mining on streaming data. *ICDM Conference*, 2005.
[58] G. Manku, R. Motwani. Approximate frequency counts over data streams. In *VLDB Conference*, 2002.
[59] J. Xu Yu, Z. Chong, H. Lu, A. Zhou. False positive or false negative: Mining frequent itemsets from high speed transaction data streams. *VLDB Conference*, 2004.
[60] Y. Chi, H. Wang, P. Yu, R. Muntz. Moment: Maintaining closed frequent itemsets over a stream sliding window. In *ICDM Conference*, 2004.
[61] C. Gianella, J. Han, J. Pei, X. Yan, P. Yu. Mining frequent patterns in data streams at multiple time granularities. In *NSF Workshop on Next Generation Data Mining*, 2002.
[62] J.H. Chang, W.S. Lee. Finding recent frequent itemsets adaptively over online data streams. In *ACM KDD Conference*, 2003.
[63] C. Aggarwal. A framework for diagnosing changes in evolving data streams. In *ACM SIGMOD Conference*, 2003.

[64] T. Dasu, S. Krishnan, S. Venkatasubramaniam, K. Yi. An information-theoretic approach to detecting changes in multi-dimensional data streams. Duke University Technical Report CS-2005-06, 2005.

[65] D. Kifer, S.-B. David, J. Gehrke. Detecting change in data streams. In *VLDB Conference*, 2004.

[66] C. Aggarwal. An intuitive framework for understanding changes in evolving data streams. In *IEEE ICDE Conference*, 2002.

[67] Y. Chen, G. Dong, J. Han, B.W. Wah, J. Wang. Multi-dimensional regression analysis of time-series data streams. In *VLDB Conference*, 2002.

[68] G. Dong, J. Han, J. Lam, J. Pei, K. Wang. Mining multi-dimensional constrained gradients in data cubes. In *VLDB Conference*, 2001.

[69] Y. Chi, H. Wang, P. Yu. Loadstar: Load shedding in data stream mining. In *Proceedings of the International Conference on Very Large Data Bases (VLDB)*, 2005.

[70] N. Tatbul, U. Cetintemel, et al. Load shedding in a data stream manager. In *Proceedings of the International Conference on Very Large Data Bases (VLDB)*, 2003.

[71] X. Zhang, H. Andrade, et al. Implementing a high-volume, low-latency market data processing system on commodity hardware using IBM middleware. In *Proceedings of the 2009 Workshop on High Performance Computational Finance (WHPCF)*, 2009.

[72] H. Andrade, B. Gedik, K.-L. Wu, P. Yu. Scale-up strategies for processing high-rate data streams in System S. In *Proceedings IEEE International Conference on Data Engineering (ICDE)*, 2009.

[73] P. Pietzuch, J. Ledlie, et al. Network-aware operator placement for stream-processing systems. In *Proceedings of the IEEE International Conference on Data Engineering (ICDE)*, 2006.

[74] U. Srivastava, K. Munagala, J. Widom. Operator placement for inNetwork stream query processing. *PODS*, 2005.

[75] S. Seshadri, V. Kumar, B. Cooper, L. Li. A distributed stream query optimization framework through integrated planning and deployment. *IEEE Transactions on Parallel and Distributed Systems*, 2009.

[76] H.-H. Lee, E.-W. Yun, W.-S. Leeb. Attribute-based evaluation of multiple continuous queries for filtering incoming tuples of a data stream. *International Journal on Information Sciences*, 2008.

[77] B. Babcock, S. Babu, R. Motwani, M. Datar. Chain: Operator scheduling for memory minimization in data stream systems. *ACM SIGMOD*, 2003.

[78] F. Fu, D.S. Turaga, et al. Configuring competing classifier chains in distributed stream mining systems. In *IEEE Journal on Selected Topics in Signal Processing*, 2007.

[79] A. Condon, A. Deshpande, L. Hellerstein, N. Wu. Flow algorithm for two pipelined filter ordering problems. In *ACM SIGMOD-SIGACT-SIGART Symposium on Principles of Database Systems*, 2006.

[80] S. Babu, R. Motwani, et al. Adaptive ordering of pipelined stream filters. *ACM SIGMOD International Conference on Management of Data*, 2003.

[81] Z. Liu, S. Parthasarathy, A. Ranganathan, H. Yang. Near-optimal algorithms for shared filter evaluation in data stream systems. In *ACM SIGMOD International Conference on Management of Data*, 2008.
[82] N. Bansal, R. Bhagwan, et al. Towards optimal resource allocation in partial-fault tolerant applications. In *IEEE International Conference on Computer Communications (INFOCOM)*, 2008.
[83] R. Ducasse, D.S. Turaga, M. van der Schaar. Adaptive topologic optimization for large-scale stream mining. *IEEE Journal on Selected Topics in Signal Processing*, 2010.

CHAPTER 2

A Tutorial on Bayesian Networks for Systems Health Management

Arthur Choi

Lu Zheng

Adnan Darwiche

Ole J. Mengshoel

CONTENTS

2.1 INTRODUCTION

Bayesian networks have established themselves as an indispensable tool in artificial intelligence and are being used effectively by researchers and practitioners, more broadly in science and engineering [1–13]. The domain of systems health management, including diagnosis, is no exception. In fact, diagnostic applications have driven much of the developments in Bayesian networks over the past few decades [4–11]. In this chapter, we provide a gentle and accessible introduction to modeling and reasoning with Bayesian networks, with the domain of systems health management in mind.

Formally, a Bayesian network is a type of statistical model that can compactly represent complex probability distributions. Bayesian networks are particularly well-suited to modeling systems that we need to monitor, diagnose, and make predictions about, all under the presence of uncertainty. The system under consideration may be a natural one, such as a patient, where our goal may be to diagnose a disease, given the outcomes of imperfect medical tests [4–6]. The system may be artificial, such as an aerospace vehicle, where our goal may be to isolate a component failure, given a vector of unreliable sensor readings [7–11]. In aerospace, these goals and algorithms are useful in developing techniques for what is often referred to as fault detection, fault isolation, and recovery (FDIR). Bayesian networks can also model large-scale systems, even in settings where reasoning must be done in real time. For example, in an online system, we may want to detect impending failures, given access to possibly noisy observations.

Our goal in this chapter is to provide the reader with a high-level perspective on Bayesian networks, with a focus on systems health management. We also hope to give a taste for the capabilities that Bayesian networks bring, in terms of modeling, reasoning, and learning in complex systems. First, in Section 2.2, we highlight briefly the problem of systems health management, and introduce a small system that will be used as a running example in the following few sections. Section 2.3 provides an introduction to Bayesian networks, and Section 2.4 highlights the process by which Bayesian networks are constructed. Section 2.5 illustrates the reasoning capabilities that Bayesian networks provide, again through simple examples. Section 2.6 gives a brief overview of the types of algorithms, developed in the artificial intelligence community over the past three decades, available for reasoning in Bayesian networks. Section 2.7 does similarly for the task of learning Bayesian networks from data. Section 2.8 presents a real-world application of Bayesian networks in diagnosing electrical power systems, developed for a testbed located at the NASA Ames Research Center. Finally, Section 2.9 concludes this tutorial, which also includes references to relevant books and software systems, for the interested readers.

2.2 SYSTEMS HEALTH MANAGEMENT AND UNCERTAINTY

Systems health management is often needed due to uncertainty regarding a system's operation as well as in the system's environment. For example, a system might be an aerospace vehicle, and we may consider how uncertainty plays a key role during a NASA vehicle's mission. Uncertainty may be induced by (i) the varying and uncertain environments in which vehicles operate, (ii) the unpredictability of hardware and software failures that can take place in these vehicles, (iii) noisy and ambiguous sensor systems, (iv) incomplete and uncertain knowledge of the physics of dynamic systems, both terrestrially and extraterrestrially.

Beyond the requirement for dealing with uncertainty, there are often requirements on the ease of modeling and on real-time reasoning [10]. Ease of modeling is critical for practitioners to employ Bayesian networks in their specific domains, without becoming experts on Bayesian networks themselves. Real-time reasoning is critical for those applications that require inferences made about the system to be reported within strict time limits, so that the system is able to react to it.

In the context of ADAPT [12], an electrical power system testbed at NASA discussed in more detail in Section 2.8, we can illustrate these requirements as follows. Modeling requirements can be addressed by using procedures that take as input high-level specifications and synthesize Bayesian networks from them. The benefit of this automated approach is that it enables the modeling of real-world systems, even by practitioners who may not be well-versed in Bayesian networks. The second requirement, for real-time reasoning, may be addressed by embedding powerful and efficient reasoning algorithms into hard, real-time systems [13]. The real-time requirement has at least two facets, namely the need for predictability and the need for fast inference. Some of the reasoning algorithms we discuss in Section 2.5, for example, the compilation to arithmetic circuits, address these concerns.

2.2.1 An Electrical Circuit Example

Consider Figure 2.1, which depicts a simple electrical system, composed of a number of components and equipped with a number of sensors. First, there are three components of interest: a battery *B*, a relay *R*, and a load *L*, which may be, for example, a lamp or a fan. Second, there are three sensors: a current sensor *C*, a voltage sensor *V*, and a feedback sensor *F* on the relay. Finally, a command may be issued to the relay *R*, to open or close the electrical circuit. The system components are also summarized in Table 2.1.

In this simple example, there are two nominal modes of operation, depending on the most recent command given to the relay. Given an open command, the current and voltage sensors are expected to read low and the feedback sensor is expected to read open. Given a close command, the current and voltage sensors are expected to read high and the feedback sensor is expected to read closed. Suppose that we have an abnormal reading, for example, if the command is to close the circuit and our current sensor reads low. We can imagine a number of faults that would explain this behavior. For example, the relay could be faulty and stuck open or the sensor reading could be unreliable.

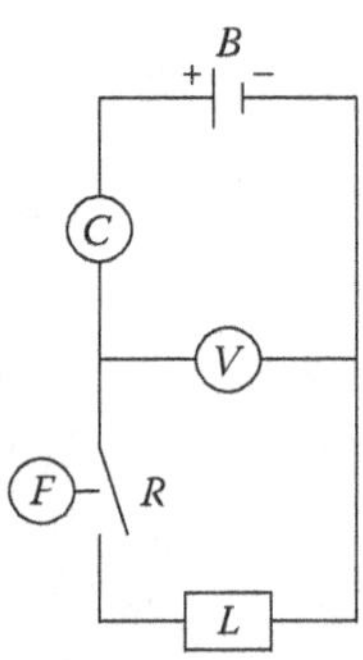

FIGURE 2.1 A diagram of a simple electrical system, with battery *B*, current censor (ammeter) *C*, voltage sensor (voltmeter) *V*, a relay (or switch) *R*, a feedback (touch) sensor *F*, and a load *L*.

TABLE 2.1 System Components

Location	Part	Mode	Health States
B	Battery	healthy	healthy
		stuck disabled	stuck-disabled
		stuck enabled	stuck-enabled
W	Voltage sensor	healthy	healthy
		reading stuck high	stuck-hi
		reading stuck low	stuck-lo
	Current sensor	healthy	healthy
		reading stuck high	stuck-hi
		reading stuck low	stuck-lo
R	Relay	healthy	healthy
		stuck open	stuck-open
		stuck closed	stuck-closed
	Feedback sensor	healthy	healthy
		reading stuck open	stuck-open
		reading stuck closed	stuck-closed
L	Load	healthy	healthy
		stuck disabled	stuck-disabled
		stuck enabled	stuck-enabled

In this chapter, our goal is to illustrate how to reason, in an automated and principled way, about such problems by means of Bayesian networks. We shall illustrate first how to model small-scale systems such as the one described above and then highlight how this process can be scaled to state-of-the-art approaches to real-world diagnostic problems. More specifically, we hope to illustrate that Bayesian networks provide an expressive and natural framework for modeling systems health management problems, both small and large. We further hope to illustrate that Bayesian networks provide an even more powerful framework for analyzing and reasoning about such problems. For illustrative purposes, we will use the simple electrical system of Figure 2.1 as a running example throughout the next few sections. First, we shall illustrate how a Bayesian network model can be constructed automatically from a formal system design (such as a schematic).

Next, we highlight some of the queries supported by Bayesian networks. For example, it is natural to ask for a given set of abnormal sensor readings:

- How likely are our observations and how likely is each individual component to be the cause of an abnormality?
- What is the likeliest explanation of our observations, i.e., what is the likeliest joint health state?
- How robust are our estimates and explanations under perturbations in the Bayesian network?

We then provide a high-level overview of the variety of algorithms available for reasoning and learning in Bayesian networks, such as those that are more suitable for embedded systems, and powerful approximation algorithms for reasoning in large and complex systems. Finally, we highlight a successful application of the Bayesian network models and techniques that we introduce for the purposes of diagnosing electrical power systems.

2.3 BAYESIAN NETWORKS

A Bayesian network is a directed acyclic graph (DAG) that is annotated with local conditional probability tables (CPTs). Nodes represent random variables, and directed edges typically denote causal relationships between variables (although causality is not a requirement). For example, an edge $C \rightarrow E$ may represent a direct cause and effect relationship between random variables C and E. More precisely, the DAG of a Bayesian network encodes a set of conditional independence statements that reads as follows: Every variable is independent of its nondescendants given its parents. When edges represent direct causal influences, these independencies read: Every variable is independent of its non-effects given its direct causes.

As a concrete example, consider Figure 2.2, which depicts a simple Bayesian network model of a voltage sensor. This model is also a submodel of a larger Bayesian network model that we shall construct for the simple electrical system in Figure 2.1 (we discuss this larger Bayesian network, in Figure 2.3, later in this chapter). In this example, we have three random variables:

- Variable P_V represents the presence of voltage on a wire. The voltage may be either high, $P_V = \text{hi}$, or it may be low, $P_V = \text{lo}$.

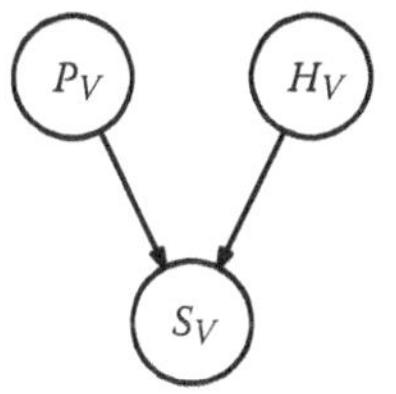

P_V	θ_{P_V}
hi	0.5
lo	0.5

H_V	θ_{H_V}
healthy	0.99
stuck-hi	0.005
stuck-lo	0.005

P_V	H_V	S_V	θ_{S_V}
hi	healthy	read-hi	1.0
hi	healthy	read-lo	0.0
hi	stuck-hi	read-hi	1.0
hi	stuck-hi	read-lo	0.0
hi	stuck-lo	read-hi	0.0
hi	stuck-lo	read-lo	1.0
lo	healthy	read-hi	0.0
lo	healthy	read-lo	1.0
lo	stuck-hi	read-hi	1.0
lo	stuck-hi	read-lo	0.0
lo	stuck-lo	read-hi	0.0
lo	stuck-lo	read-lo	1.0

FIGURE 2.2 A Bayesian network model of a voltage sensor, along with the network's CPTs.

- Variable S_V represents the reading of a voltage sensor. The sensor either reads high, S_V = read-hi, or it reads low, S_V = read-lo.
- Variable H_V represents the health of a sensor. We assume that a sensor has 3 modes: one nominal mode H_V = healthy, and two failure modes, either stuck at a high reading, H_V = stuck-hi, or stuck at a low reading, H_V = stuck-lo.

In this example, variables P_V and H_V are parents of variable S_V, as the presence of voltage and the health of a sensor dictates the value of its reading.

In a Bayesian network, we must quantify the local relationships between a variable and its parents. In particular, we specify a CPT for each variable X in the network, which is a conditional probability distribution for a variable X given its parents $\mathbf{U}$. These local conditional distributions induce a global probability distribution over all network variables $\mathbf{X}$. Let $\mathbf{x}$ denote an instantiation of these variables, where each variable $X \in \mathbf{X}$ is assigned to one of its values x. The probability associated with a variable instantiation $\mathbf{x}$ is then a product of these conditional probabilities $Pr(x|\mathbf{u})$ for each value of x that was set in $\mathbf{x}$:

$$Pr(\mathbf{x}) = \prod_{X \in \mathbf{X}} Pr(x|\mathbf{u})$$

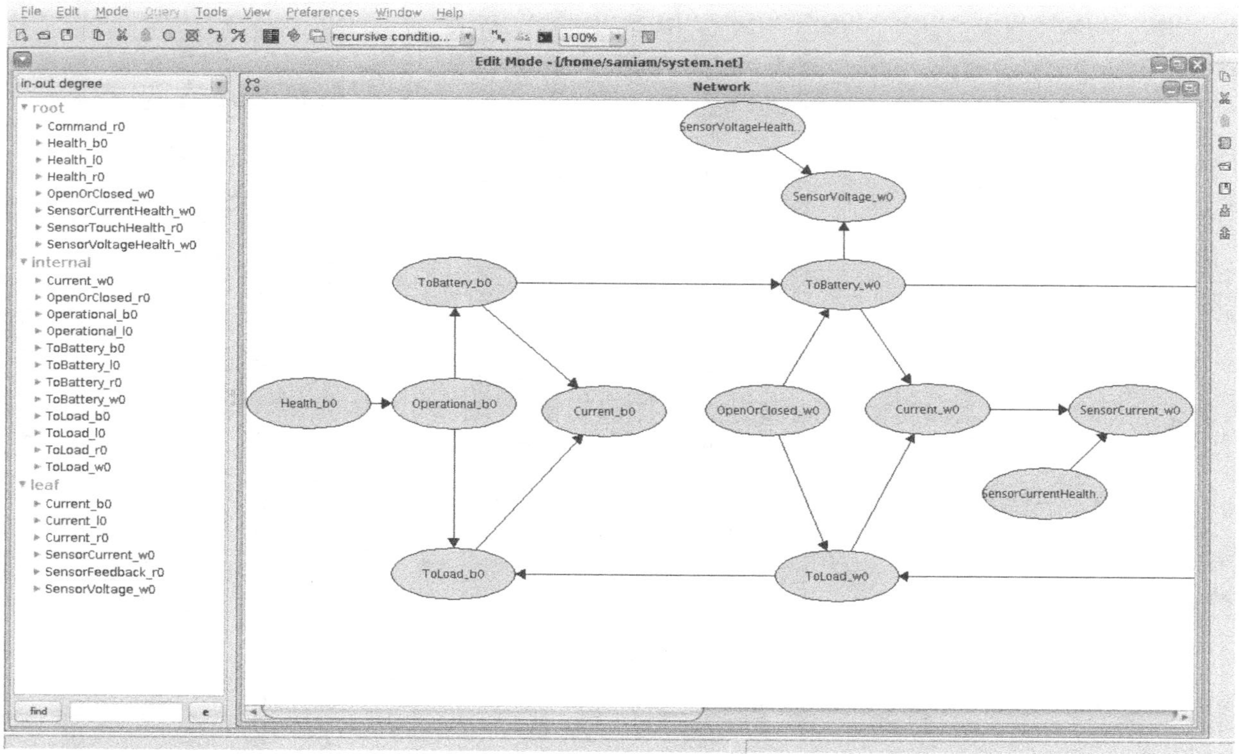

FIGURE 2.3 A partial view of a Bayesian network for the electrical system in Figure 2.1, in SamIam, a comprehensive tool for modeling and reasoning with Bayesian networks, available for download at http://reasoning.cs.ucla.edu/samiam/.

where **u** are the sub-instantiations of **x** over the parent variables **U**. One main feature of Bayesian networks is this compact representation of the joint probability distribution in terms of local conditional distributions.

In our example, we need to specify the conditional distribution $Pr(S_V|P_V,H_V)$ that a sensor has a particular reading, given the presence or absence of voltage, and the health of the sensor. In these terms, we can specify the operation of a sensor as a functional relationship: (1) if a sensor is healthy, then the sensor's reading reflects the presence or absence of voltage; otherwise, (2) if a sensor is stuck high (or stuck low), then the reading is always high (or always low). For example, we have $Pr(S_V =$ read-hi $| P_V =$ hi, $H_V =$ healthy$) = 1.0$, i.e., the sensor is sure to read high given that voltage is present and the sensor is healthy. We also refer to these conditional probabilities as parameters $\theta_{S_V|P_V,H_V}$ of the Bayesian network model. Variables P_V and H_V are root nodes in the DAG, so we need to specify the prior (unconditional) distributions $Pr(P_V)$ and $Pr(H_V)$, or equivalently, the parameters θ_{P_V} and θ_{H_V}. In our example, the probability $Pr(H_V =$ healthy$)$ that our sensor is healthy is 99.0%, and the probabilities $Pr(H_V =$ stuck-hi$)$ and $Pr(H_V =$ stuck-lo$)$ that our sensor is stuck at a high and low readings are both 0.5%. We further assumed a uniform distribution for the prior $Pr(P_V)$ on the presence of voltage. We will describe in more detail in Section 2.4 the process of constructing Bayesian networks.

As we described before, the network's parameters (i.e., the three CPTs θ_{P_V}, θ_{H_V}, and $\theta_{S_V|P_V,H_V}$) define a global probability distribution over the three variables P_V, H_V, and S_V, which can be decomposed as follows:*

$$Pr(P_V, H_V, S_V) = Pr(P_V)Pr(H_V)Pr(S_V | P_V, H_V) = \theta_{P_V}\theta_{H_V}\theta_{S_V|P_V,H_V} \quad (2.1)$$

We can visualize this joint distribution by enumerating over all possible assignments of variables to values. Consider the following table,

* This decomposition follows from the chain rule, and the property that every variable is independent of its nondescendants given its parents (thus, root variables are marginally independent).

where we list all such assignments along with their corresponding probabilities:

P_V	H_V	S_V	$Pr(P_V,H_V,S_V)$
hi	healthy	read-hi	49.50%*
hi	healthy	read-lo	0.00%
hi	stuck-hi	read-hi	0.25%*
hi	stuck-hi	read-lo	0.00%
hi	stuck-lo	read-hi	0.00%
hi	stuck-lo	read-lo	0.25%*
lo	healthy	read-hi	0.00%
lo	healthy	read-lo	49.50%*
lo	stuck-hi	read-hi	0.25%*
lo	stuck-hi	read-lo	0.00%
lo	stuck-lo	read-hi	0.00%
lo	stuck-lo	read-lo	0.25%*

where we have starred the scenarios with nonzero probability. In this submodel, we find that the two nominal scenarios,

$$\{P_V = \text{hi}, H_V = \text{healthy}, S_V = \text{read-hi}\} \ \& \ \{P_V = \text{lo}, H_V = \text{healthy}, S_V = \text{read-lo}\}$$

are the most likely, both with probability of 49.5%. There are four other possible scenarios (that have nonzero probability), although they are relatively unlikely scenarios that correspond to sensor failures.

In general, the number of entries in a joint probability table is exponential in the number of variables in the model. In contrast, a Bayesian network is defined in terms of local conditional distributions, i.e., its CPTs, which are relatively compact. In particular, a CPT has a size that is exponential in the number of variables that appear in it: the variable itself and its parents in the DAG. This local representation (coupled with a small number of parents for all nodes) is what allows us to compactly represent what would otherwise be an intractably large distribution.

As we shall see in Section 2.4, having to define only local CPTs greatly simplifies the process of constructing Bayesian network models. Moreover, the structure of a Bayesian network model, and the conditional independencies that we are able to infer from it, can be exploited for the purposes of efficient and effective inference and learning. In Section 2.5, we will highlight the utility of Bayesian networks in terms of the queries one can pose with respect to the joint probability distributions that they encode.

2.4 MODELING WITH BAYESIAN NETWORKS

There are three basic approaches for building Bayesian networks [2]: (1) they can be constructed by expert knowledge, (2) they can be synthesized automatically from formal system designs, and (3) they can be learned automatically from data. Considering the domain of interest here is in systems health management, it is likely that we already have available a formal system design (such as a schematic) from which we can construct a Bayesian network, so we will focus in this section on this particular approach.

Speaking more generally, there are three basic steps for specifying a Bayesian network.

1. Define the network variables and their values. Certain variables represent events that we are interested in estimating, such as the health of components. Such variables are often referred to as *query* variables. Other variables represent observables or Bayesian networks inputs, such as the readings of sensors. Such variables are often referred to as *evidence* variables. A Bayesian network may also involve *intermediary* variables, which are neither targets for queries nor observable events. They may aid the modeling process by relating query variables to evidence variables.

2. Define the network structure. That is, we need to specify the edges of a Bayesian network. In most domains, a causal interpretation of the network structure will be the most natural, where edges indicate a cause and effect relationship between variables. In this case, deciding what variables point to a variable X reduces to identifying its *direct* causes.

3. Define the network CPTs. How to specify a network's CPTs varies from domain to domain. In our example from the previous section, our sensor model was composed largely of functional relationships between variables, and we can rely on our domain knowledge to guide us. In other cases, we can rely on data to learn these parameters automatically, as we shall see in Section 2.7.

Consider again Table 2.1, which depicts a number of components, and Figure 2.1, which depicts how these components are connected to each other. To induce a Bayesian network model from this design, we can start by specifying a procedure for constructing Bayesian network fragments from components of interest. For example, we have already seen a submodel for a voltage sensor, which we constructed based on the knowledge of how such sensors operate. We may incorporate this submodel in a larger model, say by feeding the appropriate outputs of the larger model (e.g., variables that represent voltage on a wire) to the inputs of our sensor model. By connecting the submodels of components, we can construct subsystems, and by connecting subsystems, we can construct a complete Bayesian network that can then be used for systems health management. When the system of interest is well-understood and well-defined, as it typically is for electrical power systems for instance, this construction process can be completely or largely automated [14].

2.5 REASONING WITH BAYESIAN NETWORKS

In this section, we illustrate the reasoning capabilities enabled by a Bayesian network model. In particular, we use as an example the Bayesian network depicted in Figure 2.3, which we analyze using SAMIAM (Sensitivity Analysis, Modeling, Inference, and More), a comprehensive tool for modeling and reasoning with Bayesian networks, available for download at http://reasoning.cs.ucla.edu/samiam/. The network used in our example is also available at http://reasoning.cs.ucla.edu/samiam/tutorial/system.net. Our network in this example was automatically constructed from formal design [14], where we modeled each component Table 2.1 (including the voltage sensor model from Figure 2.2), as well as their interactions as suggested by Figure 2.1.

Suppose in our electrical circuit that we issued a command for the relay to close, but after doing so, we observed unexpected sensor readings: although the current sensor reports a high current, we find that the voltage sensor reads low and the feedback sensor reads open. We denote the command by $\{R = \text{close}\}$ and our sensor readings by $\{S_C = \text{read-hi}, S_V = \text{read-lo}, S_F = \text{open}\}$. In such a situation, we are interested in inferring the failures that led to the abnormal sensor readings that we observed.

In Figure 2.3, we modeled the health of our components using variables H_B (for the battery), H_R (for the relay), and H_L (for the load). We also modeled the health of our sensors using variables H_C (for the current sensor), H_V (for the voltage sensor), and H_F (for the feedback sensor). These random variables are the focus of the queries that we will pose in our Bayesian network, with the goal of acquiring actionable health information about our electrical system.

2.5.1 Posterior Marginals

Let $\mathbf{e} = \{R = \text{close}, S_C = \text{read-hi}, S_V = \text{read-lo}, S_F = \text{open}\}$ denote the assignment of variables to values, which we call the observed *evidence*. One of the most common queries that we pose in a Bayesian network is for the posterior probability $Pr(X|\mathbf{e})$ of a network variable X under evidence $\mathbf{e}$.

For example, we may ask what is our belief that current is present in the circuit. In this case we have probability $Pr(P_C = \text{hi}|\mathbf{e}) = 98.51\%$. Surprisingly, although two of our sensors suggest that the circuit is open and that there is no voltage present in the circuit, our model suggests that current is indeed flowing in our circuit. When we query for the marginal probabilities that our components are healthy, we find

$$Pr(H_B = \text{healthy}|\mathbf{e}) = 98.51\%$$

$$Pr(H_R = \text{healthy}|\mathbf{e}) = 98.51\%$$

$$Pr(H_L = \text{healthy}|\mathbf{e}) = 99.49\%$$

which is consistent with our belief that current is in fact flowing in the circuit. Consider now the marginal probabilities of the sensor health nodes:

$$Pr(H_C = \text{healthy}|\mathbf{e}) = 98.01\%$$

$$Pr(H_V = \text{stuck-lo}|\mathbf{e}) = 99.01\%$$

$$Pr(H_F = \text{stuck-open}|\mathbf{e}) = 99.01\%$$

which indicates that the voltage sensor is stuck at a low reading, and independently, that the feedback sensor is stuck at an open reading.

Note, however, that it may still be unlikely that the voltage sensor and the feedback sensor are both stuck at the same time. Indeed, observing two otherwise independent faults is typically considered much less likely than observing a single fault. As we shall soon see, observing these two faults is much more likely than any of the alternative scenarios in this particular example.

2.5.2 Explanations

In this section, we illustrate how Bayesian networks are capable of providing explanations for our observations. In our example, the Bayesian network model has suggested that current is flowing in our circuit despite two sensors suggesting otherwise. To consider why this may be, we can compute the maximum a posteriori (MAP) explanation.

Let H = $\{H_B, H_R, H_L, H_C, H_V, H_F\}$ denote the set of our component and sensor health variables and let **h** denote an assignment of these variables to values. Given a piece of evidence **e**, the MAP explanation is then the assignment **h** having maximum probability $Pr(\mathbf{h}|\mathbf{e})$. In our running example, the MAP explanation is the assignment:

$$\mathbf{h} = \{H_B = \text{healthy}, H_R = \text{healthy}, H_L = \text{healthy},$$
$$H_C = \text{healthy}, H_V = \text{stuck-lo}, H_F = \text{stuck-open}\}$$

with probability $Pr(\mathbf{h}|\mathbf{e})$ = 96.54%. According to this explanation, there are two faults: the voltage sensor is stuck at a low reading and the feedback sensor is stuck at an open reading. This helps to explain why we observed that the probability that current is present in the circuit was 98.51%. If we look further, the *next* most likely instantiations of our variables H assume that there are *three* simultaneous faults, for example,

$$\mathbf{h} = \{H_B = \text{disabled}, H_R = \text{stuck-open}, H_L = \text{healthy},$$
$$H_C = \text{stuck-hi}, H_V = \text{healthy}, H_F = \text{healthy}\}$$

which is significantly less likely with probability $Pr(\mathbf{h}|\mathbf{e})$ = 0.4878%.*

There are two notable classes of MAP explanations that are worth mentioning here. If the variables of interest consist of all network variables X (excluding those evidence variables E that are already observed), the MAP explanation is also called the Most Probable Explanation (MPE). In addition, the MPE is generally not as difficult to compute as MAP explanations. On the other hand, the MPE will mention any (and all) auxiliary variables that may not be of immediate interest. Moreover, the MPE and the MAP explanation may not agree on the variables of interest. On the other extreme, it is common to seek the most likely state

* Note if we had assumed instead that no current is flowing in the network: (1) the current sensor is stuck at a high reading, (2) we issued a close command to the relay, so either the feedback sensor is stuck or the relay is stuck, and (3) even if the circuit is open, the voltage sensor should still be able to read a voltage, so either the voltage sensor is stuck or the battery is disabled.

of an individual variable X, as it can easily be identified from its marginal probabilities $Pr(X|\mathbf{e})$. Again, MAP explanations are in general more difficult to compute than marginals [15]. On the other hand, maximizing two variables independently may not lead to a good explanation. As we already mentioned, a Bayesian network may find that two components are likely to be faulty, yet it may still be unlikely that both are faulty at the same time.

2.5.3 Sensitivity Analysis

The question we ask now: can we quantify the degree to which we are confident in the conclusions made by a Bayesian network? Suppose, for example, that we are in the position to make a decision based on a probability computed using a Bayesian network. For example, we may decide to take one action if the probability $Pr(P_C = \text{hi}|\mathbf{e})$ were greater than some threshold T and another action if this probability were less than T. Can we quantify the robustness of such a decision under perturbations in the Bayesian network model?

In this case, we can appeal to sensitivity analysis where we are interested in the sensitivity of probabilistic queries with respect to changes in the network parameters of a Bayesian network. Suppose, for example, that we want to perform an action that depends on whether current is flowing in our circuit, and that we make a decision to do so if we believe that current is flowing with probability greater than 50%. In our running example, this probability was $Pr(P_C = \text{hi}|\mathbf{e}) = 98.51\%$, which significantly surpasses our threshold. However, would a decision we make here be robust under changes in the network parameters? From a different perspective, we can ask how much do we need to change our network parameters so that $Pr(P_C = \text{hi}|\mathbf{e}) < 50.0\%$, contradicting our current decision.

These types of queries are supported by Bayesian networks [3,16], and in this case, one finds that significant changes in the model would be called for to contradict our current decision. For example, the parameter $\theta_{S_V=\text{stuck-lo}}$, which is the prior probability that the voltage sensor is stuck on a low reading, would have to be changed from 0.5% to less than 0.0051%. This corresponds to a sensor that is around two orders of magnitude more reliable, which is a fairly extreme assumption. We find that a similar change is required in the parameter $\theta_{S_F=\text{stuck-open}}$, the prior probability that the feedback sensor is stuck open. We could also change, for example, the parameter $\theta_{S_C=\text{healthy}}$, the prior probability that the current sensor is healthy. In this case, we need to change the value from 99.0%

to less than 33.78% to contradiction our decision, which is also quite an extreme change.

We can also perform sensitivity analysis on the MPE and ask what parameter changes are required to change the likeliest scenario [17]. The MPE in our running example also identifies two faults: that the voltage sensor is stuck at a low reading and that the feedback sensor is stuck open. Again, we find that to flip the most likely scenario calls for extreme changes to our parameters. For example, we would need to change the value of the parameter $\theta_{S_C=\text{healthy}}$ from 99.0% to less than 33.33%.

2.6 REASONING ALGORITHMS

There are several classes of algorithms for reasoning in Bayesian networks, computing marginal probabilities, and finding explanations. Each algorithm has its own advantages, which we shall highlight here.

Elimination algorithms include variable elimination [18,19] and jointree algorithms [20,21]. These are among the most common algorithms implemented in practice. These types of algorithms take advantage of the global structure of a Bayesian network to perform inference efficiently. In particular, these types of algorithms are exponential only in the *treewidth* of a Bayesian network, which is a graph theoretic property that measures, intuitively, the degree to which a graph resembles a tree [2]. Inference in a tree-structured Bayesian network, for example, can be performed efficiently in time that is only linear in the diameter of the network [1].

Conditioning algorithms include cutset conditioning [1] and recursive conditioning [22]. Intuitively, these are approaches based on decomposing the inference task into simpler cases, where each case can be solved easily, and then aggregating the results. These algorithms can have attractive computational properties compared with elimination-based algorithms. Recursive conditioning, for example, allows a natural anyspace-inference framework, where one is allowed to trade space with time [22].

Compilation algorithms are generally based on compiling a Bayesian network into arithmetic circuits [23,24]. The aim of these algorithms is to push most of the computational overhead into an offline phase, while producing secondary structures (arithmetic circuits) that can be processed quite efficiently in the online inference phase. Compilation-based algorithms can be based on elimination or conditioning techniques and seek to take advantage of the global structure of a network, in addition to the local structure that may be present in the network CPTs. Exploiting local structure can be quite time consuming in general, but since this is done only offline, the overhead is not incurred during the critical online phase.

All of the inference approaches discussed above lead to exact results and are therefore known as exact inference methods. In the case exact inference is not tractable, we can appeal to approximation algorithms, which have been influential in their own right in numerous applications.

"Loopy" belief propagation and other message-passing algorithms inspired by it have in the past decade been particularly successful and have enabled new and influential applications in fields as diverse as information theory [25], computer vision, satisfiability, and more broadly in artificial intelligence [1,26]. More recent generalizations of belief propagation also provide a means by which computational accuracy can be traded for computational resources. For example, the perspective given by [27] formulates belief propagation as a way to compensate for structural relaxations, which yields a spectrum of approximation. On one end, with a coarse, fully disconnected approximation, we have loopy belief propagation. On the other end, with the original unrelaxed model, we have exact inference. Given constraints on the computational resources available, we can then to try to identify an approximation along this spectrum that is as accurate as possible.

Stochastic sampling and search algorithms, such as importance sampling, Gibbs sampling, and stochastic local search, have the attractive property that they tend to give increasingly accurate estimates as they are given more time to run and can also work well on Bayesian networks with high treewidth. Stochastic sampling algorithms are typically used to compute marginals [28,29], while stochastic local search algorithms are more successful when computing MAP and MPE [15,30,31]. The disadvantage of these algorithms is that an impractical number of samples may be needed to produce accurate estimates, and convergence may be hard to diagnose.

Variational algorithms have also been successful, particularly in applications such as information retrieval and extraction, as well as text analysis [32]. Variational approaches reduce the inference problem to an optimization task. For example, we may specify an approximation with a tractable structure and optimize its parameterization so as to minimize the distance between the original and approximate distributions (typically the Kullback-Leibler divergence). Another attractive property of variational approaches is that they tend also to yield bounds on quantities of interest [33].

2.7 LEARNING BAYESIAN NETWORKS

As we have discussed before, Bayesian networks can be constructed by human domain experts or they can be synthesized automatically from design. In this section, we highlight how Bayesian networks can also be learned automatically from data.

Consider again our voltage sensor model from Figure 2.2 and the following data set:

Case	P_V	H_V	S_V
1	hi	healthy	read-hi
2	lo	healthy	read-lo
3	hi	healthy	read-hi
4	hi	stuck-hi	read-hi
5	lo	?	read-hi
6	?	healthy	read-lo
⋮	⋮	⋮	⋮

which consists of a number of cases where we have observed the state of a voltage sensor, which consists of three values: the presence of voltage P_V, the health of the sensor H_V, and the sensor reading S_V. In this example, we see that in some of these cases, particular values may be missing, which are marked with a "?"*

One can use such a data set to learn the network parameters given its structure or learn both the structure and its parameters. Learning parameters only is an easier task computationally. Moreover, learning either structure or parameters always becomes easier when the data set is complete—that is, the value of each variable is known in each data record.

Since learning is an inductive process, one needs a principle of induction to guide the learning process. The two main principles for this purpose lead to the maximum likelihood and Bayesian approaches to learning. The maximum likelihood approach favors Bayesian networks that maximize the probability of observing the given data set. The Bayesian approach on the other hand uses the likelihood principle in addition to some prior information that encodes preferences on Bayesian networks.

We remark here that the term "Bayesian network" does not necessarily imply a commitment to the Bayesian approach for learning networks. This term was coined by Judea Pearl [34] to emphasize three aspects: the

* Typically, the health of a sensor is unobservable. However, we can learn Bayesian networks with hidden or latent variables, where a variable's value will always be missing in data.

often subjective nature of the information used in constructing these networks; the reliance on Bayes conditioning when reasoning with Bayesian networks; and the ability to perform causal as well as evidential reasoning on these networks, which is a distinction originally underscored by Thomas Bayes.

The above learning approaches are meant to induce Bayesian networks that are meaningful, independently of the tasks for which they are intended. Consider, for example, a network that models a set of diseases and a corresponding set of symptoms. This network may be used to perform diagnostic tasks, by inferring the most likely fault given a set of observed sensor readings. It may also be used for prediction tasks, where we infer the most likely sensor readings given some faults. If we concern ourselves with only one of these tasks, say diagnostics, we can use a more specialized induction principle that optimizes the diagnostic performance of the learned network. In machine learning jargon, we say that we are learning a *discriminative* model in this case, as it is often used to discriminate among failures according to a predefined set of classes (e.g., stuck or healthy). This is to be contrasted with learning a generative model, which is to be evaluated based on its ability to generate the given data set, regardless of how it performs on any particular task.

2.7.1 Learning a Bayesian Network's Parameters

Suppose we are given a data set $\mathcal{D}$ consisting of N cases $\mathcal{D}_i$ and that we have the graph structure G of a Bayesian network whose parameters we want to learn from data set $\mathcal{D}$ (the graph structure may have, for example, been given to us by an expert, or synthesized from design).

Let θ denote a parameterization of this Bayesian network and let Pr_θ denote the probability distribution induced by graph structure G and parameters θ. Under the maximum likelihood approach to learning, the maximum likelihood parameters θ^* that we seek are those that maximize the likelihood of the data:

$$\theta^* = \underset{\theta}{\operatorname{argmax}} \prod_{i=1}^{N} Pr_\theta(\mathcal{D}_i),$$

assuming that the cases in our data $\mathcal{D}_i$ are independent and identically distributed. In the case our data is complete, computing the maximum likelihood parameters is a simple task. In the case our data is not complete and has missing values, the computation is not as straightforward, but there are a number of effective algorithms that we can employ in this case.

2.7.2 Learning with Complete Data

In the case where our data set $\mathcal{D}$ is complete, i.e., it has no missing values, then the maximum likelihood parameter estimates have a simple closed form, that is based on the "empirical" probability distribution that the data set itself induces.

In the case of Bayesian learning, instead of seeking point estimates θ^*, we seek a posterior distribution (density) over network parameterizations $\rho(\theta|\mathcal{D})$, conditioned on the data. This posterior can in turn be used to identify point estimates (such as the mean or the mode), or one can otherwise average over all possible parameterizations. We remark that in the complete data case, we again have a simple closed form, when a suitable prior over parameters is assumed.

2.7.3 Learning with Incomplete Data

In the case our data set $\mathcal{D}$ is incomplete and has missing values, the parameter learning task is more difficult since, in general, there is no tractably computable closed form. However, there are still effective learning algorithms in this case, such as gradient ascent, Gibbs sampling, and expectation-maximization (EM). Such algorithms are often iterative, and intuitively, make inferences about the data to complete the missing values, which in turn are used to update the model parameters.

The EM algorithm, for example, is an efficient iterative procedure that searches for maximum likelihood estimates, in the presence of missing data. At a high level, each iteration of the EM algorithm consists of two steps: the E-step and the M-step. We start with some initial guess of the network parameters. In the expectation, or E-step, we compute expected counts, based on our current estimates. In the M-step, we treat the expected counts like complete data and compute the maximum likelihood estimates from them. We typically repeat this iterative process until it no longer improves the likelihood.

For more on learning Bayesian networks from data, including how to learn the structure of a Bayesian network, we recommend [2,3].

2.8 APPLICATIONS AND SCALABILITY EXPERIMENTS

We now discuss recent, large-scale applications of Bayesian networks and arithmetic circuits in the area of electrical power systems. These studies demonstrate our successful probabilistic approach to diagnosis of an electrical power system (EPS) known as the Advanced Diagnostics and Prognostics Testbed (ADAPT) [12]. ADAPT facilitates the benchmarking of different technologies, and we have developed of a probabilistic approach, called ProDiagnose, to model-based diagnosis and applied it in this setting [10,11,14,35–39]. In the industrial tracks of two international diagnostic competitions, which were based on ADAPT, ProDiagnose achieved the best scores in three of the four industrial tracks. In this section, we discuss ProDiagnose and related work, illustrating how many of the techniques discussed earlier scale up in a real-world application.

2.8.1 Electrical Power Systems

Our society relies on electrical power in many ways. Obviously, there is the electrical power grid, which currently is undergoing a major upgrading effort in the context of creating a smart grid. In addition, electrical power plays a key role in vehicles, including in cars and aerospace vehicles. Spacecraft and aircraft, for example, contain multiple subsystems including navigation systems, power systems, and propulsion systems, and it is crucial to keep them all operational, even under adverse conditions. A real-world (hardware) EPS that is representative of EPSs in aerospace vehicles, namely ADAPT, has been developed and is located at the NASA Ames Research Center [12]. ADAPT includes capabilities for power storage, distribution, and consumption, and serves as a platform for development and proof-of-concept demonstrations of systems health management concepts in a realistic environment. ADAPT also provides a controlled environment in which to inject failures, either through software or hardware, as well as for benchmarking of diagnostic and prognostics software in a repeatable manner.

For experimentation using real-world data, EPS scenarios were generated using ADAPT [12]. These scenarios cover component, sensor, or both component and sensor faults. A scenario is nominal or contains one, two, or three faults, and faults may be discrete or (parametric) continuous. In 2009 and 2010, international diagnostics competitions DXC-09 and DXC-10 were arranged with ADAPT as the real-world testbed (see https://c3.ndc.nasa.gov/dashlink/projects/36/ and https://www.phmsociety.org/competition/dxc/10 for information and data sets). In DXC-09, all inserted faults were abrupt and persisted until the end of the scenario. DXC-10, on the other hand, also included drift (nonabrupt) and intermittent (nonpersistent) faults. In the following, we discuss our development of BNs for ADAPT as well as the results of experiments with these BNs on nominal and fault scenarios.

2.8.2 Bayesian Network Models

Based on the ADAPT EPS, several Bayesian networks have been developed, compiled into arithmetic circuits, and used in diagnosis experiments. While several variants of this Bayesian network exist [10,11,14,36,39], we here focus mainly on the one that had the best performance in the industrial track of the diagnostic competition DXC-09. This Bayesian network has 671 discrete nodes and 789 edges; domain cardinalities range from 2 to 16, with an average of 2.86 [11,36]. An ADAPT Bayesian network variant from 2008 contains 434 nodes and 482 edges, with node cardinalities ranging from 2 to 4, with a mean of 2.27 [10,14].

Before we briefly discuss experimental results, we would like to highlight that this approach satisfies two important requirements that often arise in systems health management, mentioned in Section 2.2, namely the modeling and real-time performance requirements [10,14,35]. To address the modeling requirement, we have developed a systematic way of representing electrical power systems as Bayesian networks, supported by an easy-to-use specification language and an autogeneration algorithm that generates Bayesian networks. The specification language and autogeneration algorithm are based on the principles laid out in Sections 2.3 through 2.5. To address the real-time requirement, Bayesian networks are compiled into arithmetic circuits. We used the ACE system to compile an ADAPT Bayesian network into an arithmetic circuit and to evaluate that arithmetic circuit [23,24] (see http://reasoning.cs.ucla.edu/ace/). Arithmetic circuits support real-time diagnosis by providing faster and more predictable inference trials. In addition, it is an exact reasoning system, with deterministic behavior.

We now briefly discuss experimental results that illustrate how our approach scales to real-world systems such as ADAPT, as well as how it compares to other approaches.

2.8.3 Experiments with Real-World Data

Using ADAPT data sets (see Section 2.8.1), experiments were used to benchmark the diagnostic quality of the ADAPT Bayesian networks and the ProDiagnose algorithm (see Section 2.8.2). Since our probabilistic models do not contain continuous random variables, experiments with continuous faults are especially challenging, and one key contribution of this work was how to handle continuous faults. Continuous data, including continuous sensor readings, were discretized before being used for clamping the appropriate discrete random variables in the ADAPT Bayesian network.

In early experiments, using 16 fault scenarios, each with one to three faults, a correct diagnosis was computed in a majority of the scenarios [10]. In fact, there was an exact match between our estimated diagnosis and the actual fault(s) inserted in 10 of the 16 scenarios, giving an accuracy of 62.5%. Even in cases where there was not exact agreement, our diagnosis was either partly correct or at the very least reasonable.

Later experiments were conducted as part of diagnostics competition DXC-09.* ProADAPT, an adaptation of ProDiagnose to the ADAPT setting of DXC-09's industrial track, achieved the best performance [36] in both the Tier 1 and Tier 2 tracks of DXC-09. ProADAPT had the highest-ranking scores in both the Tier 1 (among nine international competitors) and Tier 2 (among six international competitors) categories. On the Tier 1 data set, ProADAPT's score was 72.8, while the second-best competitor's score was 59.85. On the more difficult Tier 2 data set, ProADAPT's score was 83.20, while the second-best competitor's score was 81.50. ProADAPT's Bayesian network and other diagnostic parameters were later further improved [11]. In DXC-10, ProADAPT was further extended to include novel feature transformation algorithms—including the use of cumulative sum (CUSUM) techniques from statistics [39]—for handling offset faults, drift faults, and intermittent faults using discrete Bayesian networks. In DXC-10, ProADAPT was declared the winner in one of the two industrial tracks.

2.8.4 Experiments with Synthetic Data

In addition to experiments with real-world data from ADAPT, we have also performed experiments with synthetic data generated from ADAPT BNs [10,14]. The purpose of these experiments was to understand the performance of arithmetic circuit evaluation (using ACE) versus alternative Bayesian network reasoning algorithms, and variable elimination (VE) and jointree propagation (JTP) in particular. Synthetic data was created by a procedure that (i) generated a set of failure scenarios according to the probabilities of the ADAPT Bayesian network's health nodes and

* See https://dashlink.arc.nasa.gov/topic/diagnostic-challenge-competition/ for information about the competition.

(ii) generated evidence by performing stochastic simulation for each failure scenario. These evidence sets were then used as evidence in the three different inference systems. Both MPEs and marginals were computed for 200 simulated evidence sets generated from an ADAPT Bayesian network.

The following timing measurements are for a PC with an Intel 4 1.83-GHz processor with 1-GB RAM and running Windows XP. Considering means only, MPE computation took on average 17.79 ms for VE and 0.2370 ms for ACE, with standard deviations of 1.513 and 0.2137, respectively. Computation of posterior marginals, on the other hand, required on average 10.02 ms for JTP and 0.6981 ms for ACE, with standard deviations of 4.451 and 0.6669, respectively [10].

In summary, we have observed strong performance for our probabilistic approach in these synthetic ADAPT experiments, and would like to make two main points. First, all three approaches are relatively fast when using this ADAPT BN, with average computation time always less than 18 ms. Second, both marginals and MPEs were computed in less than 1 ms by ACE, while the competing approaches (JTP and VE) needed 10 ms or more.

2.8.5 Discussion

Empirically, we have shown that the ProDiagnose diagnostic algorithm combined with arithmetic circuits compiled from Bayesian networks provides high-quality diagnostic results for a real-world EPS called ADAPT. We believe that these ADAPT case studies clearly demonstrate the power of our Bayesian network modeling principles and how arithmetic circuits offer a scalable inference technique with potential for real-time evaluation in aircraft and spacecraft, thereby enabling substantial improvements in aviation safety in the future. This technology has potential impact on systems health management beyond EPSs in aerospace. While we discussed EPS diagnosis above, it is clear that our approach supports real-time, online diagnosis and prognosis for many types of large-scale systems more generally. Consequently, these empirical results should be of interest to the broader systems health management community interested in fast and exact computation with predictable inference times.

2.9 CONCLUSION

Over the past few decades, the Bayesian network research community has developed powerful tools and algorithms for modeling, reasoning, and learning complex systems. These tools and algorithms have been leveraged to solve challenging problems in fields as diverse as artificial intelligence and machine learning, information theory [25], computer vision, computational biology, text analysis and information extraction [32,40], and in diagnosis, prognosis, and system reliability [41]. The ADAPT electrical power system, which we highlighted in this chapter, is one of many examples where Bayesian networks have been applied to real-world diagnostic problems.

In this chapter, we have sought to illustrate that Bayesian networks provide a simple and natural language for modeling problems in systems health management. Moreover, as we highlighted, they support a variety of probabilistic queries, which provide a means to qualitatively and quantitatively reason about systems health and reliability. Further, there are a variety of effective and principled approaches to inducing Bayesian networks from data, with or without prior expert knowledge.

There are several popular software systems for modeling, and reasoning with Bayesian networks that are freely available. In this chapter, we employed UCLA's SamIam system for performing the variety of probabilistic queries we considered.* ACE, also developed at UCLA, is an advanced system for reasoning in Bayesian networks that was successfully used in the ADAPT system described in the previous section.† Other popular alternatives include Kevin Murphy's Bayesian Network Toolbox for use in MATLAB® computing environments‡ and the University of Pittsburgh's GeNIe & SMILE system, which includes specialized features for Bayesian networks used in diagnostic applications.§ There are also a variety of commercial systems for modeling and reasoning in Bayesian networks.

There are several excellent textbooks available covering Bayesian networks. A good undergraduate-level introduction is provided in Russel and Norvig's popular AI textbook [42], now in its third edition. We recommend Darwiche's recent book [2], which provides an accessible introduction to modeling with Bayesian networks, as well as a thorough treatment of both classical and advanced algorithms for reasoning and learning. Another recent book by Koller and Friedman [3] provides a thorough and comprehensive treatment on probabilistic graphical models (including Bayesian networks), which we recommend for readers with a background in machine learning. We also suggest Judea Pearl's seminal book on Bayesian networks [1].

* Available for download at http://reasoning.cs.ucla.edu/samiam/.

† Available for download at http://reasoning.cs.ucla.edu/ace/.

‡ Available for download at http://code.google.com/p/bnt/.

§ Available for download at http://genie.sis.pitt.edu/.

REFERENCES

[1] J. Pearl. *Probabilistic Reasoning in Intelligent Systems: Networks of Plausible Inference*. Morgan Kaufmann, San Mateo, CA, 1988.

[2] A. Darwiche. *Modeling and Reasoning with Bayesian Networks*. Cambridge University Press, Cambridge, UK, 2009.

[3] D. Koller, Nir Friedman. *Probabilistic Graphical Models: Principles and Techniques*. MIT Press, 2009.

[4] S. Andreassen, M. Woldbye, B. Falck, S.K. Andersen. MUNIN—A causal probabilistic network for interpretation of electromyographic findings. In *Proceedings of the Tenth International Joint Conference on Artificial Intelligence*, pp. 366–372, August 1987.

[5] M.A. Shwe, B. Middleton, D.E. Heckerman, M. Henrion, E.J. Horvitz, H.P. Lehmann, G.F. Cooper. Probabilistic diagnosis using a reformulation of the INTERNIST-1/QMR knowledge base: I. The probabilistic model and inference algorithms. *Methods of Information in Medicine*, 30(4):241–255, 1991.

[6] T.S. Jaakkola, M.I. Jordan. Variational probabilistic inference and the QMR-DT database. *Journal of Artificial Intelligence Research*, 10:291–322, 1999.

[7] U. Lerner, R. Parr, D. Koller, G. Biswas. Bayesian fault detection and diagnosis in dynamic systems. In *Proceedings of the Seventeenth national Conference on Artificial Intelligence (AAAI-00)*, pp. 531–537, 2000.

[8] I. Rish, M. Brodie, S. Ma. Accuracy vs. efficiency trade-offs in probabilistic diagnosis. In *Eighteenth National Conference on Artificial intelligence (AAAI-02)*, pp. 560–566, Edmonton, Canada, 2002.

[9] C. Romessis, K. Mathioudakis. Bayesian network approach for gas path fault diagnosis. *Journal of Engineering for Gas Turbines and Power*, 128(1):64–72, 2006.

[10] O.J. Mengshoel, A. Darwiche, K. Cascio, M. Chavira, S. Poll, S. Uckun. Diagnosing faults in electrical power systems of spacecraft and aircraft. In *Proceedings of the Twentieth Innovative Applications of Artificial Intelligence Conference (IAAI-08)*, pp. 1699–1705, Chicago, IL, 2008.

[11] B.W. Ricks, O.J. Mengshoel. Methods for probabilistic fault diagnosis: An electrical power system case study. In *Proc. of Annual Conference of the PHM Society. 2009 (PHM-09)*, San Diego, CA, 2009.

[12] S. Poll, A. Patterson-Hine, J. Camisa, D. Garcia, D. Hall, C. Lee, O.J. Mengshoel, C. Neukom, D. Nishikawa, J. Ossenfort, A. Sweet, S. Yentus, I. Roychoudhury, M. Daigle, G. Biswas, X. Koutsoukos. Advanced diagnostics and prognostics testbed. In *Proceedings of the 18th International Workshop on Principles of Diagnosis (DX-07)*, pp. 178–185, Nashville, TN, 2007.

[13] D. Musliner, J. Hendler, A.K. Agrawala, E. Durfee, J.K. Strosnider, C.J. Paul. The challenges of real-time AI. *IEEE Computer*, 28:58–66, January 1995.

[14] O.J. Mengshoel, M. Chavira, K. Cascio, S. Poll, A. Darwiche, S. Uckun. Probabilistic model-based diagnosis: An electrical power system case study. *IEEE Transactions on Systems, Man, and Cybernetics*, 40(5):874–885, 2010.

[15] J.D. Park, A. Darwiche. Complexity results and approximation strategies for MAP explanations. *Journal of Artificial Intelligence Research (JAIR)*, 21: 101–133, 2004.
[16] H. Chan, A. Darwiche. When do numbers really matter? *Journal of Artificial Intelligence Research*, 17:265–287, 2002.
[17] H. Chan, A. Darwiche. On the robustness of most probable explanations. In *Proceedings of the 22nd Conference on Uncertainty in Artificial Intelligence (UAI)*, pp. 63–71, AUAI Press, Arlington, Virginia, 2006.
[18] N. L. Zhang, D. Poole. Exploiting causal independence in Bayesian network inference. *Journal of Artificial Intelligence Research*, 5:301–328, 1996.
[19] R. Dechter. Bucket elimination: A unifying framework for probabilistic inference. In *Proceedings of the 12th Conference on Uncertainty in Artificial Intelligence (UAI)*, pp. 211–219, 1996.
[20] F.V. Jensen, S.L. Lauritzen, K.G. Olesen. Bayesian updating in recursive graphical models by local computation. *Computational Statistics Quarterly*, 4:269–282, 1990.
[21] S.L. Lauritzen, D.J. Spiegelhalter. Local computations with probabilities on graphical structures and their application to expert systems. *Journal of Royal Statistics Society, Series B*, 50(2):157–224, 1988.
[22] A. Darwiche. Recursive conditioning. *Artificial Intelligence*, 126(1–2):5–41, 2001.
[23] M. Chavira, A. Darwiche. Compiling Bayesian networks with local structure. In *Proceedings of the 19th International Joint Conference on Artificial Intelligence (IJCAI)*, pp. 1306–1312, 2005.
[24] A. Darwiche. A differential approach to inference in bayesian networks. *Journal of the ACM*, 50(3):280–305, 2003.
[25] B.J. Frey, D.J.C. MacKay. A revolution: Belief propagation in graphs with cycles. In *NIPS*, pp. 479–485, 1997.
[26] J.S. Yedidia, W.T. Freeman, and Y. Weiss. Understanding belief propagation and its generalizations. In G. Lakemeyer, B. Nebel, editors, *Exploring Artificial Intelligence in the New Millennium*, chapter 8, pp. 239–269. Morgan Kaufmann, 2003.
[27] A. Choi, A. Darwiche. An edge deletion semantics for belief propagation and its practical impact on approximation quality. In *Proceedings of the National Conference on Artificial Intelligence (AAAI)*, 2006.
[28] M. Henrion. Propagating uncertainty in Bayesian networks by probabilistic logic sampling. In *Uncertainty in Artificial Intelligence 2*, pp. 149–163. Elsevier, Amsterdam, 1988.
[29] R. Shachter, M. Peot. Simulation approaches to general probabilistic inference on belief networks. In *Uncertainty in Artificial Intelligence 5*, pp. 221–231, Amsterdam, 1990. Elsevier.
[30] O.J. Mengshoel. Understanding the role of noise in stochastic local search: Analysis and experiments. *Artificial Intelligence*, 172(8–9):955–990, 2008.
[31] O.J. Mengshoel, D. Roth, D.C. Wilkins. Portfolios in stochastic local search: Efficiently computing most probable explanations in Bayesian networks. *Journal of Automated Reasoning*, 46(2):103–160, 2011.

[32] D.M. Blei, J.D. Lafferty. Topic models. In A. Srivastava, M. Sahami, editors, *Text Mining: Classification, Clustering, and Applications*, chapter 4, pp. 71–93. Chapman & Hall/CRC, 2009.

[33] T. Jaakkola. Tutorial on variational approximation methods. In D. Saad, M. Opper, editors, *Advanced Mean Field Methods*, chapter 10, pp. 129–160. MIT Press, 2001.

[34] J. Pearl. Bayesian Networks: A model of self-activated memory for evidential reasoning. In *Proceedings of the Cognitive Science Society*, pp. 329–334, 1985.

[35] O.J. Mengshoel. Designing resource-bounded reasoners using Bayesian networks: System health monitoring and diagnosis. In *Proceedings of the 18th International Workshop on Principles of Diagnosis (DX-07)*, pp. 330–337, Nashville, TN, 2007.

[36] B.W. Ricks, O.J. Mengshoel. The diagnostic challenge competition: Probabilistic techniques for fault diagnosis in electrical power systems. In *Proc. of the 20th International Workshop on Principles of Diagnosis (DX-09)*, Stockholm, Sweden, 2009.

[37] O.J. Mengshoel, S. Poll, T. Kurtoglu. Developing large-scale Bayesian networks by composition: Fault diagnosis of electrical power systems in aircraft and spacecraft. In *Proc. of the IJCAI-09 Workshop on Self-* and Autonomous Systems (SAS): Reasoning and Integration Challenges*, 2009.

[38] W.B. Knox, O.J. Mengshoel. Diagnosis and reconfiguration using Bayesian networks: An electrical power system case study. In *Proc. of the IJCAI-09 Workshop on Self-* and Autonomous Systems (SAS): Reasoning and Integration Challenges*, 2009.

[39] B.W. Ricks, O.J. Mengshoel. Diagnosing intermittent and persistent faults using static Bayesian networks. In *Proc. of the 21st International Workshop on Principles of Diagnosis (DX-10)*, Portland, OR, 2010.

[40] A. McCallum. Information extraction: Distilling structured data from unstructured text. *ACM Queue*, 3(9):48–57, 2005.

[41] H. Langseth, L. Portinale. Bayesian networks in reliability. *Reliability Engineering and System Safety*, 92(1):92–108, 2007.

[42] S. Russell, P. Norvig. *Artificial Intelligence: A Modern Approach*. Prentice Hall, 3rd edition, 2009.

CHAPTER 3

Anomaly Detection in a Fleet of Systems

Nikunj Oza

Santanu Das

CONTENTS

"Most users of complex statistical procedures have problems which can and should be handled with simple techniques," said Andrews, 1981 (Chatfield, 1985).

3.1 WHAT IS A FLEET?

A fleet is a group of systems (e.g., cars, aircraft) that are designed and manufactured the same way and are intended to be used the same way. For example, a fleet of delivery trucks may consist of 100 instances of a particular model of truck, each of which is intended for the same type of service—almost the same amount of time and distance driven every day, approximately the same total weight carried, etc. For this reason, one may imagine that data mining for fleet monitoring may merely involve collecting operating data from the multiple systems in the fleet and developing some sort of model, such as a model of normal operation that can be used for anomaly detection. However, one then may realize that each member of the fleet will be unique in some ways—there will be minor variations in manufacturing, quality of parts, and usage. For this reason, the typical machine learning and statistics algorithm's assumption that all the data are independent and identically distributed is not correct. One may realize that data from each system in the fleet must be treated as unique so that one can notice significant changes in the operation of that system.

The reality of data mining for fleet monitoring lies between these two extremes. Even though each system must be monitored on its own so that one can observe significant changes in its unique operations, data from all systems in the fleet are helpful in determining how unique each system really is relative to the remaining systems in the fleet and when fleet personnel should expect to perform various maintenance actions. One can begin with the assumption that each system in the fleet is comparable to a sample drawn from some distribution, so that all the systems in the fleet are independent and identically distributed. However, one can then adjust the methods being used to account for changes related to any information that is known. For example, if one knows which parts were made at different times, one can make adjustments to the model to account for differences in the qualities of those parts. That way, to evaluate subsystems related to that part, we can choose to use only the other systems in the fleet that have the same part to extract information on maintenance needs and remaining useful life.

In the next section, we will describe some past work in fleet monitoring. We will describe the drawbacks of this past work that motivated the work done recently in this area. We will also discuss what we see as the role of data mining in fleet health monitoring. In the following section, we will describe the key issues of heterogeneity and efficiency issues (large data volume and distributed data) that serve as challenges to effective fleet monitoring. The following two sections will give overviews of methods designed to handle heterogeneity and efficiency issues. In each section, we will give one example algorithm and its use in anomaly detection in more detail to give an idea of how these algorithms work. We will end with a summary of this chapter.

3.2 BACKGROUND

In this section, we describe some past work in the area of fleet health monitoring and then discuss the role of data mining within this area. This will serve as background material for the following sections which will provide more details.

3.2.1 Past Work in Fleet Health Monitoring

In the next two subsections, we briefly describe several systems that have been used to monitor fleets by analyzing numeric data and some other attempts at extracting useful information from text data, respectively.

3.2.1.1 Numeric Data

In this section, we briefly describe six systems for fleetwide health management. Two systems monitor each individual unit of the fleet separately and in an online manner. Information about how other units are currently behaving is not utilized in the monitoring of any unit. Three other systems we discuss collect data from all units in one place and analyze that data to monitor the fleet as a whole and units behaving abnormally are identified. Unlike the first two systems, baselines are not developed for each individual unit or for any subpopulations. The last system attempts to create baselines for the global population and for several subpopulations.

MineFleet [1] is a system that monitors individual vehicles in a fleet by collecting and processing data on-board. It collects data that are produced by many sensors present in most modern vehicles. The sensor data streams are continuously monitored by an on-board computing device for any new emerging patterns. If necessary, these patterns are sent over a low-bandwidth wireless network to a central location for possible further analysis. The system provides the ability to calculate numerous statistical aggregates such as correlations and distance matrices efficiently on the vehicle itself on embedded computing platforms such as PDAs, with the idea that these statistical aggregates form the basis of algorithms to detect anomalies in a vehicle's operation. The system also monitors changes in these statistical aggregates in real time and in an efficient manner. This system is intended to monitor changes in the way the vehicle is operating and is being operated in a setting where the available computing power and communication bandwidth are low. Because of this, each vehicle is only monitored relative to its own baseline—data and statistics from the various vehicles are not communicated to one another automatically. Any knowledge gained from analysis of the entire fleet's data would have to be manually added to each vehicle to enable it to assess itself relative to a global baseline during operation, and that global baseline is not updated in real time. The system allows data from vehicles to be uploaded manually to a central repository for offline analysis and generation of a global baseline and global knowledge to facilitate manual updating of vehicles' baseline definitions.

The Quantitative Condition Alerting and Analysis Support (QCAAS) system [2] is an on-board software system that uses sensors currently available on most commercial aircraft to monitor for several events of interest such as severe load encounters, landing gear speed exceedances, and hard landings. This system additionally maintains cumulative load statistics for the aircraft on which it is installed and indicates when the allowable level of cumulative load between inspections has been exceeded. The system additionally uses both onboard sensors and physics-based models to estimate loads in the tail of the aircraft, where current commercial aircraft are not instrumented to take direct measurements. QCAAS checks each aircraft independently relative to a global baseline; however, this global baseline is fixed. That is, the global baseline is not learned from data but is rather set by regulations. QCAAS does not currently have the capability to detect anomalies relative to an individual vehicle's normal behavior and is, therefore, unable to adapt to typical vehicle-to-vehicle variations.

The Aviation Performance Measuring System (APMS) [3] was a NASA program to move analysis of data collected during commercial flights, known as Flight Operational Quality Assurance (FOQA) data, beyond identification of single-parameter exceedances. The program identified three major goals—analyzing data beyond simply looking for exceedances of typical ranges of single parameters, focused analysis of higher-risk phases of flight, and looking for potential precursors to aviation safety incidents and accidents. As part of APMS, numerous tools were developed such as tools to search for patterns in particular flight parameters within multiple flights' data, document the distribution of key parameters related to standard operating procedures, report key descriptive statistics grouped by phase of flight, link flight data to corresponding weather data or air traffic data, and multivariate clustering to group flights based on flight signatures derived from parameter values to identify atypical flights. APMS was clearly oriented toward analyzing the data from multiple flights together and did not look at individual flights to assess them based on their own unique baseline. Additionally, APMS's tools analyze the flight data as static data, i.e., the data at each point in time is examined independently rather than as a sequence. The sequential nature of the data is only considered indirectly—by adding parameters such as slopes and flight signatures.

Google [4] collects data from all of its systems and saves them in a repository for offline analysis. Such data is clearly too large to store in a single place; therefore, this data is stored on multiple machines using a distributed file system and Google's well-known Mapreduce framework is used to allow large-scale data analysis programs to analyze that data in a distributed manner and collect the results. However, the distributed storage of the data is not necessarily related to the distributed nature of the Google system being monitored. Therefore, the data analysis on the collected data is necessarily an offline analysis. No attempt is made to analyze individual servers in the system to obtain a server-specific baseline.

The Morning Report (MR) [5] was designed for individual airlines to use to analyze their flight data in a manner much like APMS—that is, collect all the data together in one place and analyze them as if they came from multiple independent flights drawn from the same distribution. The subsequent System Level Morning Report (SLMR) attempted to address the problem we described earlier of balancing between analyzing all the data together the way MR does versus analyzing each flight as a unique entity and setting each flight's normal baseline separately. SLMR allowed users to analyze flight data in the context of individual airlines and the context of all the airlines. In particular, each airline participating in the system would analyze its data and send summary information on the typical patterns observed in their data to a central site—only summary information is sent in order to preserve each airline's privacy. The combined information is analyzed to find global patterns in various flight phases. Information on these global patterns is then sent back to each airline. Each airline then can analyze its data not only within the context of its own flights, but also within the context of the global patterns. This system was only designed for offline analysis; therefore, after each flight, data is uploaded to that airline's system, and the data is filtered, derived parameters are calculated, single-parameter exceedance checks are performed, flights are partitioned into phases, and each flight parameter's signature is calculated (capturing characteristics like rate of change and variability). These flight signatures are then clustered, enabling identification of flights with anomalous signatures. Linear discriminant analysis is then used to devise rules for each of these clusters, which are then used to "classify" new flights. The individual airlines' clusters are then sent to a central site and combined to form a global set of clusters that represent global flight patterns. SLMR has the notion of examining flights relative to the entire global set and an individual airline's flights, but does not examine each flight independently. Additionally, just like with APMS, the sequential nature of the data within a flight is only considered indirectly through the flight signatures, which only give a local indication of how parameters are changing—SLMR is not able to examine a full flight as a sequence. Additionally, the metric by which normality is measured is such that a flight can be normal or abnormal with respect to that airline's data and normal or abnormal with respect to the the global pattern—i.e., all four combinations are possible. This means that in order to assess whether a flight is globally normal, assessing its normality within one or even multiple airlines' data is not sufficient—it has to be examined with respect to all the airlines' data at once. For the sake of efficiency, we would like to have that only locally anomalous flights are globally anomalous. This way, local anomaly detection algorithms can identify relatively few candidates that can then be evaluated for anomalousness at the global level. We will later review two algorithms for which this guarantee has been mathematically proven.

3.2.1.2 Text

In this section, we describe issues in working with text. A more comprehensive survey of text classification and text mining in general can be found in [6]. In the area of text classification, there are two key decisions that have to be made: representation of the text and the classification method used. The documents in their raw form are not amenable to classification, so another representation is needed. The simplest document representation is the bag of words (BOW). In this representation, each document is represented by a vector consisting of as many elements as there are unique words in the document repository. Each vector element represents the number of times that a given word occurs in that document. The repository can be represented by a BOW matrix in which each document's vector is a row in the matrix. In machine learning terminology, each document is an example or instance and each word frequency is a feature. In principle, one could then apply one of many possible classification algorithms to document repositories after converting them to BOW matrices. One difficulty with this scheme is the weighting of words. Both rare words and very common words should be given low weight because both types of words are not useful for classification, where one generally wants features that come as close as possible to splitting the set of examples into equal-sized bins. However, very common words are given very high weight in BOW matrices due to their high word frequency. For this reason, weighting schemes such as term frequency inverse document frequency (TFIDF) are often used. TFIDF multiplies the term frequency by the inverse document frequency, which is related to the reciprocal of the fraction of documents in which the word appears. This reduces the weights of very common words. Term frequency and even TFIDF give excessive weight to long documents, so these are often normalized so that all documents' vectors have equal length. Stop words, which are very common words such as articles (e.g., *the*, *a*, *an*), are often eliminated from documents using a stopword list.

The bag of words and its weighted variants have a major weakness—they lose all semantic information because the order of words is lost. A given document is treated exactly the same way no matter how its words are reordered. Semantic information is clearly very important in human understanding of text. Natural language processing (NLP) methods attempt to maintain semantic information in documents while making the representation more amenable to machine learning. Examples of what NLP methods do include expanding acronyms and collapsing phrases so that different expressions with exactly the same meaning (e.g., FL vs. flight level) are reduced to one word or phrase, and setting nouns to their singular form and verbs to their infinitive conjugation. Others have attempted to use pairs or triples of words as features rather than individual words.

Full parse trees can be constructed in which the subject and object for each sentence are identified and sometimes broken down further into nouns, adjectives, prepositional phrases, etc. The chief drawback of such methods is that they are typically computationally very intensive. Also, the BOW representation does maintain information on the collocation of terms within a document and the frequency with which pairs, triples, and larger groups of words appear. This appears to often be sufficient for classification, because NLP methods have so far yielded too little improvement in classification performance (at least in aviation safety [7]) to justify their computation time.

Many of the same methods used for nontext classification, such as decision trees, random forests, and support vector machines (SVMs), are used for text [6,8]. Past experiments have shown that SVMs seem to be especially suitable because they are naturally suited to dealing with very large numbers of features, which typically exist in text classification problems [6].

Topic extraction is the act of examining the text repository and identifying the topics represented. Topics are typically in the form of vectors of words that tend to co-occur frequently. Examples of methods that do this include dirichlet allocation [9] and nonnegative matrix factorization (NMF) [10]. The representation issues for topic extraction algorithms are the same as for text classification algorithms. However, instead of necessarily relying on an existing document classification system, topic extraction algorithms operate in an unsupervised manner and identify topics based just on the documents themselves.

One example of a text repository is the Aviation Safety Reporting System (ASRS), which was started 30 years ago. It is a joint effort by the FAA and NASA [11]. The program was established to collect data and information about aviation events which could lead to unsafe situations, or nonstandard procedures, and use the data and information to identify deficiencies in the National Airspace (NAS) so appropriate solutions could be implemented. ASRS reports are publicly available and are written by pilots, flight controllers, technicians, flight attendants, and others including passengers. In general, the reports are filed in response to a specific event, but general concerns and complaints are also filed. ASRS reports include factual information about the aircraft, location, parties involved and a narrative in which the author describes the event(s) and/or situation. Since these narratives are specific to aviation they are filled with acronyms and abbreviations that only those with a working knowledge of the industry are likely to understand. Each report is read by at least two aviation experts who identify hazardous conditions and underlying causes of the reported events and classify the report into the appropriate categories. There are now over 700,000 documents in the ASRS data base with more than 3000 reports added each month. Many air carriers also have their own internal safety reporting system under the Aviation Safety Action Plan (ASAP) which may or may not be linked to the ASRS. Different airlines devised different categorizations for their documents, which necessitated the development of a master version of the categorization which all organizations could reference. The archive consisting of these documents recategorized using this master version is known as the Distributed National ASAP Archive (DNAA).

The ASRS documents and their classifications can be used by text classification algorithms to learn models that can be used to classify new documents. However, aviation safety experts do not completely trust the classification system because many problems are unique and do not fit well within the classification scheme. Therefore, topic extraction algorithms that either do not use the text classifications at all or only use them as a guide are of interest with the goal of extracting new topics that represent operationally significant problems that have been previously unknown and/or unstudied.

3.2.2 Role of Data Mining in Fleet Health Monitoring

Gathering data, transforming data into information, and discovering useful knowledge from the information have been identified as the three major steps in extracting valuable resources from the operations of all kinds of processes or systems. Organizations have recognized the importance of historical data that have been collected over decades and seek ways to utilize this knowledge in improving organizational effectiveness. For this, we turn to the field of data mining. By data mining we mean the art and science of analyzing a large collection of observations. More specifically, data mining is the process of sifting through data and extracting relevant knowledge about the system being described by the data. Such knowledge can then be used to make decisions relevant to the system, such as decisions involving appropriate utilization and maintenance of the system. Data mining is highly multidisciplinary in nature and uses sophisticated techniques from various areas like statistics, pattern recognition, information retrieval, machine learning, dynamic programming, high-performance computing, data visualization, etc. The whole purpose is to develop an automated process to manage information, discover previously unknown knowledge, predict trends and behaviors, comprehend the underlying relationships in large data and information sets, and finally provide a means to aid decision-making activities.

3.2.2.1 Anomaly Detection Problem

Anomaly detection is an example of data mining. Since the theme of this chapter is anomaly detection, we will mostly discuss data mining problems and applications that involve anomaly detection, which is also known as outlier detection or surprise pattern detection. Outlier or anomaly detection refers to the task of identifying new or unknown patterns which, in many cases, are abnormal or inconsistent relative to most of the data. In the past decade or so, researchers have found tremendous potential in applications of anomaly detection algorithms in various disciplines. Some of the thrust areas include medical applications, airspace safety, fraud and intrusion detection, and business analytics. The problem of outlier detection has been extensively studied using several approaches. The two broadest categories of anomaly detection problems are supervised and unsupervised. In the supervised approach a model is built from training data containing inputs for which the class labels are known. The resulting model can classify new data into one or more of the classes, which may correspond to normal or anomalous categories. Supervised anomaly detection is typically implemented using standard machine learning methods for classification, such as SVMs. However, class labels are often not easily available and, if they are available, they are typically expensive to obtain—especially for historical data representing systems which are no longer available in the precise configuration used to generate the data. Additionally, it is often impossible to know all possible classes of normal and anomalous operation and corresponding data. Unsupervised techniques do not require knowledge of normal and anomalous operation modes/classes. Unsupervised anomaly detection methods typically assume that all or most of the training data represent normal behavior and attempt to model this data. When new data arrives, the model is used to identify them as normal if they fit the model and anomalous otherwise. We encourage interested readers to go through [12–15] which have extensively reviewed and studied various methods for anomaly detection.

3.2.2.2 Some Preliminaries

Figure 3.1 gives an overview of various popular anomaly detection techniques and some of these techniques have been extensively discussed in

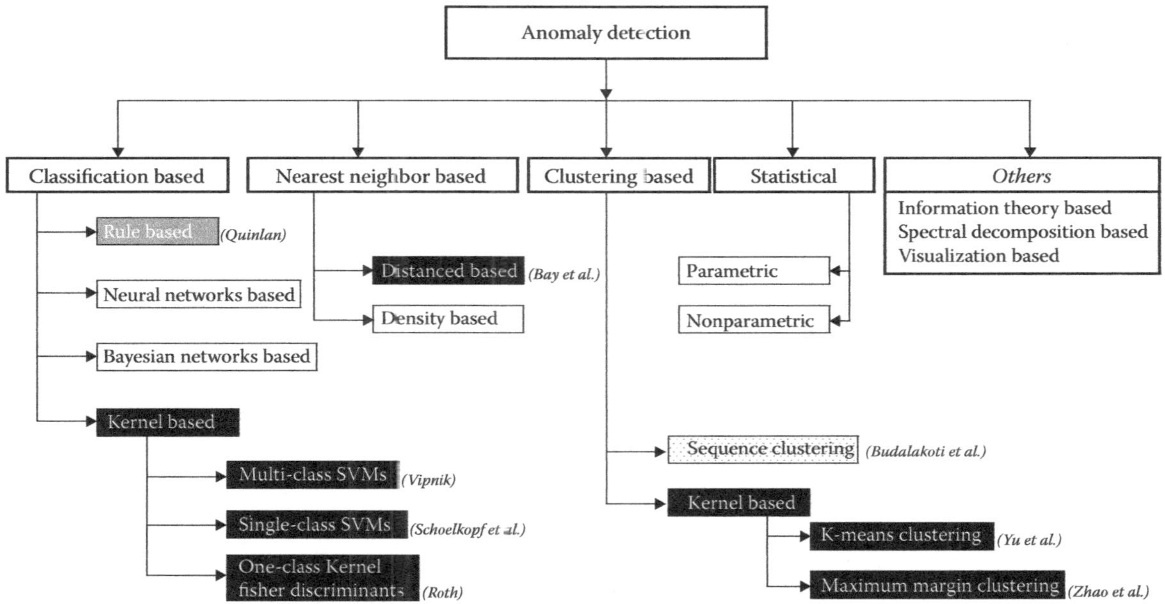

FIGURE 3.1 Systematic chart representing some popular approaches to anomaly detection and their subcategories.

a survey on outlier detection [15]. Typical classification based techniques such as Bayesian inference, decision trees, SVMs, and neural network models are built on previously labeled instances of both normal and abnormal data instances. In classification based outlier detection, the objective is to assign a new test object to one of the existing classes, i.e., label previously unseen data points either as a normal class or any of the abnormal classes. On the other hand, there are some kernel-based classification methods like one-class SVMs and one-class kernel Fisher discriminants, which fall under the anomaly detection category and are unsupervised in nature. In these detection techniques a solution is offered by modeling normal data and deriving a threshold for determining how far data can be from the model and still be considered normal. In the nearest neighbor based approach, the aim is to infer the outliers based on the data itself—for example, by finding those points which are at a greater distance from most of the other data points or by finding those points which are in low density regions. In most cases, k-nearest neighbor (k-NN)–based solutions have quadratic time complexity since a comparison between every two points is needed to find the nearest neighbors. Researchers such as Angiulli and Pizzuti [16], Angiulli and Fassetti [17], Ramaswamy et al. [18], and Bay and Schwabacher [19] have addressed this problem by introducing some promising techniques with improved run time complexities. We will elaborate on some of the above methods in the sections to follow.

In some of the popular clustering techniques such as the ones based on Gaussian mixture model (GMM) using EM, C-means fuzzy clustering, k-means (or median), and centroid linkage hierarchical approach, the basic approach is to partition the entire data into a number of clusters, where each data point will be assigned to each cluster with certain degree of association, also called "degree of membership." Here the anomaly detection problem is framed as a method of finding the odd member (data point) whose association to any of the existing clusters is very low—i.e., the sample does not belong to any of the existing clusters.

In statistical approaches, the models are based on the statistical properties of the training data themselves. In the testing phase the job boils down to estimating whether the sample under investigation could have been generated by the same distribution that generated the training data. There exist two main approaches, of which one makes some assumptions about the data distribution and the other derives the distribution along with the distribution parameters from the data. The former is known as the parametric approach, which assumes that the density function belongs to one of the standard and well-known distribution families, such as the normal distribution, binomial distribution, and Poisson distribution. On the other hand, the latter approach, which does not assume a particular form for the underlying distribution of the data in advance, is widely known as the nonparametric approach.

Most of the algorithms described above have been mostly used for data containing only continuous real-valued data attributes. However, many applications, such as aviation safety, have heterogeneous data sources (e.g., discrete, continuous, text), and therefore, there is a need to develop intelligent knowledge refinement and integration techniques that work with heterogeneous data sources. It is important to note that in order to enable knowledge discovery, algorithms in general require an integrated and merged view of the data available from various sources. Knowledge extraction from multiple heterogeneous data sources remains a challenge. Another challenging problem is to keep the methodologies flexible to the heterogeneous nature of the data sources so that problem reformulation is not required whenever there is a change in information content or data structure. In the rest of the chapter, we will describe different frameworks for data integration and knowledge discovery that can utilize heterogeneous data sources. Our notion here is to demonstrate how to extract and merge information contained in heterogeneous data sources and how the combined information can be effectively used to achieve the needs of the analysis. More recently, much research has been conducted to extend the scope of some of the state-of-the-art techniques to include information from continuous data, as well as knowledge created from other structured data sources, such as discrete data which may be binary, ordinal, categorical, or sequential. Here we will discuss a few of them. Some of these techniques can be even further extended for various unstructured data sources as well.

3.3 KEY ISSUES IN FLEET HEALTH MONITORING

In this section, we examine two challenges in many modern fleet data sets: heterogeneity and large data volume. We follow these in the next section with methods designed to address these challenges.

3.3.1 Heterogeneity

> "Heterogeneity is to be expected in a meta-analysis: it would be surprising if multiple studies, performed by different teams in different places with different methods, all ended up estimating the

same underlying parameter. From the standpoint that heterogeneity is inevitable in a meta-analysis, we are left with the question of whether there is an 'acceptable' degree of heterogeneity. My own view is that any amount of heterogeneity is acceptable, providing both that the predefined eligibility criteria for the meta-analysis are sound and that the data are correct. The challenge is then to decide on the most appropriate way to analyze heterogeneous studies, and this will depend on the aims of the synthesis. . . ."

JULIAN P. T. HIGGINS [20]

In the light of the above statement, it seems clear that we must expect the presence of varying degrees of heterogeneity across different problems. It is fairly well-accepted among practitioners that mixtures of observations described using multiple attributes (parameters, features, etc.) of diverse nature, measured with varying degrees of uncertainty results in heterogeneity in the data. Sometimes these heterogeneous data sets can reveal enriched information that can help in making critical decisions. These decisions can be in the form of detection, classification, prediction, forecasting, etc. The fact that this heterogeneous nature of the data influences the entire analysis and decision-making process is a central point of this chapter. Imagine that we are analyzing a large volume of heterogeneous data with unknown class labels and we are asked to identify those objects exhibiting strange behavior. One of the popular ways of doing this is using a data-driven anomaly detection model, which is a process of identifying abnormal or inconsistent patterns in a data set. Identifying strange events or entities can be extremely important in several applications including fraud and intrusion detection, financial market analysis, medical research, safety-critical vehicle health management, etc. These adverse events or entities are also known as anomalies or outliers which can be defined as data points that lie in low density regions or in areas far away from the majority of points. The choice of the appropriateness of the algorithm is dependent on the nature of the input features and the goal that needs to be achieved. In this chapter, we study the requirements of anomaly detection methods that take into account the heterogeneous nature of the data set to accurately find anomalies. This chapter summarizes studies on such anomaly detection techniques that have appeared in the technical literature during the past two decades. The primary purpose of this study is to bring together and compare a set of methods for anomaly detection on heterogeneous data sets. We start off by discussing various heterogeneous sources within data sets. We examine some of the existing state-of-the-art algorithms for anomaly detection in data containing both discrete and continuous variables. Furthermore, we demonstrate that some of the benchmark algorithms may be motivated by the problem of working with more complex types of data where sequential dependencies exist between points in the data set. We focus on techniques that can extract information from multivariate heterogeneous data sources.

Over the past few decades, we have seen a tremendous increase in information flow, in terms of the volume and complexity, in several disciplines. Today, we are left with the challenge of dealing with a vast number of heterogeneous information sources in a variety of semantic structures. Knowledge discovery from these heterogeneous resources is still a challenging task. By knowledge discovery we mean the process of extraction of implicit, either known or unknown, potentially useful information that can be used to infer particular events or general characteristics. Knowledge discovery is composed of many computer science subfields such as machine learning, pattern recognition, and expert systems, as well as other fields like statistics, information theory, and stochastic machines. With this increased complexity of data sources there is a potential need for building intelligent refinement and integration frameworks, focusing on information content and semantics.

The key aspect of any data analysis method is how the input data were measured and transformed into information. The two major categories of data are structured data and unstructured data. Unstructured data are data with limited restrictions on their format. A typical example of unstructured data is free text data; for example, reports or scripts describing some events, and experts' feedback on a process. Structured data have much more restricted formats. One typical example is numeric data, in which the attributes may be either continuous or discrete. Continuous data can be defined as numerical responses measured on an infinitely divisible scale. Continuous variables can be used to describe the properties of any process or the assets related to that process. For example, the measured variables can be used to express parameters like height, weight, length, temperature, pressure, etc., of a wooden box or it can express the time needed to build the box or the cost (in terms of money) of the box. Discrete data are numeric attributes measured on a finitely divisible scale. Discrete attributes can be either binary or categorical, and they can either be independent measurements or part of a sequence. An example of binary data is the switching (on/off, presence/absence, etc.) characteristics in a process. An example of categorical data is a satisfaction rating on a scale of 1 to 5. In contrast to the previous two categories, discrete sequences preserve the order (time aspect) of the discrete events. For example, the sequence *ABC* is different from *CBA* as the order information has been reversed in the later case, where *A*, *B*, and *C* are categorical values or events.

In any application, heterogeneity is the measure of how different the attributes of the data are from one another. There are a wide range of sources from which heterogeneity may be introduced in data attributes. Heterogeneity may occur due to the presence of multiple attributes as compared to a single attribute in each data point. The latter is termed as *univariate data*, while the former is *multivariate data*. In multivariate data, the attributes may or may not belong to the same data type. For example each attribute can be either continuous or discrete, independent of every other attribute. The other source of heterogeneity can be the behavioral or functional properties of these attributes. Some typical relationships that are present among data instances are spatial, temporal, spatiotemporal, graphical, etc. A detailed description of the nature of input data can be obtained in [15]. Another type of example is the logical ordering in sequential data. Either it defines a relationship among observations or it may define the switching behaviors of the associated features. As a simple illustration we can take the example of the Fibonacci numbers which are an integer sequence where the first two Fibonacci numbers are 0 and 1, and each subsequent number is the sum of the previous two [21]. An interesting way to characterize the switching behavior of a group of variables is to generate a vector where each item in the vector represents a combination of a switch attribute and the direction in which it was flipped. Later in this chapter, we will further discuss situations where discrete switching patterns may have some causal relationship with some continuous attributes.

3.3.2 Efficiency: Large Data Volume, Distributed Data

Often, data representing the operations of fleets are very large. For example, major commercial airlines have databases containing tens to hundreds of terabytes of flight-recorded data. Simply reading the data from a disk may take months. Therefore, anomaly detection methods need to be judicious in deciding how large a subset of the data they need to read in order to derive useful results. Often, these data are distributed over many storage devices. This happens not only because of the substantial amount of data which is more than typically can be stored at one location, but also because of the geographically distributed nature of the data generation, which makes storing the data at locations close to where they were generated convenient. For example, one can imagine storing the data from a commercial aviation fleet at the destination airports of the corresponding flights. That is, every time an aircraft lands, its data may be uploaded to a storage device located at that airport. Consolidating the data may be impractical given how much data there is. In addition, the fact that the data are distributed should be beneficial—methods should be able to process each storage location in parallel, which would clearly be faster than processing all the data sequentially on a single computer. Therefore, anomaly detection algorithms that can operate on the data even though they are distributed, but give the same results as what a traditional algorithm would give after centralizing all the data, are clearly useful. We will discuss such methods later in this chapter.

3.4 DEALING WITH HETEROGENEITY

As discussed earlier, we need ways to integrate and merge knowledge of the data available across various resources to discover abnormal patterns or outliers in the data. One approach, which we call the *hybrid approach*, involves multiple model development, where we choose, for each modality of data that is present, one or more models appropriate for extracting useful information from that data. The results of these models can then be combined to yield useful information about the entire data set. This is depicted in Figure 3.2. The basic idea is to complement the deficiencies of one algorithm with the strength of the other algorithm or algorithms. A decision is made taking into account the combined output of all the supporting models. For example, if the objective is to discover abnormal patterns in a heterogeneous numerical data set with both continuous and discrete values, we can choose method A that works well on the continuous data and another method B that works in the discrete domain. We can then use a suitable way to amalgamate the findings of both the methods to generate a set of anomalies in that data set. Here A and B may be any machine learning techniques (see Figure 3.2).

An alternative way to approach this problem is to examine the possibilities of knowledge integration, followed by the development of a single model on the combined knowledge. This is depicted in Figure 3.3. In general, authors approach this problem by using a group of complementary metric functions to extract useful information from various data sources and thus, in a way, thus maximizing the knowledge extraction. Each metric

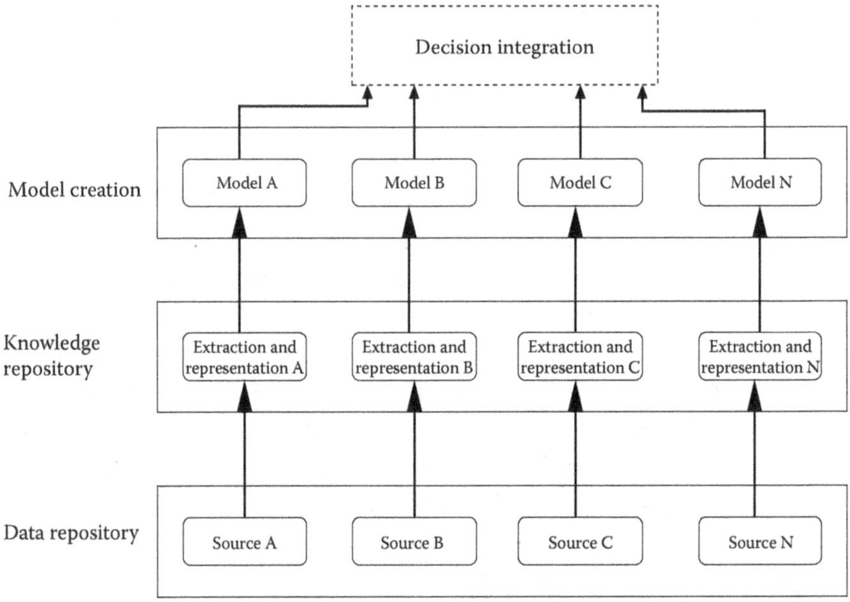

FIGURE 3.2 Two-stage conceptual model for knowledge discovery. The first layer represents heterogeneous data sources. The second layer is the knowledge management process layer where the data is appropriately processed (preprocessing, refinement, etc.) to knowledge. The third layer represents how the knowledge was communicated, corroborated, or shared to build models. In this framework individual models are created on different knowledge bases and a decision is made at the final stage corroborating all the outputs of different models.

is chosen based on its performance on some information content or data types. Once the knowledge integration is complete, the remaining task is to build a suitable model to attain the goal of the analysis.

Choosing the right metric and integrating knowledge from multiple sources must be done judiciously. We will later discuss a method that calculates a weighted combination of such functions to perform anomaly detection on heterogeneous data. Identifying the best combination of weights to integrate knowledge sources is itself an optimization problem and is another interesting research area which is not within the scope of this chapter.

In the following subsections, we will describe some anomaly detection techniques that can easily integrate information from multiple heterogeneous data sources. Some of these are shown in dark shaded blocks in Figure 3.1. The current state-of-the-art algorithms have both strengths and

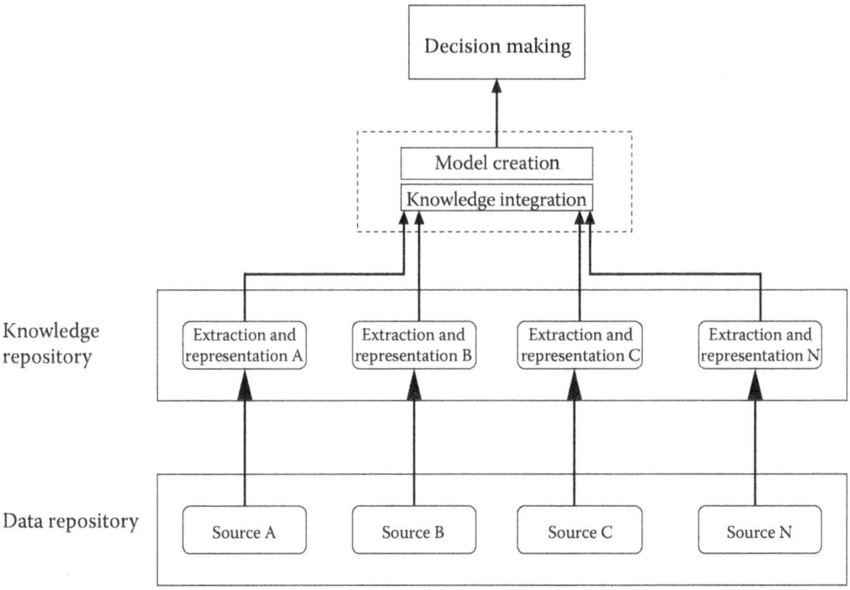

FIGURE 3.3 This is an alternative knowledge discovery scheme compared with the two stage conceptual model shown in Figure 3.2. The first two layers represent heterogeneous data sources and knowledge management process. The third layer represents how the knowledge was integrated, communicated, collaborated, or shared to a single model. In this framework, a single model is created on the fused knowledge and a decision is made at the final stage.

shortcomings in detecting a variety of anomalous conditions. The following subsections will present methods that aim to combine both strengths into a single approach to allow for detection of a variety of anomalies. The benchmark anomaly detection algorithms discussed in this chapter are Orca, GritBot, and one-class SVMs.

3.4.1 Orca

Orca [19] is a method used for detecting anomalies in both continuous and discrete (binary format) data in vector space, using a nearest neighbor based approach to detect anomalous points. For continuous data, Orca takes a reference data set and calculates some function over each data point's nearest neighbors (using the Euclidean distance) in the original vector space. For binary attributes the Hamming distance [22] is used. Orca has a nested loop structure to calculate pairwise distances between data points but uses randomization and a simple pruning rule to keep the algorithm's actual time complexity significantly less than the square of the number of data points. In fact, the pruning used in this algorithm helps to achieve near linear time performance with high dimensional data. This makes the algorithm scalable for analyzing large data sets. The advantages of distance-based anomaly detection algorithms are that no explicit distribution needs to be defined to determine a priori whether data points are normal or anomalous, and that they can be applied to any feature space for which we can define a distance measure. In this algorithm, each data point is scored independently and therefore it is unsuitable for sequential data where the order of data points is operationally significant. The pseudocode of Orca is shown in Algorithm 1.

3.4.2 GritBot

GritBot [23] is a commercially available machine learning tool that performs anomaly detection on discrete and/or continuous attributes. Gritbot uses the C4.5 decision tree algorithm [24] to detect outliers. GritBot was developed by RuleQuest Research. GritBot is more like a classifier which defines a simple boundary that separates the normal patterns from the rest and tests whether the new data falls inside or outside the boundary. The outlierness is based on values of discrete variables or ranges of continuous variables, for which the target variable holds a specific value. A data point is considered to be an anomaly when the target variable corresponding to that point is significantly different from the rest of the target values of the other points in that subset. GritBot does not provide an overall score, or even monitor the score for each data point, but instead ranks anomaly scores according to their statistical significance. Since GritBot is based on decision trees, it requires no prior knowledge of the data unlike many machine learning techniques that may require parameters or distribution models derived from the data set.

Algorithm 1. Orca Algorithm

1. Input

   A_{pq}: Matrix with q dimensional data set having p instances arranged in a random order

   k: Number of nearest neighbors (default 5)

   n: Number of outliers used (default: $n = p$)

2. Output:

   $O(n)$: Set of outliers

   S_{global}: Global scores

3. Argument:

   a: Entries in A

B: Block of examples from A
b: Entries in B
C: Cutoff threshold
w_d: Weight of discrete parameters

4. Definitions:
$x \in (x_c, x_d)$
$D \in (D_c, D_d)$
$$H_d(x_d, D_d) = w_{d_1}(x_{d_1} \neq D_{d_1}) + w_{d_2}(x_{d_2} \neq D_{d_2}) + \ldots + w_{d_n}(x_{d_n} \neq D_{d_n})$$
$$d(x,D) =$$
$$\sqrt{(x_{c_1} - D_{c_1})^2 + (x_{c_2} - D_{c_2})^2 + \ldots + (x_{c_n} - D_{c_n})^2} + H_d(x_d, D_d)$$
$d(x,D)$: maximum distance between x and an example in D
$M^k_{x,D}$: k closest example in D to x
$S(D,x) = \frac{1}{m}\sum_{i=1}^{m} d(x,D)$ distance based score

5. Initialize:
Let p instances of A_{pq} be divided in N_B blocks and $K_{nn}(x)$ be the matrix that keeps track of the nearest neighbors/examples of x. And $C = 0$ and $O = \Phi$, where Φ is a null vector
6. For *block* = 1: N_B {
$B = A(:,block)$; $K_{nn}(b) = \Phi$
7. For each a in A, {
8. For each b in B and $b \neq$ a {
9. If $Length(K_{nn}(b)) < k$ or $d(b,a) < dm(b,K_{nn}(b))$ {
$K_{nn}(b) \leftarrow M^k_{b,K_{nn}(b)\cup a}$
10. If $S(K_{nn}(b),b) < c$ {
Remove example b from set B
}}}}
11. $O = O \cup B$
12. $S_{global} = score(O)$
13. $C \leftarrow min(S_{global}(0))$
}

3.4.3 Kernel-Based Anomaly Detection

In this section, we describe kernel-based methods for anomaly detection in more detail than we devoted to other algorithms because of recent results demonstrating the strong promise of kernel methods for anomaly detection in heterogeneous data sets. In general, a kernel function, which is the heart of any kernel method, can be informally thought of as a measure to calculate the similarity between two data points. The use of certain kernel functions turns out to be equivalent to mapping n-dimensional input data into a high dimensional (possibly infinite dimensional) feature space and then using linear methods within that feature space. This is possible because vectors in the high-dimensional feature space are only present as part of dot products, which are always scalars. Tasks like anomaly detection, classification, clustering, or regression are then performed in this high-dimensional feature space. This method of mapping the data into a high-dimensional space and using linear methods rather than using nonlinear methods in the original input space yields the benefits of linear methods (well-established theory, guarantee of achieving the optimal solution given a fixed training set) with the representational benefits of a nonlinear method. Support vector–based classification, regression, and anomaly detection are classical examples of kernel based methods. Some popular kernel based techniques are shown in shaded (solid color) blocks in Figure 3.1.

3.4.3.1 Kernel Theory and Operations

Often, we find that linear methods for anomaly detection, classification, and regression are not sufficient. In such cases, we may like to use methods that are nonlinear in the input space. However, such methods' theory is not as well developed as linear methods. Additionally, methods for finding nonlinear models, such as gradient descent methods, are typically not able to guarantee that they find the optimal solution given the training data supplied. Kernel methods avoid this problem by instead mapping the data to a much higher dimensional space (called the *feature space*) and then using linear methods within the new space. The conceptual diagram shown in Figure 3.4 depicts the data mapping. The left side shows the representation of the data in the space of the original features (depicted here as a two-dimensional space) whereas on the right hand side the same data has been represented in the feature space, which is depicted here as three-dimensional. The feature space is almost always of much higher dimension than the input space. In the input space $\mathcal{R}$, suppose we are given the data $\mathcal{D}=\{(\vec{x}_i)\}_{i=1}^{n}$, where $\vec{x}_i \in \mathcal{R}^d$ (d = 2 in Figure 3.4). We define a feature space $\mathcal{F}$, and assume that there exists a function ϕ that can be used to map any variable x from the input space to the feature space i.e., $\phi : \mathcal{R}^d \rightarrow \mathcal{F}$. The feature space is often of infinite dimension. However, the feature space is never explicitly defined, but rather is induced by a *kernel function*. That is, a kernel function k is defined such that it induces a

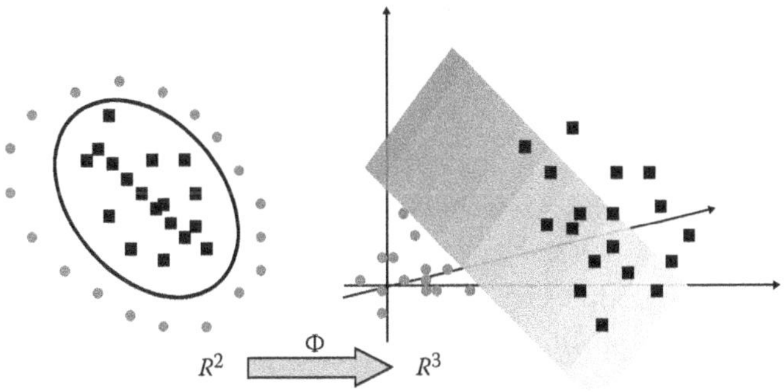

FIGURE 3.4 Higher-dimensional mapping of the data. The data is nonlinearly separable in 2-D data space, whereas it is linearly separable in the 3-D feature space.

mapping ϕ. That is, $k(\vec{x}_i,\vec{x}_j)=\phi(\vec{x}_i)\cdot\phi(\vec{x}_j)$. For example, the Gaussian kernel $k(\vec{x}_i,\vec{x}_j)=exp(-\|\vec{x}_i,\vec{x}_j\|)/2\sigma^2$ induces a function ϕ that maps the input space into an infinite-dimensional feature space, but one need not know what that feature space is because the feature-space vectors are only represented as dot products which can be replaced by $k(\vec{x}_i,\vec{x}_j)$. This implicit use of very high or infinite-dimensional vectors is commonly referred to as the *kernel trick* in the machine learning literature.

3.4.3.2 Is It a Mercer Kernel?

Now the question arises how to choose a kernel function that is "appropriate?" The answer to this question is illustrated using a combination of two very different views. The first and definitely the most immediate requirement to be an "appropriate" kernel function $k(x_i,x_j)$ is to prove that the function satisfies Mercer's conditions. This means that the function must be continuous, symmetric, and positive definite. Any function that satisfies these three conditions is known as a Mercer kernel. In general, a Mercer kernel can be used because the kernel trick can be applied to it.

$$\textit{Continuous}\;\; \forall\varepsilon > 0\; \exists\delta > 0\; s.t.\; |k(x_i,x_j) - k(x_i \pm \delta,x_j)| < \varepsilon \tag{3.1}$$

$$\textit{Symmetric}\;\; k(x_i,x_j) = k(x_j,x_i) \tag{3.2}$$

$$\textit{Positive definite}\;\; \sum_{i,j} a_i a_j (k(x_i,x_j)) \geq 0 \tag{3.3}$$

The continuous (Equation 3.1) and symmetric (Equation 3.2) properties are fairly easy to understand. The positive definiteness property of function k means that it must satisfy the following mathematical condition. For all n, $i \in \{1,2,\ldots,n\}$ and $\vec{x}_i \in \mathcal{R}^d$, the function k results in a kernel matrix (Gram matrix) K such that $K_{i,j} := k(x_i, x_j)$ which is positive definite as shown in Equation 3.3.

Apart from mathematical convenience, another criterion for selecting a kernel function typically depends on the purpose of the study. Using an appropriate function will help in extracting useful knowledge from the underlying data. The kernel function will be more effective if it is sensitive to the knowledge the user is looking for. This implies that the choice of the kernel is based on how the kernel function matches the application. The kernel function should be a measure of similarity appropriate to the application.

3.4.3.3 Problem-Specific Kernel Functions

There are several classes of kernels that satisfy Mercer's conditions. In this section, we will elaborate on a few interesting candidates out of many popular kernel functions which are greatly used in kernel based learning algorithms. A very popular example of such kernel is the Gaussian kernel, also known as radial basis function (RBF), which takes the form of,

$$k(\vec{x}_i, \vec{x}_j) = e^{-\|\vec{x}_i - \vec{x}_j\|^2 / 2\sigma^2} \tag{3.4}$$

where $\|.\|$ denotes the Euclidean norm and σ defines the width of the Gaussian distribution, also known as kernel width. RBF is one of the most widely used kernel functions due to its translation invariance property—the value of the kernel at any two points does not depend on their absolute positions but only on the distance between them in the input space. The RBF kernel width (σ) must be greater than zero. The choice of the kernel width plays an important role in effectively capturing the spread among the data and thus affects the performance of algorithms. There is active research on techniques to automatically choose the optimal value of the kernel width (σ) from the data.

Another commonly used kernel is the polynomial kernel. There are two distinct features of the polynomial function, namely the degree or power (d) of the variables and the offset (c). The two different types of polynomial functions are shown in Equations 3.5 and 3.6.

$$k(\vec{x}_p,\vec{x}_q)=\left\langle \vec{x}_p,\vec{x}_q \right\rangle^d \quad \textit{Homogeneous} \tag{3.5}$$

$$or=\left(\left\langle \vec{x}_p,\vec{x}_q \right\rangle + c\right)^d \quad \textit{Inhomogeneous} \tag{3.6}$$

Under specific scenarios when the user wants to model switching sequences for a given process and where the order of the switching is important, normalized longest common subsequence (nLCS)–based kernel is a good candidate. To define this, we first give some preliminary definitions. z is a *subsequence* of x if there are symbols that can be added before and/or after z to obtain x. z is a *common subsequence* of $\vec{x}_i$ and $\vec{x}_j$ if z is a subsequence of both $\vec{x}_i$ and $\vec{x}_j$. The longest such subsequence of is called the longest common subsequence (LCS) and is denoted by $LCS(\vec{x}_i,\vec{x}_j)$ and $|LCS(\vec{x}_i,\vec{x}_j)|$ is its length. Such a kernel over discrete sequences, when normalized, takes the form of

$$k(\vec{x}_i,\vec{x}_j)=nLCS(\vec{x}_i,\vec{x}_j)=\frac{|LCS(\vec{x}_i,\vec{x}_j)|}{\sqrt{l_{\vec{x}_i}l_{\vec{x}_j}}}, \tag{3.7}$$

where $l_{\vec{x}}$ is the number of symbols in sequence $\vec{x}$. Each sequence of switches is compared against other sequences by using the nLCS as the metric for comparison. Sequences that are similar are bound to hold high nLCS values, while dissimilar sequences will hold very low nLCS values.

If the data being used is text, then one may use the bag-of-words representation, which is a very simple representation of text. A bag of words is a vector whose length is the number of words M used in all the documents in a repository (minus "stop words"—very commonly used words such as *the*, *a*, *an*, and others that are found in nearly all documents and therefore have little usefulness is distinguishing unique characteristics of documents). Each entry in this vector is the number of times that this word appears in the corresponding document. Clearly, this representation does not retain information on the order of words in the document.

Given a corpus of N documents and a dictionary of M words, one possible goal is to establish "relationships" between documents. Latent semantic analysis aims to do this. A co-occurrence matrix (C) of size $M \times N$

can easily be formed. Each column in the co-occurrence matrix represents a document ($d_{\Diamond}$) with the number of occurrences of every term while each row ($t_{\Diamond}$) represents the number of times a term from the dictionary is contained in each document. Entry C_{ij} represents the frequency of the ith term in the jth document. A kernel based on the co-occurrence matrix can be defined as,

$$k(\vec{x}_p, \vec{x}_q) = d_p^T d_q \quad \forall p, q = 1,..N \tag{3.8}$$

$$or = t_p^T t_q \quad \forall p, q = 1,..M \tag{3.9}$$

Here Equation 3.8 gives the similarity between two documents over all the terms while Equation 3.9 gives the similarity between two terms over all documents.

There are many papers describing the use of kernel methods for many different types of data simultaneously with various types of features such as graphs and multiple feature types in computer vision such as color, shape, texture, and graphs based on image segmentations. Interested readers can explore literature [25–29] that looks into various other classes of kernels such as sigmoid, spline, graph-based, tree-based, mismatch-based functions, etc.

3.4.3.4 Information Fusion

In previous sections, we have demonstrated that evaluating a kernel functions on a pair of objects is equivalent to measuring the similarity between those objects. Some of these similarity measures are normalized between 1 and 0. Once an "appropriate" kernel function is chosen and the kernel matrix K is formed, it can be incorporated into any kernel based classification, regression, or anomaly detection methods where the kernel matrix is the sufficient representation of the input data. In another word, by "kernelizing" any method we encode knowledge about the data in terms of pairwise similarities. This provides us with the opportunity to incorporate vast knowledge from heterogeneous sources using multiple-kernel functions. This field of research is known as multiple-kernel learning (MKL) [30,31]. MKL takes advantage of the mathematics of kernels to derive new kernels from existing ones, thereby using multiple kernels simultaneously. For any λ, if K_λ is a Mercer kernel then it can be easily shown that for positive coefficients, a weighted combination of kernels (Equation 3.10) also preserves the Mercer kernel properties, i.e., $\hat{K}$ is also a Mercer kernel.

$$\hat{K} = \sum_{\lambda} \beta_{\lambda} K_{\lambda} \qquad \lambda \in R_{+} \tag{3.10}$$

A common practice is to use a convex combination (i.e., $\sum_{\lambda} \beta_{\lambda} = 1$) of various kernels which may be constructed on very different feature sets such as color, shape, texture, etc. Therefore, a major advantage of the multiple-kernel learning approach is its ability to incorporate more knowledge in the decision process while analyzing complex heterogeneous systems that involve various data sources and data structures.

3.4.3.5 One-Class SVMs: An Overview

Here we provide a brief overview of one-class SVMs, followed by a description of a case study for handling heterogeneous data. Let us consider situations where the users have ideas about what could be the normal behavior of the system (or data) as opposed to the nature of abnormal patterns. Unsupervised techniques like one-class SVM are a perfect fit to such an anomaly detection problem. One-class SVM is designed to estimate the density of the data. It builds a model on single (known) class of data and then finds a set of outliers using a decision boundary—a hyperplane that separates these outliers from the rest of the training examples. Schölkopf et al. [32] showed that in the high dimensional feature space it is possible to construct an optimal hyperplane by maximizing the margin between the origin and the hyperplane in the feature space by solving the following primal optimization problem,

$$\text{minimize} \quad P(\mathbf{w}, \rho, \xi_i) = \frac{1}{2} w w^T + \frac{1}{\nu \ell} \sum_{i=1}^{\ell} \xi_i - \rho$$

$$\text{subject to} \quad (\mathbf{w}.\phi(x_i)) \geq \rho - \xi_i, \quad \xi_i \geq 0, \quad \nu \in [0,1] \tag{3.11}$$

where ν is an user-specified parameter that defines the upper bound on the training error, and also the lower bound on the fraction of training points that are support vectors, ξ is the nonzero slack variable, ρ is the offset, $\phi(x_i)$ represents the transformed image of x_i in the Euclidean space and $i \in [\ell]$. The origin is the only representative of the second class with negative label. For multiple kernels, the dual form of the optimization can be written as,

$$\text{minimize} \quad Q = \frac{1}{2}\sum_{i,j} \alpha_i \alpha_j \left(\sum_{\lambda} \beta_\lambda K_{i,j}^{\lambda} \right)$$

$$\text{subject to} \quad 0 \leq \alpha_i \leq \frac{1}{\ell \nu}, \sum_i \alpha_i = 1, \sum \lambda = 1, \rho \geq 0, \quad \nu \in [0,1] \qquad (3.12)$$

where β_λ are the weights of the kernels and the α_i are Lagrange multipliers. Once this problem is solved at least $\nu \ell$ training points with nonzero Lagrangian multipliers ($\vec{\alpha}$) are obtained and these points $\{x_i : i \in [\ell], \alpha_i > 0\}$ are called *support vectors*. The selected points can be marginal $\mathcal{I}_m = \{i : 0 < \alpha_i < 1\}$ and nonmarginal $\mathcal{I}_{nm} = \{i : \alpha_i = 1\}$ support vectors. Once $\vec{\alpha}$ is obtained, SVMs compute the following decision function.

$$f(\vec{x}_z, \alpha, \beta, \rho) = sign\left(\sum_{i \in \mathcal{I}} \alpha_i \left(\sum_{\lambda} \beta_\lambda K_{i,z}^{\lambda} \right) - \rho \right) \qquad (3.13)$$

where $\mathcal{I} = \mathcal{I}_m + \mathcal{I}_{nm}$. The key aspect of this formulation is that many training examples are no longer needed—only the support vectors are used to define the function. If the decision function predicts a negative label for a given test point x_j, then it is classified as an outlier. Test examples with positive labels are classified as normal. The pseudo-code of one-class SVMs is shown below.

Algorithm 2. Single Class SVMs Algorithm

1. Input vector: $X = \{x_1, x_2 \ldots x_m, z\}$, $X \in \mathcal{R}^d$.
2. Map features: $\sum_{\lambda} \beta_\lambda K_{i,j}^{\lambda}$.
3. Solve Equation 3.12 to obtain α corresponding to support vectors (SVs).
4. Calculate bias, $\rho = \sum_{k=1}^{Ns} \alpha_k \left(\sum_{\lambda} \beta_\lambda K_{\vec{x}, \vec{x}_k}^{\lambda} \right)$.

Calculate score, $f(\vec{z})=\sum_{k=1}^{Ns}\alpha_k\left(\sum_{\lambda}\beta_\lambda K_{\vec{z},\vec{x}_k}^{\lambda}\right)$.

if $f(\vec{z})>\rho$ **then**

    return 1

else

    return 0

end if

3.4.3.6 *An Example of Multiple-Kernel Anomaly Detection*

The whole detection process is nothing but a systematic way of looking at events in the collected data, analyzing the extracted knowledge, and finally making some decisions using certain criteria. As researchers, we may gain a better understanding of why a certain event happened as it did, and what might be causing it. We give here an example that focuses on the use of data analysis techniques to identify anomalies on systems where the data is recorded in both the continuous and discrete form across its subsystems. We chose to elaborate on this example because it is the only example we are aware of that utilizes multiple-kernel learning for anomaly detection over heterogeneous data. We feel this method has great potential to improve anomaly detection for engineered systems such as commercial aircraft.

Here we consider a scenario where the order of switching is absolutely important as compared to the magnitude. Here, the sequential nature of the discrete switches influences the dynamics of the system and hence constitutes the main driving factor in the measured continuous output parameters of that system. For an algorithm to find anomalous behavior in such systems it should be able to detect anomalies in both the continuous and discrete sequences simultaneously. Anomaly types that may appear in such continuous and discrete sequences are shown in Figure 3.5. By continuous sequence we mean the quantized version of the continuous variables arranged in a sequence form. Discrete sequences are representations of the order in which the transitions of the switches happen. The details on the preprocessing and feature extraction steps are scripted in [33].

A synthetic data set was generated to simulate the scenario where the discrete switches are the driving factors in the measured continuous parameters. The data generation method allows the continuous parameters to vary directly with the state of the binary inputs. In addition to this,

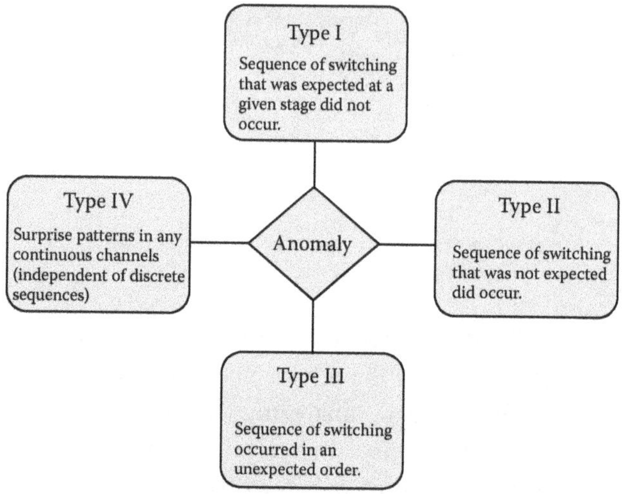

FIGURE 3.5 Summary of the different abnormality categories injected in the synthetic data. A total of 12 faults have been randomly injected, out of which 3 are continuous (Type IV) and 9 are discrete (Type I, II, and III).

Types I, II, III, and IV anomalies were randomly injected in the synthetic data. Ten binary parameters were generated with three fundamental behaviors: random flipping, constant throughout, and deliberate switching. One parameter was set to randomly switch between 0 and 1, while two parameters never changed states. For the deliberate switching six channels would hold a value at their initial state and then change to the alternate state when a separate channel toggled from 0 to 1. Apart from all the above three, abnormal patterns which are independent of discrete variables were injected arbitrarily in certain parts of the continuous data. This serves as an excellent platform where different algorithms could be tested to demonstrate their ability to detect anomalies of each type and for comparing the performances of multiple algorithms.

For comparison to MKL we choose Orca and SequenceMiner, which, with their own strengths, are strong contenders for detecting a variety of anomalous conditions. SequenceMiner [34] was originally designed to model binary switch inputs from pilots during important periods in flight. The switch transitions for a given flight are formatted into a single sequence of switches. Each sequence of switches is compared against other sequences by using the nLCS as the metric for comparison. Sequences that are similar are clustered together. Outliers are sequences that have very low nLCS values. Since SequenceMiner takes into account the order in which the switches were triggered it has the ability to identify anomalies in the temporal domain; however, it is unable to handle continuous data and therefore does not have the ability to detect anomalies in continuous parameters. Both Orca and SequenceMiner will be compared with multi-kernel one-class SVMs.

In using multi-kernel one-class SVMs, as an initial step, we preprocessed the data and then generated continuous and discrete features. We used a convex combination of two kernels with equal weights. The resultant kernel takes the form of

$$K(\vec{x}_i,\vec{x}_j)=\eta K_d(\vec{x}_i,\vec{x}_j)+(1-\eta)K_c(\vec{x}_i,\vec{x}_j), \tag{3.14}$$

where K_d is a kernel over discrete sequences constructed using the normalized nLCS metric, K_c is a kernel constructed over the symbolic aggregate approximation (SAX) [35] representation of continuous data, i.e., continuous sequences, using the same nLCS metric, and η is used to weight the two kernels. From our earlier discussions, one can understand that the constructed kernel K, as well as K_d and K_c are all symmetric positive semidefinite matrices. MKL appears to be a promising way to meet our requirement of incorporating knowledge of both discrete and continuous sequences in anomaly detection as this method aims to combine both strengths of Orca and SequenceMiner into a single approach to allow for detection of a varicty of anomalies. The different steps of multiple-kernel one-class SVMs are shown in Figure 3.6.

Figure 3.7 summarizes the outcomes. Since the actual fault injection incidents are known, we are able to evaluate the performance of all algorithms in detecting those faults. Of 12 injected faults, Orca was able to find the three continuous anomalies. Even though Orca can handle both discrete (binary) variables and continuous variables, the algorithm is unable to detect sequential anomalies where the ordering of transitions is embedded in some form. SequenceMiner, using one standard deviation threshold calculated from the reference set, was able to detect most of the discrete anomalies and clearly missed all the continuous anomalies, whereas the multiple-kernel one-class SVMs stand out among all the algorithms since it was able to identify all twelve fault types (both discrete and continuous).

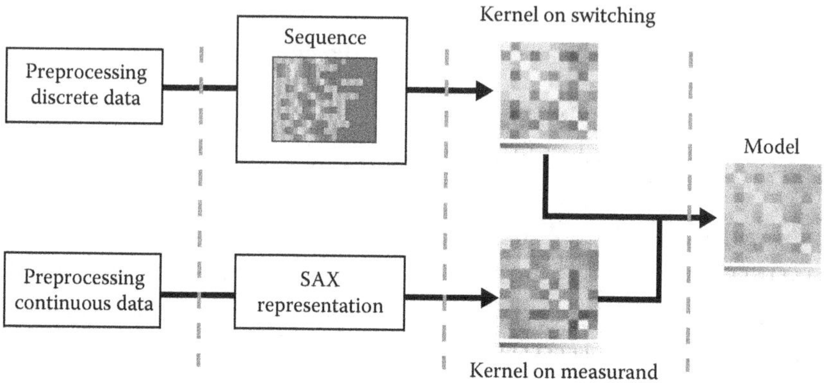

FIGURE 3.6 The three stages of kernel-based anomaly detection technique. The first stage is the data preprocessing and refinement, the second stage is information fusion, followed by the final stage which is model development.

3.4.4 Text Mining

3.4.4.1 Data and Algorithms

As mentioned in Section 3.2.1.2, our example text repository, ASRS, has two relevant problems: classification and topic extraction. We now discuss two algorithms developed and implemented to analyze and extract useful information from ASRS reports. The first is Mariana [36], an SVM

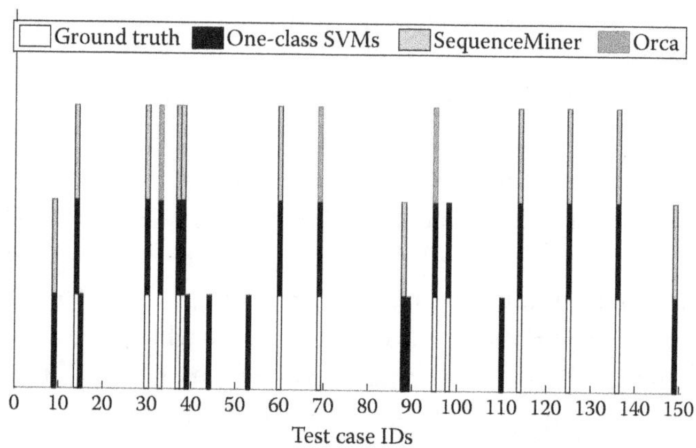

FIGURE 3.7 Summary of the performance of all three algorithms in detecting the abnormalities in the synthetic data for each abnormality category. A total of 12 abnormal candidates have been randomly injected, and these abnormal cases are represented by the ground truth. Multiple-kernel one-class SVMs was the only algorithm to detect all fault types.

with simulated annealing, which is a search method used to find the best hyperparameter settings for the model. Since SVMs were explained earlier, we will not do so here. The second method is classification built on top of nonnegative matrix factorization (NMF), which attempts to find a document model that represents document features that add up in various combinations to form documents. NMF has the potential to provide basis vectors that are interpretable and indicative of different topics present within the repository. This is critical to discovering new, previously undiscovered problem areas within ASRS and other repositories.

Nonnegative matrix factorization (NMF) [37] is a variation on the host of mathematically motivated techniques for factoring large vector-valued data sets into basis and distribution matrices. Suppose we have d documents and t terms.* The general approach is to seek a relatively small set of k basis vectors represented by the $t \times k$ matrix W, and a corresponding set of distribution weight vectors represented by the $k \times d$ matrix H, such that the transposed bag-of-words matrix X (for NMF, we assume X is $t \times d$ rather than the usual convention of being $d \times t$) is factored according to $X \approx W * H$ by minimizing some measure of the difference, $X - W * H$. The hope is that the basis vectors in W will correspond to some fundamental properties of the data set, with the distributions in H combining those properties to form the data.

The convention in NMF-based text analysis is that the jth column of X represents the term weights of document j, each column of W is a basis vector over the term set, and each column of H is a set of weights over the basis vectors. Thus, if we use $H_{\bullet j}$ to denote the jth column of H and $H_{i\bullet}$ to denote the ith row of H, then for the jth document, $W * H_{\bullet j} \approx X_{\bullet j}$. If each basis vector $W_{\bullet b}$ is individually L1 normalized to a single term weight, and the rows $H_{b\bullet}$ are inversely scaled to maintain the product $W_{\bullet b}H_b$, then H_{bj} is the approximate number of terms, from basis vector $W_{\bullet b}$, found in the jth document. If the columns $H_{\bullet j}$ are then $L1$-normalized, they give the relative basis vector weights, independent of document size.

In NMF applications, the data values are nonnegative, typically counts or scalar measurements, and the factorization is constrained to keep both W and H nonnegative. With nonnegativity, the basis vectors may be thought of as components, and the distributions as recipes for adding

* In general, NMF can be used in the case where there are d examples that have t features each, but we present NMF as used for text classification.

components to match the data. Both aspects seem more natural than the alternatives, particularly so for intrinsically nonnegative data.

While nonnegativity is an appealing property for factorization methods, constraining conventional difference minimization algorithms to maintain nonnegativity has mostly been difficult. This changed in 2002, with spreading recognition of the potential of Lee and Seung's multiplicative update approach [37,38]. For the squared Frobenius norm, the standard sum of squared matrix values $\|X-WH\|_F^2$, Lee and Seung's original paper gives the minimizing reestimation equations for iteration as:

$$W_{ab} \leftarrow W_{ab} \frac{(X*H^T)_{ab}}{(W*H*H^T)_{ab}} \tag{3.15}$$

$$H_{bi} \leftarrow H_{bi} \frac{(W^T*X)_{bi}}{(W^T*W*H)_{bi}} \tag{3.16}$$

where *a* and *i* index over the attributes and instances of *X*, respectively, and *b* indexes over the basis vectors. This is actually a reformulation of the standard gradient driven norm minimizing search, augmented with a conceptually simple step size computation that maintains the nonnegativity constraint.

Starting with nonnegative *W* and *H*, and applied alternately, these reestimation equations are proven to monotonically lower the norm toward a local stationary point, while maintaining the nonnegative properties of *W* and *H*. Lee and Seung also provided an alternate multiplicative minimization for the Kullback-Liebler divergence of probability matrices, and [39,40] describe schemes for other matrix norms.

In factoring the prepared data, there is the fundamental choice of what function of *X* – *WH* shall be minimized with consequences that are not yet well understood. A basis size must be chosen, or a range of sizes searched over and evaluated. Algorithmic details, particularly factor initializations, may have significant effects. Since the multiplicative NMF algorithms are gradient driven, they approach their stationary points at exponentially decreasing rates. This requires stopping criteria that balance the opposing requirements of computational efficiency and numerical accuracy.

The columns of W form vectors of words, each of which can be seen as a topic represented in the text repository. If one wants to use NMF for classification, then a linear model is fit using the criterion that the product of the H matrix and the document categorization matrix is the same for the training and the test set. Since the H matrices for the training and test data are known and the categorization matrix for the training set is known, one can solve for the categorization matrix for the test data.

3.4.4.2 Results

We first discuss text classification results. The authors of [7] tested Mariana with raw text in BOW format and the same text after being processed by an NLP system called PLADS to assess how useful PLADS and the simulated annealing within Mariana would be. Mariana's process of choosing better hyperparameters improves the area under the ROC curve by 10% or more in some of the categories on raw text. It also gives equal improvement on text that has been processed by the PLADS system. As shown in Figure 3.8, there is little difference in performance when this process is used on raw text and PLADS processed text. Others have also found that NLP methods do not give enough performance improvement to justify their high running time.

The authors of [7] then evaluated the performance of Mariana by auto-classifying 100 randomly selected reports and having the results reviewed by a problem report expert. The reviewer agreed with the top classification by Mariana 73% of the time. The reviewer agreed with one or both of the top two classifications 86% of the time, and with the top three classifications 90% of the time. A separate review of the 100 reports was done by another subject expert. Then the first reviewer reviewed the second reviewer's classifications, and agreed with their top (and only) classification 89% of the time. These results indicate that Mariana is classifying

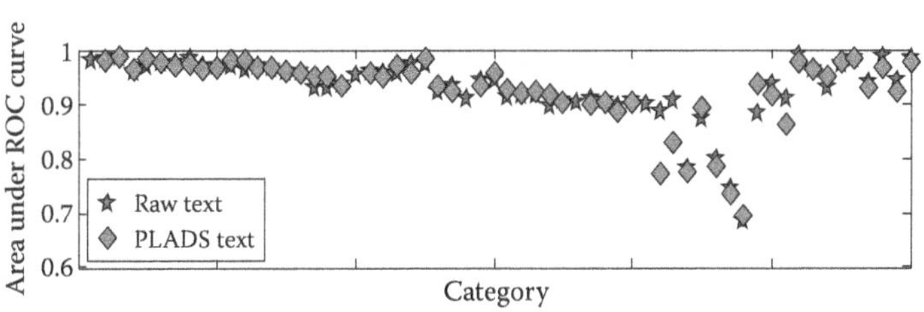

FIGURE 3.8 Comparison of raw text and PLADS processed text using optimum hyperparameters.

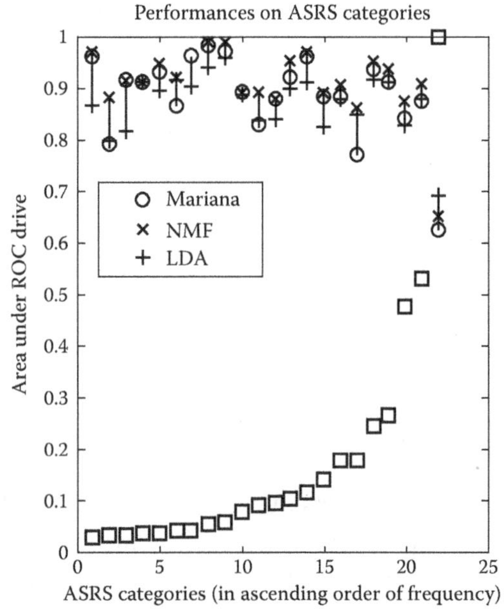

FIGURE 3.9 Performances on ASRS categories.

documents reasonably well and in a manner reasonably consistent with human experts given that reviewers (including presumably those who provided the original classifications to the documents that they used in training) disagree among themselves.

Figure 3.9 shows the areas under the ROC curves for Mariana, NMF, and a baseline method (linear discriminant analysis) applied to the ASRS categories, as well as the fraction of documents in each category (the boxes), normalized so that the category with the largest number of documents has a value of 1. The categories on the x-axis are sorted in ascending order of this normalized fraction. The vertical lines connect the highest area to the lowest area for each category and are meant to facilitate comparisons within each category. Overall, Mariana outperforms linear discriminant analysis on 16 of the 22 categories and NMF outperforms linear discriminant analysis on 21 of the 22 categories.

We now discuss NMF's results for topic extraction. Table 3.1 gives examples of two basis vectors from three runs of NMF (each with a different training and test set mix drawn from the original data set) on the ASRS repository. One basis vector was drawn from each of the three runs such that the resulting three basis vectors were the closest such triple in

TABLE 3.1 Examples of ASRS Basis Vectors

FUEL	FUEL	FUEL
TANK	TANK	TANK
POUND	POUND	POUND
GALLON	GALLON	GALLON
GAUGE	GAUGE	GAUGE
PUMP	PUMP	PUMP
FUELTANK	BURN	BURN
BURN	FUELTANK	FUELTANK
FUELER	FUELER	FUELER
FUELQUANTITY	FUELQUANTITY	FUELQUANTITY
CENTER	CENTER	CENTER
MAINTANK	DISPATCH	FUELGAUGE
FUELGAUGE	FUELGAUGE	MAINTANK
IMBAL	MAINTANK	IMBAL
REFUEL	IMBAL	REFUEL
CROSSFEED	REFUEL	PLAN
QUANTITY	QUANTITY	CALCULATE
BALANCE	PLAN	CROSSFEED
CALCULATE	CROSSFEED	BALANCE
EMPTY	CALCULATE	EMPTY
INSTALL	INSTALL	INSTALL
INSPECT	INSPECT	REMOVE
REMOVE	REMOVE	REPLACE
REPLACE	MECHANIC	ENGINEER
MECHANIC	REPLACE	MANUAL
FOUND	PART	INSPECT
WORK	MANUAL	WORK
MANUAL	WORK	SHIFT
REPAIR	REPAIR	FOUND
PART	FOUND	ASSEMBLE
ENGINEER	SIGN	TECHNICIAN
TEST	ENGINEER	REPORT
CHECK	NUMBER	PANEL
SHIFT	SHIFT	REPAIR
SIGN	MAINTAIN	JOB
ASSEMBLE	TEST	XYZ
MAINTAIN	ASSEMBLE	BOLT
SERVE	AIRCRAFT	CARD
CARD	XYZ	LEAK

Note: After performing three runs of NMF (each with a different initialization of the W and H matrices), we selected the three basis vectors (one column of W from each run) that were closest to each other in terms of $L1$ norm—these constitute the first three columns of this table. The next three columns are the second closest such triple of basis vectors.

terms of $L1$-norm—these form the three columns of Table 3.1 above the line. The three columns below the line are the basis vectors corresponding to the second closest such triple. Within each column, the words are given from top to bottom in order of weight. One can see from these examples that the three runs came up with bases that are relatively close to one another. The two triples are also quite different from each other, indicating that the basis vectors represent different topics (combinations of categories). However, the 20 words within a basis vector are clearly quite related. The first basis set clearly seems to be describing some fuel tank related problem and the second basis set clearly describes an issue related to a repair. It appears that NMF is finding the significant collocations that are indicative of topics and keeping them together.

3.5 DEALING WITH EFFICIENCY ISSUES: DISTRIBUTED ANOMALY DETECTION

Often, data representing the operations of fleets are distributed over many storage devices. This happens not only because of the substantial amount of data which is more than typically can be stored at one location, but also because of the geographically distributed nature of the data generation process, which makes storing the data at locations close to where they were generated convenient. For example, one can imagine storing the data from a commercial aviation fleet at the destination airports of the corresponding flights. That is, every time an aircraft lands, its data may be uploaded to a storage device located at that airport. Consolidating the data may be impractical given how much data there is. Therefore, anomaly detection algorithms that can operate on the data even though they are distributed, but give the same results as what a traditional algorithm would give after centralizing all the data, are clearly useful.

We briefly give overviews of two algorithms for which there are mathematical proofs that they yield the same anomalies with the same ranking of anomalousness on any distributed data set as what a corresponding centralized algorithm would give after centralizing all the data. Additionally, these algorithms are designed to be efficient in that they require much less communication than what would be required to actually copy all the data from the different nodes to one location. One algorithm is a distributed expectation-maximization (EM) algorithm for learning GMM parameters. The second algorithm is a distance-based anomaly detection algorithm.

3.5.1 Gaussian Mixture Model

Here, we discuss a density-based anomaly detection algorithm. This is an example of a class of algorithms that learn a density function over the training data set. This density function can clearly be used for many purposes, but when used for anomaly detection, the user assumes that normal data tend to be in higher density regions whereas anomalous data tend to be in lower density regions. The density may be learned using nonparametric methods, such as histogram methods or methods to construct Voronoi tesselations of the data [41]. A parametric method may also be used, whereby a model structure is assumed and the learning task involves calculating the parameters of the model that best fit the training data. Either way, when new data arrive, the learned density function is calculated on these points and those for which the probability density is lower than a certain threshold may be judged to be anomalous data. The algorithm we discuss in this section is an example of a parametric method.

This algorithm, called PeDEM [42], learns the parameters of a GMM on the entire distributed data set, monitors whether the GMM's parameters are still valid given new data that has arrived at each of the nodes of the distributed system, and, if necessary, updates the GMM's parameters to reflect the inclusion of the new data in the distributed repository.

EM is an iterative optimization technique to estimate unknown parameters Θ (typically sufficient statistics for the probability distribution being used to model the data) given data D. The EM algorithm alternates between two steps to maximize the posterior probability distribution of Θ given D. The algorithm utilizes hidden variables H, which often are assumed to represent hidden state information. The two steps are

- E-step: Calculate the expected value of H given Θ and D.
- M-step: Calculate Θ to maximize the likelihood of the data D given the expected value of H.

For GMMs, the generative model for a vector $\mathbf{x} \in R^d$, where $\mathbf{x}$ is a d-dimensional vector of real numbers, is

$$P(\mathbf{x}) = \sum_{j=1}^{M} \pi_j P(\mathbf{x} \mid j),$$

where M is the number of Gaussians in the mixture, π_j is the prior probability of the jth mixture being used to generate a data point, and the jth mixture's density is

$$P(\mathbf{x}\mid j)=\frac{1}{(2\pi)^{d/2}\left\|\mathbf{C}_j\right\|^{1/2}}exp-(\mathbf{x}-\mu_j)^T\mathbf{C}_j^{-1}(\mathbf{x}-\mu_j)/2.$$

Each density is parameterized by its mean vector $\mu_j = [\mu_{j.1}\mu_{j,2}\ldots\mu_{j,d}]$ and its covariance matrix $\mathbf{C}_j = (\mathbf{x} - \mu_j)(\mathbf{x} - \mu_j)^T$. Assuming that we have a training data set $X = \{\mathbf{x}_1,\mathbf{x}_2,\ldots,\mathbf{x}_n\}$, EM attempts to estimate the model parameters such that they maximize the likelihood of the parameters given the data. Normally, the log-likelihood is maximized instead, which gives the same result since logarithm is a monotonically increasing function.

A centralized version of EM for GMMs would have the following E-step:

$$q_{j,i}=\frac{\pi_j N(\mathbf{x}_i;\mu_j,\mathbf{C}_j)}{\pi_k N(\mathbf{x}_i;\mu_k,\mathbf{C}_k)},$$

and the following M-step:

$$\pi_j=\frac{\sum_{i=1}^{n}q_{j,i}}{n}$$

$$\mu_j=\frac{\sum_{i=1}^{n}q_{j,i}\mathbf{x}_i}{\sum_{i=1}^{n}q_{j,i}}$$

$$\mathbf{C}_j=\frac{\sum_{i=1}^{n}q_{j,i}(\mathbf{x}_i-\mu_j)(\mathbf{x}_i-\mu_j)^T}{\sum_{i=1}^{n}q_{j,i}},$$

for each of the M Gaussians in the Gaussian mixture. $N(\mathbf{x}_i;\mu_j,\mathbf{C}_j)$ denotes the probability density function at data point $\mathbf{x}_i$ under a normal distribution with mean μ_j and covariance matrix $\mathbf{C}_j$.

The corresponding version of EM in PeDEM has the following E-step:

$$q_{a,j,i} = \frac{\pi_j N(\mathbf{x}_{\mathbf{a,i}};\boldsymbol{\mu}_{\mathbf{j}},\mathbf{C}_{\mathbf{j}})}{\sum_{k=1}^{M} \pi_k N(\mathbf{x}_{\mathbf{a,i}};\boldsymbol{\mu}_{\mathbf{k}},\mathbf{C}_{\mathbf{k}})},$$

and the following M-step, where B is the number of nodes in the distributed system:

$$\pi_j = \frac{\sum_{a=1}^{B} \sum_{i=1}^{n} q_{a,j,i}}{\sum a = 1^{B} N_a}$$

$$\boldsymbol{\mu}_{\mathbf{j}} = \frac{\sum_{a=1}^{B} \sum_{i=1}^{n} q_{u,j,l}\mathbf{x}_{\mathbf{a,l}}}{\sum_{a=1}^{B} \sum_{i=1}^{n} q_{a,j,i}}$$

$$\mathbf{C}_{\mathbf{j}} = \frac{\sum_{a=1}^{B} \sum_{i=1}^{n} q_{a,j,i}(\mathbf{x}_{\mathbf{a,i}} - \boldsymbol{\mu}_{\mathbf{j}})(\mathbf{x}_{\mathbf{a,i}} - \boldsymbol{\mu}_{\mathbf{j}})^T}{\sum_{a=1}^{B} \sum_{i=1}^{n} q_{a,j,i}}.$$

The E-step computation for each node involves only data and parameters available at that node. However, the M-step computation at each node requires information from all the other nodes. Obviously, the distributed anomaly detection problem would be trivial if all the computations at each node only involved information available at that node.

The goal of the exercise in [42] was to continually check that the current parameters of the GMM are valid with respect to the global data set even as new data are added and to update the parameters when they are incompatible with the data. In PeDEM, four monitoring problems are solved—one each for the log-likelihood ($\overline{L}$), prior probabilities of the mixture models, the means of the mixture models, and the covariances. In particular, they monitor for

$$\bar{L}(\hat{\Theta}|G) > \varepsilon$$

$$\left|\pi_j - \hat{\pi}_j\right\| < \varepsilon_1$$

$$\left\|\mu_j - \hat{\mu}_j\right\| < \varepsilon_2$$

$$\left\|\left\|\mathbf{C}_j\right\|_F - \hat{\mathbf{C}}_j\right| < \varepsilon_3.$$

In case of log-likelihood, clearly one would want the log-likelihood to be high—if it drops below a threshold, then that means that the new data is too different from the data seen so far for the current model to be valid, so model updating is needed. For the remaining three conditions, if the difference between the last stable value calculated based on the global data and what is estimated at a node based on current data at a given node is too high, then model updating is needed. More details on the algorithm and the proof that the distributed algorithm returns the same results as the centralized EM algorithm are shown in [42]. However, in summary, the proof relies on a notion of covering the global data with convex regions where if each node's knowledge (the data stored at that node plus any data communicated to it by other nodes) are contained within one such convex region, then the results that each node returns would be the same as what would be returned by a centralized algorithm. This condition means that the nodes do not need to communicate until the condition is violated. The authors of [42] demonstrated experimentally that the level of communication required for the algorithm is quite low assuming that the data is not changing wildly at every node.

3.5.2 Distance Based

In distance-based anomaly detection methods, data points for which neighboring data points are relatively nearby are judged to be normal, whereas those with relatively distant neighbors are judged to be anomalous. A variety of distance functions is used, such as distance to the *k*th nearest neighbor for some chosen *k*, and average distance to the *k* nearest neighbors. Given a data set, one can calculate this distance function for every data point and rank order the data points in decreasing order of the distance function. The data points with high values for the distance function can be returned as examples of anomalies.

A naive algorithm for this task would require calculation of the distance between all pairs of data points. In particular, for a given data point x, its distances to all the other data points seemingly need to be calculated in order to determine its k nearest neighbors, which in turn can be used to calculate the distance function. However, there is one restricted scenario in which all pairs of points do not need their distances calculated. This idea has been implemented within Orca [43]. The algorithm takes as input a data set and a number L and returns the top L anomalies in the data set by one of several metrics such as distance to the kth nearest neighbor or average distance to the k nearest neighbors—that is, the points having the L highest values of the chosen metric are returned as anomalies. Orca starts running in the naive way by calculating the distance from the first data point to all the remaining data points, calculating the chosen metric for that data point, calculating the distance from the second data point to the following data points, etc. However, Orca keeps track of the top L anomalies and their values for the metric. Note that while a data point is being examined, its value for the metric can only decrease.* Therefore, if the value of the metric for the current data point being examined drops below the value of the metric for the Lth most anomalous data point found so far, then the current data point cannot be an anomaly, so we can stop calculating distances for the current point and move onto the next one. This happened because we have already determined that there are at least L data points with higher values for the metric than the current point being considered. Orca has been demonstrated experimentally to have a running time that is, on average, slightly greater than linear in the number of data points submitted, although its worst-case running time is quadratic in the number of data points just like the naive algorithm.

For distance-based distributed anomaly detection algorithms to return the same anomalies as the centralized algorithm discussed above, we depend on this property that adding new data points can only reduce a given data point's value for a metric. This is known as the antimonotonicity property [44]. More formally, assume that the anomaly detection algorithm A takes as input a finite set of points P, a function $R : D \times 2^D \rightarrow R^+$,

* For example, suppose we are calculating the metric for the first data point, have checked its distances to the next 20 data points, and are finding points based on distance to the 10th nearest neighbor. In this case, the 22nd data point can either be closer than the current 10th nearest neighbor or farther away. If the 22nd data point is farther away, then the distance to the 10th nearest neighbor will not change. If it is closer, then either the 22nd point or the current 9th nearest neighbor will become the 10th nearest neighbor and the distance to the 10th nearest neighbor will decrease.

and a positive integer L, and returns the data points having the L highest values of R. Given a data point $x \in D$ and two sets of data points $P_1 \subseteq P_2 \subseteq D$, we require the following two properties:

- Antimonotonicity: $R(x,P_1) \geq R(x,P_2)$.
- Smoothness: if $R(x,P_1) > R(x,P_2)$, then there exists $y \in P_2/P_1$ such that $R(x,P_1) > R(x,P_1 \cup y)$.

According to the theorems and proofs in [44], only the antimonotonicity property is needed to prove that the distributed algorithm will terminate in finite time with all the nodes agreeing on the anomalous points and their supports (the nearest neighbors needed to calculate the metric). Both antimonotonicity and smoothness properties are needed to prove that the agreed upon anomalies are the globally correct ones—that is, the ones that would have been identified by the corresponding distance-based anomaly detection acting on all the data after being collected at one site.

3.6 CONCLUSIONS

It may be noticed that in this chapter, the detection problem was formulated with an emphasis on how to handle heterogeneous knowledge sources to obtain a meaningful solution in a more robust way. The data-driven concept for developing heterogeneous anomaly detection models outlined in this chapter is an attempt to provide a broad sample of recent techniques with detailed coverage on a few algorithms to illustrate how they work but obviously, we are unable to cover all approaches in a single chapter. Readers are advised to consider that there is no single universally accepted approach that can perform extremely well on all possible data sets and anomalous conditions. The goodness of any algorithm depends a lot on its applicability which in most cases is very application specific. So far authors have developed a wide variety of machine learning tools and techniques and have demonstrated their applicability on various complex systems. Often it is seen that a particular approach is well suited to address certain problem. Also one cannot ignore the importance of data refinement, cleaning, and transformation techniques in effective data mining and knowledge discovery. The nature of the algorithm decides how well and in what form this knowledge can be integrated in the core algorithm. We have already seen in previous examples that the ability to represent the data in the form of discrete and continuous sequences and to integrate this knowledge, with the notion of similarity between sequences, in the optimization problem of the multiple-kernel based technique, provides us with an opportunity to detect all different types of anomalies simultaneously. Moreover since the discrete and continuous sequences can be transformed into a single sequence of switches and are representative of "system level" information over time, another important feature of the multiple-kernel anomaly detection algorithm is its ability to do "systemwide" analysis to detect anomalies. Even though improvements to the multiple-kernel learning-based technique that we discussed may be found, it does point in the direction of the types of methods that we want—those that can handle heterogeneous sequence data.

REFERENCES

[1] H. Kargupta, V. Puttagunta, M. Klein, K. Sarkar. On-board vehicle data Stream monitoring using minefleet and fast resource constrained monitoring of correlation matrices. *Next Generation Computing*, 25(1):5–32, 2006.

[2] P. Robinson. A real-time quantitative condition alerting and analysis support system for aircraft maintenance. *NASA SBIR Phase II Abstract*. (from: http://www.atr-usa.com/documents/AeroTech_QCAAS_Narrative_Brief.pdf). 2007.

[3] T. Chidester. Understanding normal and atypical operations through analysis of flight data. In *Proceedings of the 12th International Symposium on Aviation Psychology*, pp. 239–242, May 1997.

[4] E. Pinheiro, W.-D. Weber, L.A. Barroso. Failure trends in a large disk drive population. In *Proceedings of the 5th USENIX Conference on File and Storage Technologies (FAST'07)*, Feb 2007.

[5] B.G. Amidan, T.A. Ferryman. APMS SVD methodology and implementation. Technical report PNWD-3026, Battelle, 2000.

[6] A. Hotho, A. Nurnberger, G. Paab. A brief survey of text mining. *Zeitschrift fuer Computerlinguistik und Sprachtechnologie (GLDV-Journal for Computational Linguistics and Language Technologie*, 20(2):19–62, 2005.

[7] J.P. Castle, J.C. Stutz, D.M. McIntosh. Automatic discovery of anomalies peported in aerospace systems health and safety documents. In *AIAA Infotech@Aerospace*. American Institute of Aeronautics and Astronautics. 2007.

[8] T. Hastie, R. Tibshirani, J. Friedman. *The Elements of Statistical Learning: Data Mining, Inference, and Prediction*. Springer, 2001.

[9] D. Blei, A. Ng, M. Jordan. Latent dirichlet allocation. *Journal of Machine Learning Research*, 3:993–1022, 2003.

[10] D.D. Lee, H.S. Seung. Algorithms for non-negative matrix factorization. In *NIPS*, pp. 556–562, 2000.

[11] NASA. Aviation Safety Reporting System (ASRS) program overview.

[12] M. Markou, S. Singh. Novelty detection: A review—Part 1: Statistical approaches. *Signal Processing*, 83: 2003.

[13] M. Markou, S. Singh. Novelty detection: A review—Part 2: Neural network based approaches. *Signal Processing*, 83:2499–2521, 2003.

[14] J.A. Quinn, C.K.I. Williams. Known Unknowns: Novelty detection in condition monitoring. In *Proceedings of 3rd Iberian Conference on Pattern Recognition and Analysis, Lecture Notes in Computer Science 4477*. Springer-Verlag, pp. 1–6. Springer LNCS. 2007.

[15] A. Banerjee, V. Chandola, V. Kumar. Anomaly detection: A Survey. *ACM Computer Survey*, 41:1–58, 2009.

[16] F. Angiulli, C. Pizzuti. Outlier mining in large high-dimensional data sets. *IEEE Transactions on Knowledge and Data Engineering*, 17(2):203–215, 2005.

[17] F. Angiulli, F. Fassetti. Very efficient mining of distance-based outliers. In *Proceedings of CIKM '07*, pp. 791–800, 2007.

[18] S. Ramaswamy, R. Rastogi, K. Shim. Efficient algorithms for mining outliers from large data Sets. *SIGMOD Rec.*, 29(2):427–438, 2000.

[19] S. D. Bay, M. Schwabacher. Mining distance-based outliers in near linear time with randomization and a simple pruning rule. In *Proceedings of KDD'03*, pp. 29–38, 2003.

[20] J.P.T. Higgins. Commentary: Heterogeneity in meta-analysis should be expected and appropriately quantified. *International Journal of Epidemiology*, 37(5):1158–1160, 2008.

[21] Wikipedia. Fibonacci Number, The Free Encyclopedia. 2004.

[22] Wikipedia. Hamming Distance, The Free Encyclopedia. 2010.

[23] RuleQuest Research. Gritbot. 2007.

[24] J.R. Quinlan. *C4.5: Programs for Machine Learning*. Morgan Kaufmann, San Francisco, CA, 1993.

[25] B. Schölkopf, A. Smola. *Learning with Kernels*. MIT Press, 2002.

[26] H. Kashima, K. Tsuda, A. Inokuchi. Kernels for Graphs. *Kernel Methods in Computational Biology*, 39(1):101–113, 2004.

[27] O. Chapelle, P. Haffner. Support vector machines for histogram-based classification. *IEEE Transactions on Neural Networks*, 10(5):1055–1064, 1999.

[28] H. Zhang, A.C. Berg, M. Maire, J. Malik. Svm-knn: Discriminative nearest neighbor classification for visual category recognition. In *Computer Vision and Pattern Recognition*, pp. 2126–2136, 2006.

[29] Z. Harchaoui, F.R. Bach. Image classification with segmentation graph kernels. In *Computer Vision and Pattern Recognition*, pp. 1–8, 2007.

[30] F.R. Bach, G.R.G. Lanckriet, M.I. Jordan. Multiple kernel learning, conic duality, and the SMO algorithm. In *International Conference on Machine Learning*, 2004.

[31] G.R.G. Lanckriet, N. Cristianini, L. El Ghaoui, P. Bartlett, M.I. Jordan. Learning the kernel matrix with semidefinite programming. *Journal of Machine Learning Research*, 5:27–72, 2004.

[32] B. Schölkopf, J.C. Platt, J.C. Shawe-Taylor, A.J. Smola, R.C. Williamson. Estimating the support of a high-dimensional distribution. *Neural Computing*, 13(7):1443–1471, 2001.

[33] S. Das, B.L. Matthews, A.N. Srivastava, N.C. Oza. Multiple kernel learning for heterogeneous anomaly detection: Algorithm and aviation safety case study. In *Proceedings of the 16th ACM SIGKDD international conference on Knowledge Discovery and Data Mining*, KDD '10, pp. 47–56, New York, NY, USA, ACM. 2010.
[34] Suratna Budalakoti, Ashok N. Srivastava, and Matthew E. Otey. Anomaly detection and diagnosis algorithms for discrete symbol sequences with applications to airline safety. *Trans. System Man Cyber Part C*, 39:101–113, January 2009.
[35] P. Patel, E. Keogh, J. Lin, S. Lonardi. Mining motifs in massive time series databases. In *Proceedings of the IEEE International Conference on Data Mining (ICDM-02)*, 370–377, 2002.
[36] N.C. Oza, J.P. Castle, J. Stutz. Classification of aeronautics system health and safety documents. *IEEE Transactions on Systems, Man, and Cybernetics, Part C: Applications and Reviews*, 39(6):670–680, 2009.
[37] D.D. Lee, H.S. Seung. Algorithms for non-negative matrix factorization. In T.K. Leen, T.G. Dietterich, V. Tresp, editors, *Advances in Neural Information Processing Systems-13*, pp. 556–562. MIT Press. 2001.
[38] L.K. Saul, D.D. Lee. Multiplicative updates for classification by mixture models. In T.G. Dietterich, S. Becker, Z. Ghahramani, editors, *Advances in Neural Information Processing Systems-14*, pp. 897–904. MIT Press, 2002.
[39] I.S. Dhillon, S. Sra. Generalized nonnegative matrix approximations with Bregman divergences. In Y. Weiss, B. Schölkopf, J. Platt, editors, *Advances in Neural Information Processing Systems*, pp. 283–290. MIT Press, 2006.
[40] Y. Wang, Y. Jia, C. Hu, M. Turk. Fisher non-negative matrix factorization for learning local features. In *Asian Conference on Computer Vision*, 2004.
[41] J.D. Scargle. A novel approach to novelty detection: Voronoi Tesselation, Machine Learning Seminar, NASA Ames Research Center. 2004.
[42] K. Bhaduri, A.N. Srivastava. A local scalable distributed expectation maximization algorithm for large peer-to-peer networks. In *Proceedings of the IEEE International Conference on Data Mining*, pp. 31–40, 2009.
[43] S.D. Bay, M. Schwabacher. Mining distance-based outliers in near linear time with randomization and a simple pruning rule. In *Proceedings of The Ninth ACM SIGKDD International Conference on Knowledge Discovery and Data Mining*. 2003.
[44] C. Giannella, R. Wolff, J. Branch, B. Szymanski, H. Kargupta. In-network outlier detection in wireless sensor networks. In *International Conference on Distributed Computing Systems*, pp. 51–58, 2006.

CHAPTER 4

Discriminative Topic Models

Hanhuai Shan
Amrudin Agovic
Arindam Banerjee

CONTENTS

4.1 INTRODUCTION

The Aviation Safety Reporting System (ASRS) [1] is used to collect voluntarily submitted aviation safety reports from pilots, controllers, and others. Each report is manually assigned to anomaly classes such as "maintenance problems" or "operation problems." Given these documents, one can consider two important tasks: First, although we already have the manually labeled reports in the database, it would still be interesting to develop an understanding of the content of "topic" from the reports. These topics could be the subcategories of predefined classes, hence become a finer level of anomaly categories for analysis. Second, given the labeled reports in the database, it would be interesting if we can automatically categorize the new coming reports.

The first task above is topic modeling and the second is text classification. Both tasks are important for aviation health management systems. For example, for the first task, given the aviation safety reports under the class of "maintenance problems," it would be interesting to automatically discover subcategories (i.e., topics) such as power system problems, communication system problems, monitoring system problems, etc. The experts on aviation system health management could then focus on the specific problems and related reports. For the second task, since a large number of aviation safety reports are generated everyday, if we can label them automatically, it would greatly facilitate the analysis on these reports, based on which we can improve the system health management.

In recent years, several models for analyzing topics from document collections have been proposed [2–4]. Among these algorithms, latent Dirichlet allocation (LDA) [2] is one of the most widely used topic models. Such models provide a succinct representation of long documents, and they also generate interpretable word lists for the topics. Despite the prominent advantages, one important restriction of most existing topic models is that they are unsupervised models and cannot leverage class label information for classification. In terms of the ASRS example, LDA and related models cannot take advantage of the labeled reports in the database to classify the new coming reports. On the other hand, while popular classification algorithms, such as support vector machines (SVMs) [5] and logistic regression (LR) [6], perform well on classification, the classifier itself is often hard to interpret. The above observations motivate our current work on designing accurate discriminative classification algorithms while leveraging topic models for interpretability.

Supervised latent Dirichlet allocation (SLDA) [7] is such a topic modeling algorithm that takes response variables into account. However, the response variable in SLDA is a real number, which is assumed to be generated from a normal linear model, so it is different from categorical labels in the context of classification. In principle, the authors proposed a general framework to extend SLDA to deal with other types of response variables, including categorical labels, based on generalized linear models (GLM) [8]. However, efficient inference in the general case is difficult without the good properties of Gaussian distribution.

In this chapter, we discuss discriminative* latent Dirichlet allocation (DLDA), which is a discriminative topic model combining multiclass logistic regression with LDA. Just like LDA, DLDA is able to generate interpretable word lists for the topics. The mixed-membership representation generated by DLDA is biased by class labels and can be viewed as a supervised dimensionality reduction. Further, since DLDA allows the number of components k in the mixed membership to be different from the number of classes c, the model often discovers additional latent structure beyond what is implied by the class labels with a larger k. To learn the model, we propose two families of variational inference algorithms: one is based on ideas originally proposed in [2] and the other is more efficient in space and time complexity by using a significantly less number of parameters. Unlike Taylor expansion-based approximations suggested in [7], the proposed inference algorithms maintain the lower bound maximization strategy used in variational inference.

Recently, there has been an increasing interest in mixed-membership models combining supervision information. Based on SLDA [7], maximum entropy discrimination latent Dirichlet allocation (MedLDA) [9] combines supervised topic models with maximum margin learning. Labeled LDA [10] incorporates functional annotation of known genes to guide gene clustering. DiscLDA [11] determines document position on

* "Discriminative" here does not mean a discriminative model, but a generative model used for classification instead of clustering.

topic simplex with guidance of labels. Discriminative topic models [12] combine topic modeling with manifold learning on a neighborhood graph, which is constructed with the help of labels. In addition, [13] proposed a Dirichlet multinomial regression, which accommodates different types of metadata including labels, and [14] proposed a correlated labeling model for multilabel classification. Also, [15] extends SLDA for image classification and annotation. The DLDA discussed in this chapter belongs to this broad class of models but is specifically geared toward scalable text analysis and classification.

The rest of the chapter is organized as follows: in Section 4.2, we give a brief overview of LDA and SLDA. In Section 4.3, we propose DLDA. In Section 4.4, two variational approaches for learning DLDA are discussed. We present the experimental results on the ASRS corpus as well as standard benchmarks in Section 4.5, and we conclude in Section 4.6.

4.2 LATENT DIRICHLET ALLOCATION AND SUPERVISED LATENT DIRICHLET ALLOCATION

In this section, we give an overview on two topic models—LDA and SLDA.

4.2.1 Latent Dirichlet Allocation

LDA [2] is one of the most widely used topic modeling algorithms. It is capable of extracting topics from documents in an unsupervised fashion. In LDA, each document is assumed to be a mixture of topics, whereby a topic is defined to be a distribution over words. LDA assumes that each word in a document is drawn from a topic z, which in turn is generated from a discrete distribution discrete(π) over topics. Each document is assumed to have its own distribution discrete(π), whereby all documents share a common Dirichlet prior α. The generative process for each document with N words $x_{1:N}$ is as follows (Figure 4.1):

1. Draw $\pi \sim$ Dirichlet(α).
2. For each of N words x_n in document $x_{1:N}$:
   (a) Draw a topic $z_n \sim$ Discrete(π).
   (b) Draw x_n from $p(x_n|\beta_{1:k}, z_n)$.

where $\beta_{1:k} = \{\beta_i, [i]_1^k\}$ ($[i]_1^k \equiv i = 1 \ldots k$) is a collection of parameters for k topic distributions over words and β_i denotes a V-dimensional discrete

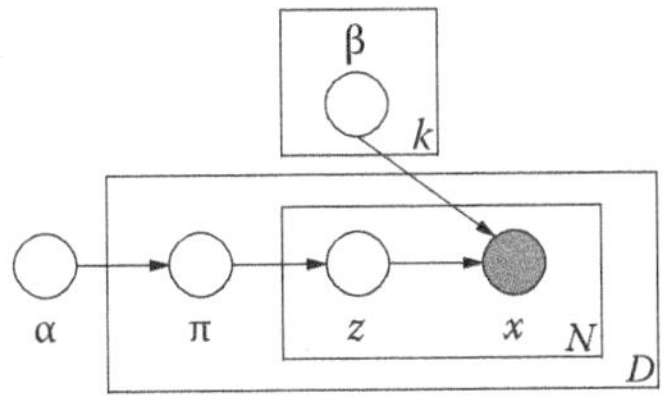

FIGURE 4.1 Graphical model representation of LDA.

distribution with V as the total number of words in the dictionary. The generative process chooses β_i corresponding to z_n. The chosen topic distribution β_i is subsequently used to generate the word x_n. The most likely words within β_i can be used as a representation for topic i.

Given the generative process, for each document $x_{1:N}$, LDA has a density of the form

$$p(x_{1:N} \mid \alpha, \beta_{1:k}) = \int_{\pi} p(\pi \mid \alpha) \left(\prod_{n=1}^{N} \sum_{z_n} p(z_n \mid \pi) p(x_n \mid z_n, \beta_{1:k}) \right) d\pi.$$

LDA is an extension of probabilistic latent semantic indexing (pLSI) [16]. In pLSI, each document is also represented as a distribution over the topics, but there is no generative model to generate these distributions. In LDA, a Dirichlet distribution is introduced to generate the topic distributions, which yields a complete generative model for the documents.

We can also view LDA as an extension of multinomial principal component analysis (PCA). If we represent each document as a V-dimensional word occurrence vector y, following multinomial PCA [17], the generative process for a document is given by

1. Draw $\pi \sim \text{Dirichlet}(\alpha)$.

2. Draw $y \sim \text{Multinomial}(\beta\pi, N)$, where β is a matrix with β_i, $[i]_1^k$ on columns, i.e., we are taking average of $\beta_{1:k}$ according to π.

In multinomial PCA, all words in a document are generated from a same distribution $\text{Multinomial}(\beta\pi, N)$. To consider LDA as an extension of multinomial PCA, we introduce latent mixture components, such that the words in a document could be generated from different component distributions [17]:

1. Draw $\pi \sim$ Dirichlet(α).
2. Draw $c \sim$ Multinomial(π,N), where c is a k-dimensional vector and c_i denotes the number of words in the document representing the ith topic.
3. For each topic i, $[i]_1^k$, draw $x(i,\cdot) \sim$ Multinomial(β_i,c_i), where $x(i,\cdot)$ is the set of words in x representing the ith topic.

Therefore, the words representing different topics i are generated from different component distributions β_i.

4.2.2 Supervised Latent Dirichlet Allocation

SLDA [7] is an extension of LDA that accommodates the response variables other than the documents. The response variable is assumed to be generated from a normal linear model $N(\eta^T\bar{z},\sigma^2)$, where η and σ^2 are the parameters and the covariates $\bar{z} = \sum_{n=1}^{N} z_n/N$ are the empirical average frequencies of each latent topic in the document. The generative process of SLDA for generating the document $x_{1:N}$ and the response y are given as follows (Figure 4.2):

1. Draw $\pi \sim$ Dirichlet(α).
2. For each of N words x_n in document $x_{1:N}$:
   (a) Draw a topic $z_n \sim$ Discrete(π).
   (b) Draw x_n from $p(x_n|\beta_{1:k},z_n)$.
3. Draw y from $N(\eta^T\bar{z},\sigma^2)$.

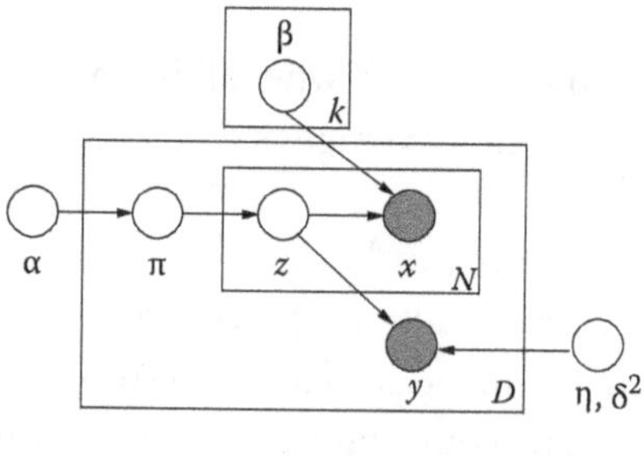

FIGURE 4.2 Graphical model representation of SLDA.

Similar to LDA, $\beta_{1:k}$ in SLDA is also a collection of k distributions over all words for k topics. Therefore, the first two steps of SLDA are exactly the same as in LDA, but SLDA has an additional step 3 to generate the response y.

The density function of SLDA is given as follows:

$$p(x_{1:N}, y \mid \alpha, \beta_{1:k}, \eta, \sigma^2) = \int_{\pi} p(\pi \mid \alpha) \sum_{z_{1:N}} \left(\prod_{n=1}^{N} p(z_n \mid \pi) p(x_n \mid z_n, \beta_{1:k}) \right)$$
$$\times p(y \mid z_{1:N}, \eta, \sigma^2) d\pi.$$

4.3 DISCRIMINATIVE LATENT DIRICHLET ALLOCATION

As indicated by its name, SLDA is a supervised topic model that is able to incorporate the response variable. However, it has an important limitation that prevents it from being used as a discriminative text classification algorithm: The response variable in SLDA is a univariate real number, which is assumed to be generated from a normal linear model, whereas the response variables, i.e., labels, are discrete categories in the classification setting. Although the authors pointed out that the response variables can be of various types obtained from generalized linear models, variational inference is difficult in the general case. While a Taylor expansion is recommended [7] to obtain an approximation of the log-likelihood, such an approach forgoes the lower bound guarantee of variational inference.

Therefore, in this section, we propose DLDA. It is built on top of LDA and is a variant of SLDA to incorporate categorical response variables. On the one hand, DLDA is a classification algorithm like logistic regression or SVM; on the other hand, it inherits the advantages of topics models hence is able to generate interpretable results, such as low dimensional representation for the documents or word lists for the topics.

Assuming there are c classes and the number of topics we choose is k, the graphical model for DLDA is given in Figure 4.3, where α is a k-dimensional parameter of a Dirichlet distribution, $\beta_{1:k}$ are the parameters for k topic distributions over the words, and $\eta_{1:c} = [\eta_1,\cdots,\eta_c]^T$ is a matrix with c k-dimensional logistic regression parameters as the rows, where η_c is a zero vector by default, so we only use $\eta_{1:c-1}$ as the parameter to be estimated. Assuming there are totally V words in the dictionary, then each word x_n could be represented as a V-dimensional vector with only the *vth* dimension x_n^v being 1 and others being 0 if it denotes the vth word in the dictionary. The generative process for each document $x_{1:N}$ is given as follows:

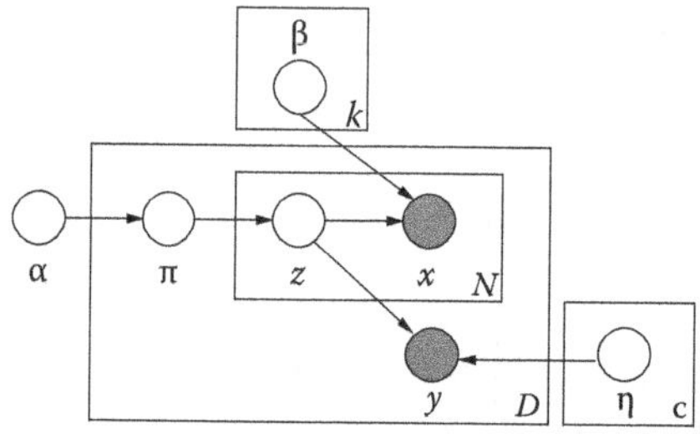

FIGURE 4.3 Graphical model representation of DLDA.

1. Choose a component proportion $\pi \sim$ Dirichlet(α).
2. For each word x_n in the document with N words $\mathbf{x} = x_{1:N}$,
   (a) Choose a component $z_n = i \sim$ Discrete(π).
   (b) Choose a word $x_n \sim$ Discrete(β_i).
3. Choose the label from a multiclass logistic regression

$$y \sim \mathrm{LR}\left(\frac{\exp\left(\eta_h^T \bar{z}\right)}{1+\sum_{h=1}^{c-1} \exp\left(\eta_h^T \bar{z}\right)}\right), [h]_1^{c-1}.$$

As in SLDA, $\bar{z}$ is an average of $z_{1:N}$ over all observed words, where each z_n is a k-dimensional unit vector with only the ith entry being 1 if it denotes the ith topic, so $\bar{z}$ is also a k-dimensional vector. The categorical response variable y can be considered as a sample generated from the discrete distribution $\left(p_1,\ldots p_{c-1}, 1-\sum_{h=1}^{c-1} p_h\right)$, where $p_h = \dfrac{\exp\left(\eta_h^T \bar{z}\right)}{1+\sum_{h=1}^{c-1}\left(\eta_h^T \bar{z}\right)}$ for $[h]_1^{c-1}$. In two-class classification, y is 0 or 1 generated from Bernoulli $\left(\dfrac{1}{1+\exp\left(-\eta^T \bar{z}\right)}\right)$, i.e., the model needs only one η in the two-class case.

In the generative process, following [7], we have used $\bar{z}$ (the mean of z for all words) as an input to logistic regression. In principle, any other transformation on $z_{1:N}$ could work, as long as it gives a reasonable representation of the original data point. We choose $\bar{z}$ due to the following two reasons. (1) Optimality: Given a set of data points, their best representative is always the mean according to a wide variety of divergence functions [18,19]. We also notice that $\eta_h^T \bar{z} = \eta_h^T E[z] = E[\eta_h^T z]$, which means that if we take the mean of $\eta_h^T z$ on each feature as the input to logistic function, it is equivalent to using $\eta_h^T \bar{z}$ as the input to logistic function. (2) Simplicity: Since z is the latent variable, if we use other complicated transformation on z such as a nonlinear function, it would greatly increase the difficulty in inference and learning.

There are two important properties of DLDA compared with LDA: (1) The k-dimensional mixed-membership $\bar{z}$ effectively serves as a low dimensional representation of the original document. While $\bar{z}$ in LDA is inferred in an unsupervised way, it is obtained from a supervised dimensionality reduction in DLDA. We give the explanation in Section 4.4. (2) DLDA allows the number of classes c and the number of components k in the generative model to be different. If k was forced to be equal to c, for problems with a small number of classes, $\bar{z}$ would have been a rather coarse representation of the document. In particular, for two-class problems, $\bar{z}$ would lie on the two-simplex, which may not be an informative representation for classification purposes. Decoupling the choice of k from c prevents such pathologies. In principle, we may find a proper k using Dirichlet process mixture models [20].

From the generative model, the joint distribution of latent and observable variables for DLDA is given by

$$p(\pi, z_{1:N}, x_{1:N}, y \mid \alpha, \beta_{1:k}, \eta_{1:c-1}) = p(\pi \mid \alpha)\left(\prod_{n=1}^{N} p(z_n \mid \pi) p(x_n \mid z_n, \beta_{1:k})\right) \times p(y \mid z_{1:N}, \eta_{1:c-1}). \tag{4.1}$$

Integrating Equation 4.1 over π and summing it over $z_{1:N}$ yields the marginal distribution of $(x_{1:N}, y)$:

$$p(x_{1:N}, y \mid \alpha, \beta_{1:k}, \eta_{1:c-1}) = \int_{\pi} p(\pi \mid \alpha) \sum_{z_{1:N}} \left(\prod_{n=1}^{N} p(z_n \mid \pi) p(x_n \mid z_n, \beta_{1:k})\right) \times p(y \mid z_{1:N}, \eta_{1:c-1}) d\pi. \tag{4.2}$$

The probability of the entire data set of D documents and their labels $(\mathcal{X} = \{x_d, [d]_1^D\}, \mathcal{Y} = \{y_d, [d]_1^D\})$ is given by

$$p(\mathcal{X},\mathcal{Y}|\alpha,\beta_{1:k},\eta_{1:c-1}) = \prod_{d=1}^{D}\int_{\pi_d} p(\pi_d|\alpha)\sum_{z_{d,1:N_d}}\left(\prod_{n=1}^{N_d} p(z_{dn}|\pi_d)p(x_{dn}|z_{dn},\beta_{1:k})\right)$$

$$\times p(y_d|z_{d,1:N_d},\eta_{1:c-1})d\pi_d. \tag{4.3}$$

4.4 INFERENCE AND PARAMETER ESTIMATION

Since DLDA assumes a generative process for both labels as well as the data points, instead of using labels directly to train a classifier, we use both $\mathcal{X}$ and $\mathcal{Y}$ as samples from the generative process to estimate the parameters of DLDA, such that the likelihood of observing $(\mathcal{X},\mathcal{Y})$ is maximized. Unlike naive Bayes [21], the parameters cannot be directly estimated from the class labels due to the latent mixed memberships. In particular, due to the latent variables, the computation of the likelihood in Equation 4.2 is intractable. In this section, we present two alternative approaches to obtain a variational approximation of the log-likelihood and propose an expectation-maximization (EM)–style algorithm to iteratively obtain better estimates of the model parameters. Finally, we show how the estimated parameters can be used to do prediction on test data.

4.4.1 Variational Approximation

For each document $x_{1:N}$, to obtain a tractable lower bound to log $p(x_{1:N},y|\alpha,\beta_{1:k},\eta_{1:c-1})$, we introduce a variational distribution $q(\pi,z_{1:N}|\Omega)$ as an approximation of the true posterior distribution $p(\pi,z_{1:N}|\alpha,\beta_{1:k},\eta_{1:c-1})$ over the latent variables, where Ω is the set of variational parameters.* By a direct application of Jensen's inequality [2], the lower bound to log $p(x_{1:N},y|\alpha,\beta_{1:k},\eta_{1:c-1})$ is given by:

$$\log p(x_{1:N},y|\alpha,\beta_{1:k},\eta_{1:c-1}) \geq E_q[\log p(\pi,z_{1:N},x_{1:N},y|\alpha,\beta_{1:k},\eta_{1:c-1})] + H(q(\pi,z_{1:N})), \tag{4.4}$$

where $H(\cdot)$ denotes the Shannon entropy. We use L to denote the lower bound. Following [2] and noticing that $x_{1:N}$, and y are conditionally independent given $z_{1:N}$, we have

* To avoid clutter, we do not show the free variational parameters of q unless necessary in the sequel.

$$L = E_q[\log p(\pi \mid \alpha)] + E_q[\log p(z_{1:N} \mid \pi)] + E_q[\log p(x_{1:N} \mid z_{1:N}, \beta_{1:k})]$$
$$- E_q[\log q(\pi)] - E_q[\log q(z_{1:N})] + E_q[\log p(y \mid z_{1:N}, \eta_{1:c-1})]. \quad (4.5)$$

We propose two different variational distributions $q(\pi, z_{1:N})$. Following [2], we consider

$$q_1(\pi, z_{1:N} \mid \gamma, \phi_{1:N}) = q_1(\pi \mid \gamma) \prod_{n=1}^{N} q_1(z_n \mid \phi_n), \quad (4.6)$$

where $q_1(\pi|\gamma)$ is a Dirichlet distribution for π and each $q_1(z_n|\phi_n)$ is a discrete distribution for z_n. Also, we propose

$$q_2(\pi, z_{1:N} \mid \gamma, \phi) = q_2(\pi \mid \gamma) \prod_{n=1}^{N} q_2(z_n \mid \phi), \quad (4.7)$$

where $q_2(\pi|\gamma)$ is a Dirichlet distribution for π and $q_2(z_n|\phi)$ is a discrete distribution for all z_n. In both q_1 and q_2, we have a k-dimensional Dirichlet(γ) for each document, but we have a k-dimensional discrete(ϕ_n) for each of N words in q_1 and only one discrete(ϕ) for all words in q_2. By keeping a substantially smaller number of parameters, q_2 is space-efficient especially for high-dimensional data. It is also time-efficient with a substantially smaller number of parameters to optimize over. q_1 and q_2 determine two different variational inference algorithms. We call the first one "standard variational inference" and the second one "fast variational inference," which accordingly yield standard DLDA (std DLDA) and fast DLDA. In the sequel, we use q to denote q_1 or q_2 unless otherwise necessary.

Given the variational distribution as in Equation 4.6 or Equation 4.7, the first five terms in the lower bound (Equation 4.5) can be easily obtained following LDA [2] or fast LDA [22]. The most difficult part is the last term, which cannot be computed exactly even after introducing the variational distribution q, so we work on the last term for further approximation. The class label y is from a multiclass logistic regression $\text{LR}\left(\frac{\exp\left(\eta_h^T \bar{z}\right)}{1 + \sum_{h=1}^{k-1} \exp\left(\eta_h^T \bar{z}\right)}, [h]_1^{k-1}\right)$, i.e., y is from a discrete distribution with $\eta_{1:c-1}\bar{z}$ the natural parameter. Therefore,

$$p(y \mid z_{1:N}, \eta_{1:c-1}) = \exp\left(\sum_{h=1}^{c-1} \eta_h^T \bar{z} y_h - \log\left(1 + \sum_{h=1}^{c-1} \exp\left(\eta_h^T \bar{z}\right)\right)\right).$$

Accordingly,

$$\begin{aligned} E_q[\log p(y \mid z_{1:N}, \eta_{1:c-1})] &= E_q\left[\sum_{h=1}^{c-1} \eta_h^T \bar{z} y_h - \log\left(1 + \sum_{h=1}^{c-1} \exp\left(\eta_h^T \bar{z}\right)\right)\right] \\ &= \sum_{h=1}^{c-1} \sum_{i=1}^{k} \eta_{hi} E_q[\bar{z}_i] y_h - E_q\left[\log\left(1 + \sum_{h=1}^{c-1} \exp\left(\eta_h^T \bar{z}\right)\right)\right]. \end{aligned} \tag{4.8}$$

The second term of Equation 4.8 could be expanded as follows:

$$\begin{aligned} -E_q\left[\log\left(1 + \sum_{h=1}^{c-1} \exp\left(\eta_h^T \bar{z}\right)\right)\right] &\geq -\log\left(1 + \sum_{h=1}^{c-1} E_q\left[\exp\left(\sum_{i=1}^{k} \eta_{hi} \bar{z}_i\right)\right]\right) \\ &\geq -\log\left(1 + \sum_{h=1}^{c-1} E_q\left[\sum_{i=1}^{k} \bar{z}_i \exp(\eta_{hi})\right]\right) \\ &= -\log\left(1 + \sum_{h=1}^{c-1} \sum_{i=1}^{k} E_q[\bar{z}_i] \exp(\eta_{hi})\right) \\ &\geq -\frac{1}{\xi} \sum_{h=1}^{c-1} \sum_{i=1}^{k} E_q[\bar{z}_i] \exp(\eta_{hi}) + 1 - \frac{1}{\xi} - \log(\xi), \end{aligned} \tag{4.9}$$

where the first inequality is from Jensen's inequality, the second inequality is also from Jensen's inequality noticing that $\bar{z}$ is actually a discrete distribution, and the third inequality is from $-\log(x) \geq 1 - \frac{x}{\xi} - \log(\xi)$ [23] by introducing a new variational parameter $\xi > 0$. Given Equation 4.9, for std LDA, we have

$$E_q[\log p(y \mid z_{1:N}, \eta_{1:c-1})] \geq \frac{1}{N}\sum_{n=1}^{N}\sum_{i=1}^{k}\phi_{ni}\left(\sum_{h=1}^{c-1}\eta_{hi}y_h - \frac{1}{\xi}\sum_{h=1}^{c-1}\exp(\eta_{hi})\right) + \left(1 - \frac{1}{\xi} - \log\xi\right), \tag{4.10}$$

and for fast DLDA, we have

$$E_q[\log p(y \mid z_{1:N}, \eta_{1:c-1})] \geq \sum_{i=1}^{k}\phi_i\left(\sum_{h=1}^{c-1}\eta_{hi}y_h - \frac{1}{\xi}\sum_{h=1}^{c-1}\exp(\eta_{hi})\right) + \left(1 - \frac{1}{\xi} - \log\xi\right). \tag{4.11}$$

Putting $E_q[\log p(y|z_{1:N},\eta_{1:c-1})]$ back to Equation 4.5 gives us the complete expression for L. By maximizing Equation 4.5 with respect to the variational and model parameters alternatively, we find the optimal value for $(\alpha,\beta_{1:k},\eta_{1:c-1})$.

4.4.1.1 Inference

Given a choice of model parameters $\left(\alpha^{(t)},\beta^{(t)}_{1:k},\eta^{(t)}_{1:c-1}\right)$, the lower bound to the log-likelihood for each data point in Equation 4.5 becomes a function of the variational parameters $L(\alpha^{(t)},\beta^{(t)}_{1:k},\eta^{(t)}_{1:c-1},\gamma,\phi,\xi)$. The goal of the inference step is to obtain the tightest lower bound to the true log-likelihood, which is achieved by maximizing $L(\alpha^{(t)},\beta^{(t)}_{1:k},\eta^{(t)}_{1:c-1},\gamma,\phi,\xi)$ with respect to γ, ϕ, and ξ. In particular, for std DLDA, we have

$$\phi_{ni} \propto \exp(\Psi(\gamma_i) - \Psi\left(\sum_{l=1}^{k}\gamma_l\right) + \sum_{v=1}^{V}x_n^v\log\beta_{iv} + \frac{1}{N}\sum_{h=1}^{c-1}(\eta_{hi}y_h - \exp(\eta_{hi})/\xi)) \tag{4.12}$$

$$\gamma_i = \alpha_i + \sum_{n=1}^{N}\phi_{ni} \tag{4.13}$$

$$\xi = 1 + \frac{1}{N}\sum_{h=1}^{c-1}\sum_{i=1}^{k}\sum_{n=1}^{N}\phi_{ni}\exp(\eta_{hi}), \tag{4.14}$$

and for fast DLDA, we have

$$\phi_i \propto \exp(\Psi(\gamma_i) - \Psi\left(\sum_{l=1}^{k} \gamma_l\right) + \frac{1}{N}\sum_{n=1}^{N}\sum_{v=1}^{V} x_n^v \log\beta_{iv}$$

$$+\frac{1}{N}\sum_{h=1}^{c-1}(\eta_{hi}y_h - \exp(\eta_{hi})/\xi)) \tag{4.15}$$

$$\gamma_i = \alpha_i + N\phi_i \tag{4.16}$$

$$\xi = 1 + \sum_{h=1}^{c-1}\sum_{i=1}^{k}\phi_i \exp(\eta_{hi}), \tag{4.17}$$

$\bar{\phi} = \left[\sum_{n=1}^{N} \phi_{n1}/N, \ldots, \sum_{n=1}^{N} \phi_{nk}/N\right]$ for std DLDA and $\phi = [\phi_1, \ldots, \phi_k]$ for fast DLDA actually give the posterior of $\bar{z}$, i.e., the low-dimension representation of each data point. Note that the last terms in both Equations 4.12 and 4.15 contain y, indicating that the low-dimension representation not only depends on $x_{1:N}$, but also depends on y, which means DLDA models achieve supervised dimension reduction. Removing the last term gives the expression of ϕ in the corresponding unsupervised settings.

4.4.1.2 Parameter Estimation

Variational parameters $(\phi^*, \gamma^*, \xi^*)$ from the inference step gives the optimal lower bound to the log-likelihood of each pair of $(x_{1:N}, y)$. Since we cannot maximize $\log p(\mathcal{X}, \mathcal{Y} | \alpha, \beta_{1:k}, \eta_{1:c-1})$ directly, we maximize the aggregate lower bound $\sum_{d=1}^{D} L(\phi_d^*, \gamma_d^*, \xi_d^*, \alpha, \beta_{1:k}, \eta_{1:c-1})$ over all data points with respect to $\alpha, \beta_{1:k}$ and $\eta_{1:c-1}$, respectively to obtain the estimated parameters. The estimations of α and $\beta_{1:k}$ for std DLDA and fast DLDA are the same as in LDA and fast LDA, respectively [2,22]. In particular, for both std DLDA and fast DLDA, the updating equation for α is given by

$$\alpha_i' = \alpha_i - \frac{g_i - u}{f_i}, \tag{4.18}$$

where

$$g_i = D\left(\Psi\left(\sum_{l=1}^{k}\alpha_l\right) - \Psi(\alpha_i)\right) + \sum_{d=1}^{D}\left(\Psi(\gamma_{di}) - \Psi\left(\sum_{i=1}^{k}\gamma_{di}\right)\right)$$

$$f_i = -D\Psi'(\alpha_i)$$

$$u = \frac{\sum_{i=1}^{k} g_i/f_i}{w^{-1} + \sum_{i=1}^{k} f_i^{-1}}$$

$$w = D\Psi'\left(\sum_{i=1}^{k}\alpha_i\right).$$

For std DLDA, the updating equation for β is given by

$$\beta_{iv} \propto \sum_{d=1}^{D}\sum_{n=1}^{N_d}\left(\phi_{dni}x_{dn}^{v}\right), \tag{4.19}$$

and for fast DLDA, the updating equation for β is given by

$$\beta_{iv} \propto \sum_{d=1}^{D}\left(\phi_{di}\sum_{n=1}^{N_d}x_{dn}^{v}\right), \tag{4.20}$$

where N_d is the total number of words in document d. As for the updating equation for η, for std DLDA, we have

$$\eta_{hi} = \log\frac{\sum_{d=1}^{D}\sum_{n=1}^{N_d} y_{dh}\phi_{dni}/N_d}{\sum_{d=1}^{D}\sum_{n=1}^{N_d}\phi_{dni}/(N_d\xi_d)},$$

and for fast DLDA, we have

$$\eta_{hi} = \log \frac{\sum_{d=1}^{D} \phi_{di} y_{dh}}{\sum_{d=1}^{D} \phi_{di} / \xi_d}.$$

4.4.2 Variational EM Algorithm

We propose an EM-style algorithm to find out the optimal model parameters alternatively. Given the model parameters $(\alpha^{(t-1)}, \beta_{1:k}^{(t-1)}, \eta_{1:c-1}^{(t-1)})$ from the initial guess or the last iteration. The algorithm alternates between the following two steps until convergence:

1. E-step: Given $(\alpha^{(t-1)}, \beta_{1:k}^{(t-1)}, \eta_{1:c-1}^{(t-1)})$, for each data point, find the variational parameters

$$(\phi_d^{(t)}, \gamma_d^{(t)}, \xi_d^{(t)}) = \underset{(\phi_d, \gamma_d, \xi_d)}{\operatorname{argmax}}\, L(\phi_d, \gamma_d, \xi_d, \alpha^{(t-1)}, \beta_{1:k}^{(t-1)}, \eta_{1:c-1}^{(t-1)}).$$

   Then $L(\phi_d^{(t)}, \gamma_d^{(t)}, \xi_d^{(t)}; \alpha, \beta_{1:k}, \eta_{1:c-1})$ gives a lower bound to log $p(\mathbf{x}_d, y_d | \alpha, \beta_{1:k}, \eta_{1:c-1})$.

2. M-step: Maximizing the aggregate lower bound yields an improved estimate of model parameters:

$$(\alpha^{(t)}, \beta_{1:k}^{(t)}, \eta_{1:c-1}^{(t)}) = \underset{(\alpha, \beta_{1:k}, \eta)}{\operatorname{argmax}} \sum_{d=1}^{D} L(\gamma_d^{(t)}, \phi_d^{(t)}, \xi_d^{(t)}; \alpha, \beta_{1:k}, \eta_{1:c-1}).$$

After t iterations, the objective function becomes $\sum_{d=1}^{D} L(\gamma_d^{(t)}, \phi_d^{(t)}, \xi_d^{(t)}; \alpha^{(t)}, \beta_{1:k}^{(t)}, \eta_{1:c-1}^{(t)})$. In iteration $t+1$,

$$\begin{aligned}\sum_{d=1}^{D} L(\gamma_d^{(t)}, \phi_d^{(t)}, \xi_d^{(t)}; \alpha^{(t)}, \beta_{1:k}^{(t)}, \eta_{1:c-1}^{(t)}) &\leq \sum_{d=1}^{D} L(\gamma_d^{(t+1)}, \phi_d^{(t+1)}, \xi_d^{(t+1)}; \alpha^{(t)}, \beta_{1:k}^{(t)}, \eta_{1:c-1}^{(t)}) \\ &\leq \sum_{d=1}^{D} L(\gamma_d^{(t+1)}, \phi_d^{(t+1)}, \xi_d^{(t+1)}; \alpha^{(t+1)}, \beta_{1:k}^{(t+1)}, \eta_{1:c-1}^{(t+1)}).\end{aligned}$$

The first inequality holds because $(\gamma_d^{(t+1)},\phi_d^{(t+1)},\xi_d^{(t+1)})$ maximizes $L(\gamma_d,\phi_d,\xi_d;\alpha^{(t)},\beta_{1:k}^{(t)},\eta_{1:c-1}^{(t)})$ in E-step. The second inequality holds because $(\alpha^{(t+1)},\beta_{1:k}^{(t+1)},\eta_{1:c-1}^{(t+1)})$ maximizes $\sum_{d=1}^{D} L(\gamma_d^{(t+1)},\phi_d^{(t+1)},\xi_d^{(t+1)};\alpha,\beta_{1:k},\eta_{1:c-1})$ in M step. Therefore, the objective function is guaranteed to be nondecreasing until convergence.

4.4.3 Prediction

Once we have the model parameters from variational EM, we can use $\eta_{1:c-1}$, the parameters for logistic regression, to do prediction. Given a document $x_{1:N}$, we have

$$E[\log p(y=h\,|\,x_{1:N},\alpha,\beta_{1:k},\eta_{1:c-1})]$$

$$=\begin{cases} \eta_h^T E[\bar{z}]-E\left[\log\left(1+\sum_{h=1}^{c-1}\exp\left(\eta_h^T\bar{z}\right)\right)\right] & [h]_1^{c-1} \\ 0-E\left[\log\left(1+\sum_{h=1}^{c-1}\exp\left(\eta_h^T\bar{z}\right)\right)\right] & h=c. \end{cases}$$

Since the second term for $[h]_1^{c-1}$ and $h = c$ are the same, we only need to compare $\left(\eta_1^T E[\bar{z}],\cdots,\eta_{c-1}^T E[\bar{z}],0\right)$. If the hth term is the largest, the predicted class is h.

The computation for $E[\bar{z}]$ is intractable, so we again introduce variational distribution $q(\pi,z_{1:N})$ and calculate $E_q[\bar{z}]$ as an approximation of $E[\bar{z}]$. In particular, $E_q[\bar{z}]=\frac{1}{N}\sum_{n=1}^{N}\phi_n$ for std DLDA and $E_q[\bar{z}]=\phi$ for fast DLDA, where ϕ_n and ϕ are obtained from running an E-step on the test documents given the model parameters from the training process.

4.5 EXPERIMENTAL RESULTS ON ASRS DATA SETS

In this section, we present experimental results for DLDA. The experiments will include two parts: in the first part, we use several small text data sets to show the properties of DLDA, as well as its performance compared with LDA and other state-of-the-art classification algorithms. In the second part, we focus on the ASRS data set and show fast DLDA's performance on it. The code for fast DLDA is available at https://c3.ndc.nasa.gov/dashlink/resources/316/.

TABLE 4.1 Text Data

	Nasa	Classic3	Cmudiff	Cmusim	Cmusame
D	4226	3893	3000	3000	3000
V	604	5923	7666	10,083	5932
Classes	Passenger	Aeronautics	Atheism	Guns	Graphics
	Flight crew	Medicine	Baseball	Mideast	Windows
	Maintenance	Information retrieval	Space	Politics	MS Windows

4.5.1 DLDA Properties

For the first part of experiments, we pick five small text data sets to run DLDA. The number of documents (D), the number of words in the dictionary (V), and the classes of documents are in Table 4.1. Nasa in Table 4.1 is a subset of ASRS database with three classes of reports. Others are commonly used benchmark data sets for text classification. We run std DLDA and fast DLDA on five small text data sets in Table 4.1 to show DLDA's performance and properties. We first compare DLDA with LDA and then compare DLDA with other state-of-the-art classification algorithms.

4.5.1.1 Discriminative Latent Dirichlet Allocation versus Latent Dirichlet Allocation

We compare DLDA and fast DLDA with $k = c$ to corresponding LDA and fast LDA [22],* where fast LDA uses a same variational inference strategy as fast DLDA. We initialize model parameters using all data points and their labels in the training set, in particular, we set the number of topics k to be the number of classes c, use the mean of documents (word occurrence vectors) in each class to initialize β, and use D_i/D to initialize α_i, where D_i is the number of data points in class i and D is the total number of data points. For $\eta_{1:c-1}$, we run a cross-validation by holding out 10% of training data as the validation set and use the parameters generating the best results on the validation set. In particular, each η_h in $\eta_{1:c-1}$ takes value of ru_h, where u_h is a unit vector with the hth dimension being 1 and others being 0, and r takes values from 0 to 100 in steps of 10. In principle, std LDA and fast LDA are not used for classification, but given the initialization we have introduced, there is a one-to-one mapping between the topic

* Just to be consistent with Std DLDA and Fast DLDA, we will use Std LDA and Fast LDA hereafter. Also, we do not compare DLDA with SLDA, since SLDA does not have a ready formulation for classification, as mentioned in the introduction.

TABLE 4.2 Accuracy for LDA and DLDA with $k = c$

	Nasa	Classic3	Cmudiff	Cmusim	Cmusame
Std LDA	0.9140 ± 0.0140	0.6733 ± 0.0254	0.9677 ± 0.0069	0.8143 ± 0.0161	0.5633 ± 0.0243
Fast LDA	0.9194 ± 0.0148	0.6748 ± 0.0242	0.9773 ± 0.0110	0.8553 ± 0.0197	0.7730 ± 0.0205
Std DLDA	0.9220 ± 0.0127	0.6710 ± 0.0256	0.9600 ± 0.0089	0.8140 ± 0.0252	0.6267 ± 0.0348
Fast DLDA	**0.9237 ± 0.0163**	**0.6756 ± 0.0234**	**0.9800 ± 0.0102**	**0.8653 ± 0.0182**	**0.7900 ± 0.0315**

and the class. Therefore, given the mixed membership on a test document, we pick the topic i with the largest probability as the predicted topic, if the corresponding class of topic i is the same with the class label, we consider the data point as correctly classified; otherwise, it is mistakenly classified. We use the percentage of correctly classified data points, i.e., the accuracy, to compare DLDA with LDA.

The results for std/fast LDA and std/fast DLDA are presented in Table 4.2. We make two observations: (1) Fast DLDA achieves a higher accuracy than std DLDA. (2) Std DLDA is not necessarily better than std LDA, but fast DLDA is always better than fast LDA in our experiment. The higher accuracy of fast DLDA demonstrates the effects of logistic regression in accommodating label information.

We compare the running time between std DLDA and fast DLDA. The results are presented in Table 4.3, where fast DLDA is about 20 to 150 times faster than the std DLDA, showing fast DLDA's superiority in terms of time efficiency. Combining the results with the accuracy comparison in Table 4.2, we can see that fast DLDA are generally more accurate and substantially faster than LDA. Such properties enable fast DLDA to be applied to large real data set such as ASRS database.

We further investigate the membership vectors generated by std/fast LDA and std/fast DLDA. In particular, we compute the Shannon entropy for the membership as a discrete distribution and compare the entropy

TABLE 4.3 Running Time (Seconds) of Std DLDA and Fast DLDA with $k = c$

	Nasa	Classic3	Cmudiff	Cmusim	Cmusame
Std DLDA	549.1762 ± 5.7491	2176.6794 ± 21.6241	1752.7828 ± 22.3689	2344.6408 ± 966.5029	1981.4625 ± 289.2406
Fast DLDA	**3.6359 ± 0.2090**	**114.3461 ± 18.1366**	**27.569 ± 0.61151**	**36.1029 ± 2.9873**	**40.1892 ± 5.8339**

among different algorithms. A low entropy implies almost a "sole membership," whereas a higher entropy implies a real mixed membership. Figure 4.4 is an example showing the histogram of membership entropy on Cmusim using four variants of LDA. We can see that for fast LDA/DLDA, almost all documents have extremely small membership entropies, while for std LDA/DLDA, the entropies fall into different ranges. The interesting observation indicates that fast variational inference actually generates "sole membership," while standard mean-field variational inference generates real "mixed membership." The fact that fast DLDA generates sole membership, as well as the previous observation that fast DLDA are better than std DLDA in terms of accuracy, shows the correlation between "sole membership" and higher classification accuracy, although we are not sure about the existence of causality between them. "Mixed membership" may be useful in various real applications, but it does not seem to help in terms of classification accuracy. Therefore, in ASRS data analysis, if we are only interested in the classification accuracy, it is safe or even better to use fast DLDA instead of std DLDA.

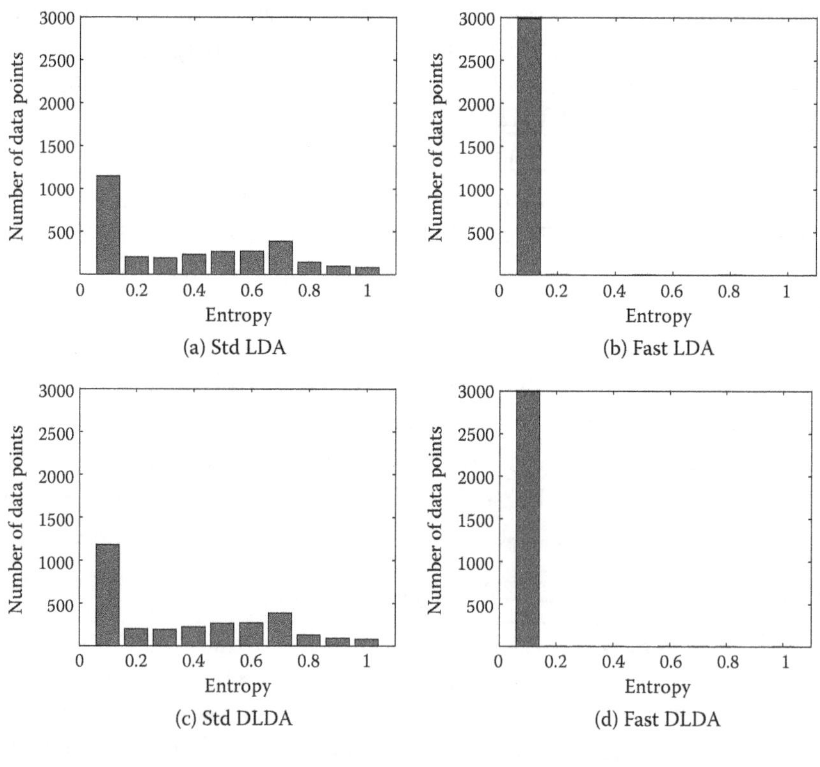

FIGURE 4.4 Histogram of membership entropy on Cmusim.

4.5.1.2 Fast DLDA versus Other Classification Algorithms

Since fast DLDA has better performance than std DLDA, in this subsection, we use fast DLDA to compare with other classification algorithms. In particular, we compare fast DLDA with support vector machine (SVM), naïve Bayes (NB), logistic regression (LR) and mixture of von Mises-Fisher (vMF) model on text data. Since fast DLDA is a combination of logistic regression and fast LDA, we also compare the results from DLDA with those from fast LDA and logistic regression in two steps sequentially.

For fast DLDA models, we run the experiments with an increasing k. In particular, we use $k = (c, c + 15, c + 30, c + 50, c + 100)$. For initialization of β, we use the mean of the training data in given classes plus some perturbation if $k > c$; for α, we set it to be $1/k$ on each dimension; and for $\eta_{1:c-1}$, we again use a cross-validation as in Section 4.1. For SVM, we use linear and RBF kernel with same cross-validation strategy on the penalty parameter and the kernel parameter (for RBF only), taking values from 10^{-5} to 10^{5} in multiplicative steps of 10, respectively.

The results are presented in Table 4.4. The top parts of the tables are the results from the generative models, and the bottom parts are the results from discriminative classification algorithms. For SVM, we report the highest accuracy of linear and RBF kernels with different parameters. We

TABLE 4.4 Accuracy for Different Classification Algorithms with Different Choices of k

	Nasa	**Classic3**	**Cmudiff**	**Cmusim**	**Cmusam**
Fast DLDA ($k = c$)	0.9237 ± 0.0163	0.6756 ± 0.0234	0.9800 ± 0.0102	0.8653 ± 0.0182	0.7900 ± 0.0315
Fast DLDA ($k = c + 15$)	0.9232 ± 0.0144	0.6858 ± 0.0216	0.9747 ± 0.0121	0.8713 ± 0.0264	0.8458 ± 0.0214
Fast DLDA ($k = c + 30$)	0.9301 ± 0.0128	0.6838 ± 0.0234	0.9817 ± 0.0099	0.8707 ± 0.0228	***0.8468 ± 0.0190***
Fast DLDA ($k = c + 50$)	0.9237 ± 0.0138	0.6854 ± 0.0211	***0.9823 ± 0.0083***	0.8700 ± 0.0230	0.8150 ± 0.0184
Fast DLDA ($k = c + 100$)	0.9261 ± 0.0102	***0.6866 ± 0.0245***	0.9760 ± 0.0108	***0.8718 ± 0.0182***	0.8347 ± 0.0187
vMF	0.9216 ± 0.0113	0.6509 ± 0.0246	0.9530 ± 0.0071	0.7447 ± 0.0214	0.7600 ± 0.0347
NB	***0.9334 ± 0.0094***	0.6766 ± 0.0230	0.9813 ± 0.0069	0.8613 ± 0.0216	0.8410 ± 0.0262
LR	0.9209 ± 0.0157	0.6396 ± 0.0252	0.9553 ± 0.0157	0.6750 ± 0.1330	0.4823 ± 0.1283
SVM	0.9192 ± 0.0146	0.6854 ± 0.0278	0.9563 ± 0.0105	0.8357 ± 0.0156	0.8120 ± 0.203

use boldface for the best results among the generative models and boldface-italic for the best results among all algorithms. Three parts of information could be read from the tables. (1) Overall, fast DLDA does better than all other algorithms, including SVM, on almost all data sets, which is a promising result although more rigorous experimentations may be needed to make a further investigation. (2) The better performance of fast DLDA compared with LR on original data sets indicates that the supervised low-dimensional representation we generate helps the classification. (3) Roughly, for DLDA, an increasing of accuracy with a larger k is observed, although the result goes up and down without a clear trend. One possible reason for the increasing accuracy is as follows: When k is too small, we are performing a drastic dimension reduction to represent each data point in a k-dimensional mixed-membership representation, which may cause a huge loss of information, but the loss may decrease when k increases.

Fast DLDA does dimensionality reduction and classification in one shot via a combination of fast LDA and logistic regression. In principle, we may also use these two algorithms sequentially in two steps, i.e., first using fast LDA to get a low-dimensional representation and then applying logistic regression on the low-dimensional representation for classification. The results with different choices of k following this two-step strategy are presented in Table 4.5. Comparing these results with Table 4.4, it is clear that DLDA models outperform the algorithm using fast and logistic regression sequentially, which means that by combining fast LDA and logistic regression, fast DLDA achieves supervised dimensionality reduction to obtain a better low-dimensional representation than fast LDA, which further

TABLE 4.5 Accuracy from Fast LDA and Logistic Regression Together with Different Choices of k

	Nasa	**Classic3**	**Cmudiff**	**Cmusim**	**Cmusame**
$k = c$	0.9194 ± 0.0148	0.5609 ± 0.0281	0.9513 ± 0.0268	0.8560 ± 0.0196	0.7733 ± 0.0339
$k = c + 15$	0.9118 ± 0.0124	0.5611 ± 0.0284	0.9756 ± 0.0112	0.8550 ± 0.0226	0.8173 ± 0.0197
$k = c + 30$	0.9080 ± 0.0143	0.5611 ± 0.0284	0.9760 ± 0.0116	0.8530 ± 0.0216	0.8183 ± 0.0168
$k = c + 50$	0.9085 ± 0.0132	0.5596 ± 0.0284	0.9746 ± 0.0123	0.8546 ± 0.0248	0.8040 ± 0.0201
$k = c + 100$	0.8926 ± 0.0942	0.6537 ± 0.0598	0.9423 ± 0.0896	0.7726 ± 0.1715	0.6726 ± 0.6726

TABLE 4.6 Extracted Topics Using Fast DLDA from ASRS Database

Communication with tower	report, aircraft, information, problem, conversation, revealed, callback, flight, air traffic control, controller
Power problem	engine, aircraft, start, gate, ground, shut, auxiliary power unit, flight, ramp, power
Fire emergency	smoke, flight, emergency, cabin, aircraft, landing, fire, runway, attendant, passenger
Navigation	flight, chart, navigation, zzz, takeoff, operations, aircraft, information, data, route
Passenger encountering turbulence	flight, turbulence, aircraft, passenger, attendants, attendant, moderate, severe, encountered, approximate
Approaching ground	approach, terrain, ground proximity warning system, warning, runway, mean sea level, aircraft, air traffic control, climb, approximate
Flight crew schedules	hours, flight, time, crew, day, duty, rest, trip, minutes, zzz
Maintenance for damage repair	maintenance, aircraft, repair, damage, zzz, inspection, manual, found, flight, item
Landing	runway, approach, tower, cleared, visual, instrument landing system, controller, told, landing, localizer
Departure	departure, turn, heading, runway, degrees, tower, takeoff, aircraft, traffic, climb

Note: The left column shows the topic, and the right column shows the top-ranked word list in this topic.

helps classification. Comparing these results with the accuracy of logistic regression on original data, we can see that there is no clear winner, which may depend on the quality of low-dimensional representation generated from LDA.

4.5.2 Fast DLDA on ASRS

For the second set of experiments, we use an ASRS data set that contains 66,309 reports labeled as anomalous events. Within these extracted reports there are 58 predefined classes. For instance, "anomaly. ground-encounters. vehicle" would denote one such class name, where "anomaly. ground-encounter" is the attribute name and "vehicle" is the value of the attribute. Other class names have the same structure.

4.5.2.1 Topics from Fast DLDA

We use fast LDA to extract topics from ASRS data. Table 4.6* shows some examples of obtained topics. The right column denotes a list of top ranked words within a given topic, and the left column contains a name that is

* "zzz" denotes specific names that cannot be revealed.

manually assigned to the topic in question. As we can see, these word lists are quite interpretable, and provide a reasonable representation of discussed topics. The results show that the reported problems include communication problems, power problems, etc. The experts could therefore focus on specific problems by looking at the related reports on certain topics. The analysis on these reports could help to improve the aviation system health management.

4.5.2.2 Relationship between Classes and Topics

Figure 4.5 shows the number of documents in each of the 58 classes. We can see that the classes are highly unbalanced with some classes containing more than 10,000 documents and others containing less than 50. Four largest classes are "anomaly. other-anomaly. other," "anomaly. nonadherence. published-procedure," "anomaly. nonadherence. clearance," and "anomaly. nonadherence. far," meaning that quite a few anomalies are the nonadherence of prescribed procedures or clearance. Four smallest classes are "anomaly. ground-encounters. gear-up-landing," "anomaly. ground-encounters. animal," "anomaly. cabin-event. galley-fire," and "anomaly. inflight-encounter. skydivers." Judging from these names, we can see that all of them are potentially dangerous accidents, hence should rarely happen, but the reports in these classes requires attention to avoid serious accident or injury.

We investigate the relationship between 58 classes and 200 topics in ASRS data. The number of topics was chosen upfront to be multiple times

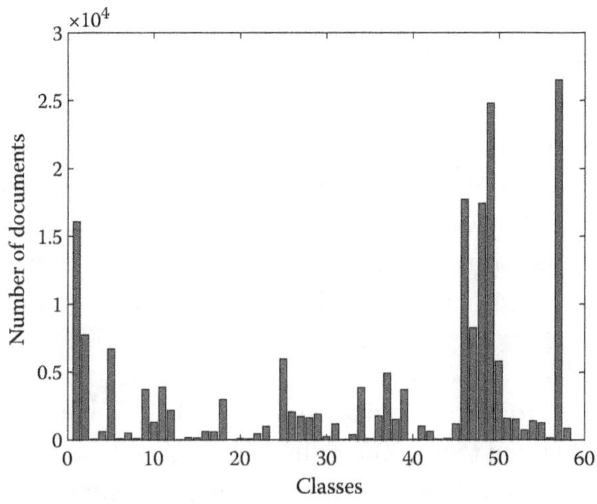

FIGURE 4.5 The number of documents in each of 58 classes.

larger than the number of predefined classes. For each document, we have a posterior over all 200 topics. We assign a document to its most likely topic. Meanwhile, each document is also assigned to multiple classes. Therefore, we can count the number of the documents falling in both class l and topic i, a higher value indicates a closer relationship. Such a strategy yields a 58 × 200 matrix M, with $M(l,i)$ denoting the approximate relationship between class l and topic i. We visualize the matrix M in Figure 4.6, where a lighter color indicates a closer relationship. As we can see, there are several bright rows in the figure, corresponding to the classes that have a close relationship with most of the topics. These classes include "anomaly. other-anomaly. other," "anomaly. nonadherence. published-procedure," "anomaly. nonadherence. clearance," and "anomaly. nonadherence. far." These classes are the largest classes in Figure 4.5. Since the size of these classes is large, they have a higher chance to co-occur with the topics. This also reflects the fact that some of the larger classes include very broad types of documents. For instance, "anomaly. other-anomaly. other" is lumping together anomalies that are not described by other predefined classes.

Given the matrix M, for each class, we can rank topics according to their closeness to the class. Some example classes with top five topics are presented in Table 4.7. Overall, the topic lists in each class appear reasonable. Topics in the same class are either similar to each other or are different but explain the class from different perspectives. For example, in class 1

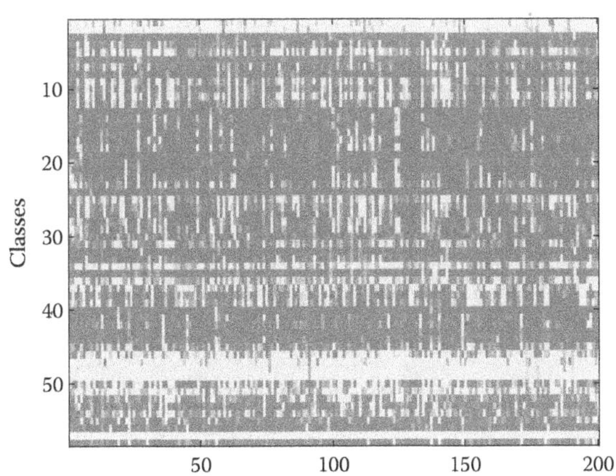

FIGURE 4.6 The relationship among 58 classes and 200 topics. A lighter color indicates a closer relationship.

TABLE 4.7 Top-Ranked Topics in Each Class

Class 1	anomaly.other-anomaly.smoke-or-fire
Topics	smoke, flight, emergency, cabin, aircraft, landing, fire, runway, attendant, passenger
	smoke, flight, passenger, aircraft, cabin, fire, attendant, gate, captain, door
	flight, smoke, smell, cabin, cockpit, captain, fire, attendant, emergency, aircraft
	report, smoke, aircraft, flight, stated, maintenance, cabin, information, emergency, revealed
	fire, engine, aircraft, warning, auxiliary power unit, light, flight, maintenance, checklist, emergency
Class 2	anomaly.inflight-encounter.turbulence
Topics	flight, turbulence, aircraft, passenger, attendants, attendant, moderate, severe, encountered, approximate
	flight, turbulence, passenger, attendant, attendants, seat, galley, aircraft, back, cabin
	aircraft, turbulence, autopilot, flight, air traffic control, moderate, severe, encountered, approximate, captain
	approach, runway, knots, aircraft, degrees, landing, tower, airspeed, approximate, winds
	deviation, air traffic control, emergency, clearance, aircraft, turbulence, captain, requested, flight, advised
Class 3	anomaly.other-anomaly.loss-of-aircraft-control
Topics	aircraft, wake, turbulence, report, approach, runway, degrees, roll, flight, encountered
	aircraft, runway, landing, approach, knots, captain, normal, touchdown, flight, gear
	runway, aircraft, braking, taxiway, landing, action, approach, turn, captain, tower
	aircraft, turbulence, autopilot, flight, air traffic control, moderate, severe, encountered, approximate, captain
	flight, turbulence, aircraft, passenger, attendants, attendant, moderate, severe, encountered, approximate
Class 4	anomaly.cabin-event.passenger-illness
Topics	passenger, flight, attendant, medical, oxygen, captain, report, attendants, aircraft, cabin
	passenger, flight, medical, emergency, attendant, landing, aircraft, dispatch, gate, air traffic control
	passenger, flight, attendant, told, captain, seat, asked, back, aircraft, sit
	flight, passenger, attendant, seat, captain, lavatory report, told, attendants, back
	flight, passenger, aircraft, attendant, attendants, crew, captain, time, told, gate

(*continued*)

TABLE 4.7 Top-Ranked Topics in Each Class (Continued)

Class 5	anomaly.incursion.landing-without-clearance
Topics	tower, approach, runway, landing, clearance, cleared, frequency, control, land, aircraft
	runway, approach, tower, cleared, visual, instrument landing system, controller, told, landing, localizer
	runway, approach, airport, visual, captain, tower, cleared, final, sight, instrument landing system
	runway, approach, tower, aircraft, landing, captain, knots, time, instrument landing system, final
	runway, tower, aircraft, landing, cleared, approach, land, taxiway, final, short
Class 6	anomaly.aircraft-equipment-problem.critical
Topics	engine, oil, landing, emergency, pressure, flight, light, checklist, declared, zzz
	hydraulic, system, landing, pressure, quantity, emergency, aircraft, gear, maintenance, runway
	engine, aircraft, runway, landing, fire, emergency, flight, takeoff, normal, captain
	tire, aircraft, brake, wheel, maintenance, tires, main, nose, installed, found
	engine, report, aircraft, stated, information, revealed, callback, conversation, maintenance, oil

"anomaly.other-anomaly.smoke-or-fire," most of the topics are related to smoke or fire. Among these some are related to passenger, and some are related to power and maintenance. In class 2 "anomaly.inflight-encounter. turbulence," three topics are directly related to the turbulence, and the other two are about the airspeed and deviation caused by the turbulence, respectively. For class 3 "anomaly.other-anomaly.loss-of-aircraft-control," we can see it also contains several topics about turbulence, In class 4 "anomaly.cabin-event.passenger-illness," the first two topics are directly related to the medical emergency of the passenger, the other three topics are also related to passenger's conduct. For the class 5 "anomaly.incursion. landing-without-clearance," there are topics about landing or clearance or both. Finally, the class 6 "anomaly.aircraft-equipment-problem.critical" contains equipment problem from different aspects, such as engine, hydraulic pressure, tire, brake, etc.

These top-ranked topics provide a finer-level overview for each anomaly class, hence are helpful to build a better understanding of the safety issues arising in the aviation domain. For example, from class 1, we can see that the main causes of smoke/fire are the power system problems or passenger misconduct, so attention should be given to these two aspects to avoid smoke/fire. From class 6, we know that the commonly reported equipment problems include engine problems, hydraulic pressure problems, and so on. The experts then have a clear target on these specific problems, and the related reports could be used for analysis. In addition, the shared topics among different classes help us to discover the relationship between classes. For example, the class 3 contains several topics about turbulence, which are shared with the class 2, indicating that loss of aircraft control are mostly caused by the turbulence. Therefore, how to deal with turbulence might be essential in aircraft control.

4.5.2.3 Classification Results

We compare the classification performance of fast DLDA with the result from LDA+SVM, i.e., using LDA and subsequently applying SVM in two steps. Again, fast DLDA integrates topic extraction and classification in one model, and LDA+SVM first generates the low-dimensional representation over the topics from LDA and then runs SVM to get the classification result. Since neither fast DLDA nor SVM is designed for multilabel classification, we use the "one-versus-rest" strategy. In particular, for each label l, we build a two-class classifier that separates the documents with label l from others. Therefore, given s labels in total for each document, we can run s classifiers for each of the labels to get an s-dimensional predicted multilabel vector.

As for the evaluation, we use five different measures: one error, precision, coverage, ranking loss, and hamming loss. Let $g(\mathbf{x},l)$ denote a real-valued function that assigns a score to label l for document $\mathbf{x}$, such that a larger score is considered better. For fast DLDA, $g(\mathbf{x},l) = E[\log p(y = l|\mathbf{x})]$, and for SVM, we use the soft prediction score from libsvm [24]. Also, let $f(\mathbf{x})$ denote the classifier whose output is the predicted multilabel vector. Further, let $L_{\mathbf{x}}$ denote a set of true labels associated with $\mathbf{x}$. The measurements we use are defined as follows:

1. One error evaluates how frequently the top ranked predicted label is not among the true labels. If $1[\cdot]$ denotes the indicator function, we have

$$OneError(g) = \frac{1}{D}\sum_{d=1}^{D} 1\left[\text{argmax}_{l\in L}\, g(\mathbf{x}_d, l) \notin L_{\mathbf{x}_d}\right]. \tag{4.21}$$

2. For true labels $l \in L_x$, *average precision* evaluates the fraction of labels in L_x that rank at least as high as l according to the scoring rule g on average. For any data point $\mathbf{x}$ and any label $l \in L_x$, let $\mathcal{R}(\mathbf{x},l) = \{l' \in L_x | rank_g(\mathbf{x},l') \leq rank_g(\mathbf{x},l)\}$, where the ranking is among all possible labels. Then, average precision is

$$AvePrec(g) = \frac{1}{D}\sum_{d=1}^{D}\frac{1}{|L_{\mathbf{x}_d}|}\sum_{l \in L_{\mathbf{x}_d}}\frac{|\mathcal{R}(\mathbf{x}_d,l)|}{rank_g(\mathbf{x}_d,l)}. \tag{4.22}$$

3. *Coverage* reflects on average how far one needs to go down in the label ranking to cover all actual labels of an instance:

$$Coverage(g) = \frac{1}{D}\sum_{d=1}^{D}\left(\max_{l \in L_{\mathbf{x}_d}} rank_g(\mathbf{x}_d,l) - 1\right). \tag{4.23}$$

4. *Hamming loss* evaluates the fraction of label instance pairs that were misclassified:

$$HammingLoss(f) = \frac{1}{D}\sum_{d=1}^{D}\frac{1}{C}\left|f(\mathbf{x}_d)\Delta L_{\mathbf{x}_d}\right| \tag{4.24}$$

where $(\cdot\Delta\cdot)$ denotes the symmetric difference between two sets.

5. *Ranking loss* reflects the average number of labels that are reversely ordered for a given instance. Let $\mathcal{T}(\mathbf{x}_d) = \{(l_1,l_2) | g(\mathbf{x}_d,l_1) \leq g(\mathbf{x}_d,l_2), (l_1,l_2) \in L_{\mathbf{x}_d} \times \bar{L}_{\mathbf{x}_d}\}$, where $\bar{L}_{\mathbf{x}_d}$ denotes the complement of $L_{\mathbf{x}_d}$. Ranking loss is defined as:

$$RankLoss(g) = \frac{1}{D}\sum_{d=1}^{D}\frac{|\mathcal{T}(\mathbf{x}_d)|}{|\bar{L}_{\mathbf{x}_d}||L_{\mathbf{x}_d}|}. \tag{4.25}$$

For OneError, Coverage, HammingLoss, and RankLoss, smaller values are considered better. In particular, for a perfect performance, *HammingLoss*(h) = *RankLoss*(g) = 0. For AvePrec, the higher, the better.

TABLE 4.8 Classification Performance from Fast DLDA and LDA+SVM

	HammingLoss	OneError	RankLoss	Coverage	AvePrec
LDA + SVM	0.073 ± 0.007	0.863 ± 0.032	0.334 ± 0.027	26.374 ± 1.557	0.169 ± 0.005
Fast DLDA	**0.049 ± 0.001**	**0.492 ± 0.007**	**0.141 ± 0.031**	**16.000 ± 2.631**	**0.507 ± 0.019**

Note: For HammingLoss, OneError, RankLoss, and Coverage, the lower, the better; for AvePrec, the higher, the better.

The classification results using fivefold cross-validation on ASRS are in Table 4.8. We can see that fast DLDA performs better than LDA + SVM across all five measurements, and the advantage is significant judging by the standard deviation. The high accuracy of fast DLDA enables us to automatically categorize the new coming report given the labeled ones in ASRS database, which means that we can quickly discover the main problem covered in the new report. It helps to identify the right expertise needed to analyze the new reports, and also helps to organize the new coming reports automatically for future reference on the aviation safety issues.

4.6 CONCLUSION

In this chapter, we have discussed DLDA, which is a combination of LDA and multilabel logistic regression. DLDA is able to generate interpretable topics while leveraging labels for classification. We proposed two variational inference algorithms to learn the model, yielding std DLDA and fast DLDA, respectively. By running experiments on multiple benchmark text data sets as well as the ASRS data, we showed that fast DLDA is competitive with several state-of-the-art classification algorithms. At the same time, it is also capable of extracting interpretable topics from the ASRS data.

Since the documents in ASRS database have multiple labels, using a one-versus-rest approach is potentially crude and neglects important relationships between classes. As future work, we intend to develop a discriminative topic model specifically for multilabel classification that is capable of modeling correlations among classes. Another useful extension would be to utilize Dirichlet process priors, to automatically learn the number of topics.

REFERENCES

[1] Aviation Safety Reporting System. http://akama.arc.nasa.gov/ASRSDBOnline/QueryWizard_Filter.aspx.

[2] D. Blei, A. Ng, M. Jordan. Latent Dirichlet allocation. *Journal of Machine Learning Research*, 3:993–1022, 2003.

[3] D. Blei, J. Lafferty. Dynamic topic models. In *Proceedings of the 23rd International Conference on Machine Learning (ICML)*, 2006.

[4] D. Blei, J. Lafferty. Correlated topic models. In *Proceedings of the 19th Annual Conference on Neural Information Processing Systems (NIPS)*, 2006.

[5] C. Burges. A tutorial on support vector machines for pattern recognition. *Data Mining and Knowledge Discovery*, 2:121–167, 1998.

[6] F. Pampel. *Logistic Regression: A Primer*. Sage, 2000.

[7] D. Blei, J. McAuliffe. Supervised topic models. In *Proceedings of the 20th Annual Conference on Neural Information Processing Systems (NIPS)*, 2007.

[8] P. McCullagh, J.A. Nelder. *Generalized Linear Models*. Chapman & Hall/CRC, 1989.

[9] J. Zhu, A. Ahmed, E. Xing. Medlda: Maximum margin supervised topic models for regression and classification. In *Proceedings of the 26th International Conference on Machine Learning (ICML)*, 2009.

[10] P. Flaherty, G. Giaever, M. Jordan, A. Arkin. A latent variable model for chemogenomic profiling. *Bioinformatics*, 21:3286–3293, 2005.

[11] S. Lacoste-Julien, F. Sha, M. Jordan. DiscLDA: Discriminative learning for dimensionality reduction and classification. In *Proceedings of the 21th Annual Conference on Neural Information Processing Systems (NIPS)*, 2008.

[12] S. Huh, S. Fienberg. Discriminative topic modeling based on manifold learning. In *Proceedings of the 17th ACM International Conference on Knowledge Discovery and Data Mining (KDD)*, 2010.

[13] D. Mimno, A. McCallum. Topic models conditioned on arbitrary features with Dirichlet-multinomial regression. In *Proceddings of the 24th Conference on Uncertainty in Artificial Intelligence (UAI)*, 2008.

[14] H. Wang, M. Huang, X. Zhu. A generative probabilistic model for multi-label classification. In *Proceedings of the IEEE International Conference on Data Mining (ICDM)*, 2008.

[15] C. Wang, D. Blei, L. Fei-Fei. Simultaneous image classification and annotation. In *Proceedings of IEEE International Conference on Computer Vision and Pattern Recogonition (CVPR)*, 2009.

[16] T. Hoffman. Probabilistic latent semantic indexing. In *Proceedings of the 15th Conference in Uncertainty in Artificial Intelligence (UAI)*, 1999.

[17] W. Buntine. Variational extensions to EM and Multinomial PCA. In *European Conference on Machine Learning (ECML)*, 2002.

[18] A. Banerjee, S. Merugu, I. Dhillon, J. Ghosh. Clustering with Bregman divergences. *Journal of Machine Learning Research*, 6:1705–1749, 2005.

[19] A. Banerjee. An analysis of logistic models: Exponential family connections and online performance. In *Proceedings of the 7th SIAM International Conference on Data Mining (SDM)*, 2007.

[20] D. Blei, M. Jordan. Variational inference for Dirichlet process mixtures. *Bayesian Analysis*, 1(1):121–144, 2006.
[21] P. Domingos, M. Pazzani. On the optimality of the simple Bayesian classifier under zero-one loss. *Machine Learning Journal*, 29:103–130, 1997.
[22] H. Shan, A. Banerjee. Mixed-membership naive Bayes models. *Data Mining and Knowledge Discovery*, 2010.
[23] T. Minka. A comparison of numerical optimizers for logistic regression. Technical report. 2003.
[24] C. Chang, C. Lin. *LIBSVM: a library for support vector machines*. 2001. Software available at http://www.csie.ntu.edu.tw/~cjlin/libsvm.

CHAPTER 5

Prognostic Performance Metrics

Kai Goebel

Abhinav Saxena

Sankalita Saha

Bhaskar Saha

Jose Celaya

CONTENTS

5.1 INTRODUCTION

We define prognostics here strictly as the predicting of remaining useful life (RUL) of a component or system. The prediction is typically performed only after the "health" of the component or system deteriorates beyond a certain threshold. Often times, that threshold is tripped because a fault occurs. A fault is a state of a component or system that deviates from the normal state such that the integrity of the component is outside of its required specification. A fault does not necessarily imply that the overall system does not operate anymore; however, the damage that characterizes the fault often grows under the influence of operations to a failure. The latter is the state at which the component or system does not meet its desired function anymore. It is the task of prognostics to estimate the time that it takes from the current time to the failed state, conditional on anticipated future usage. This would give operators access to information that has significant implications on system safety or cost of operations. Where safety is impacted, the ability to predict failure allows operators to take action that preserves the assets either through rescue operation or through remedial action that avert failure altogether. Where minimizing cost of operations is the primary objective, predictive information allows operators to avert secondary damage or to perform maintenance in the most cost-effective fashion. Often times, there is a mix of objectives that need to be optimized together, sometimes weighted by different preferences.

Predicting remaining component or system life can be accomplished in several ways. Where sufficient historical run-to-failure trajectories are available, data mining techniques can be employed to perform the predictions. Traditionally, reliability-based predictions have been used widely in the manufacturing industry to schedule preventive maintenance. In contract, the focus of this chapter is mainly on condition-based prognostic systems for a particular monitored unit under test (UUT). Instead of considering the entire population for a statistical life estimates, one can employ physics-based models to perform the predictions or a combination of models and history data. In either case, predictions are conditional on future conditions and are subject to significant amounts of uncertainty. Methods for prognostics ideally express their confidence of their own prediction based on an assessment of the various uncertainty sources. Besides uncertainty of future usage, uncertainty also comes from the current state assessment, the models used, measurement noise, etc.

Metrics can be understood as a standardized language by which technology developers and users communicate their findings and compare results. This aids in allowing the proper expression of requirements as well as the dissemination of scientific information. Two surveys on methods for prognostics, one on data-driven methods [1] and one on artificial-intelligence–based methods [2] reveal a lack of standardized methodologies for performance evaluation or a lack of performance methods altogether. The most recent ISO standard by the International Organization for Standards [3] for prognostics in condition monitoring and diagnostics of machines does not even provide a firm definition of any such method. Nonetheless, there has been recently a significant push toward crafting suitable metrics to evaluate prognostic performance [4,5]. These metrics address primarily evaluation of algorithmic performance for prognostics applications. They are mostly focused on tackling offline performance evaluation methods for applications where run-to-failure data are available and true end-of-life (EoL) is known a priori. They are therefore particularly useful for the algorithm development phase where feedback from the metrics can be used to fine-tune prognostic algorithms. It needs to be appreciated that these metrics are continuously evolving. Efforts are also underway toward designing on-line performance metrics although they have not reached a significant level of maturity.

This chapter presents a discussion on prognostics metrics. After a review of performance assessment for prediction/forecasting applications a categorization of prognostic metrics into several classes is performed. This categorization suggests that there can be various different objectives that drive improvements in prognostic performance and correspondingly a different set of metrics may be used to obtain performance feedback.

5.2 BACKGROUND

As more diverse research communities and practitioners start adopting prognostics and health management (PHM) techniques, it becomes imperative to use standardized prognostic methodologies [6] as well as to use metrics to measure performances. However, since prognostics is a developing field, the challenges in developing standards and metrics are numerous [7,8]. We start out by providing an overview of prognostic concepts that are used in a variety of domains.

5.2.1 Prediction Categorization

Prior to delineating the methods to assess prognostic performance it may be useful to provide a brief discussion about different types of applications in which predictions are employed. Based on an analysis of aerospace, medicine, nuclear energy, finance, weather, and automotive domains, it was found that one can distinguish roughly between forecasting and prognostics.

5.2.1.1 Forecasting

Forecasting is found in applications where predictions are made to describe expected future behavior without predicting a fixed target. That is, there is no notion of EoL, and consequently, there is no concept of RUL. Example application areas are weather or finance domains. The prediction format can be either quantitative (e.g., prediction of exact numbers) or qualitative (e.g., high or low demands) in nature. Furthermore, the data trends are generally nonmonotonic in such applications. Predictions may be discrete (e.g., forecasting market demand for a particular month) or continuous (e.g., variation of temperature over the period of next week).

Details and more references to such applications in various domains can be found in [29].

5.2.1.2 Prognostics

The other class of applications makes use of critical thresholds such that if the system under test crosses this threshold it is declared to have failed or lost its functional capability. This class of applications—e.g., medicine, nuclear, mechanical, and electrical industrial systems—involves predictions of RUL and involves decay or fault propagation models to capture the behavior of the system.

Predictions can be made in two forms: (1) an event prediction where the time for EoL is estimated and (2) a decay prediction where the complete future trajectory is predicted until EoL is reached. It must be noted, however, that EoL criteria need not always be a complete loss or failure. In safety critical applications, EoL is often a degraded state where performance level has deteriorated to cross a predetermined safety margin even though the component may still retain partial functionality. For example, in the electronics domain, EoL of a switching device (such as a MOSFET) is not necessarily the complete loss of the switching functionality. Instead, it could be a decrease in the switching frequency below a certain threshold level.

There are two main types of applications where predictions for system health are made: These include *predicting wear* of a system or *predicting failure* in the event of a fault.

Failure predictions: An otherwise healthy system may encounter a fault that grows due to continued usage (or exposure to adverse conditions) that may result in a failure. In such cases, it is critical to detect the presence of a fault (ideally shortly after it happens), the particular fault mode, its severity, and its rate of growth so that appropriate decisions may be taken to avoid undesired, possibly catastrophic, events. Here, the task of prognosis is to estimate expected EoL, i.e., determine when the system will no longer operate under specifications. In some cases, it is not only important to know when the system will break but also how it will approach the failure. In those cases, instead of predicting just the event of EoL, a complete trajectory may be predicted, where the end point of the trajectory also determines the EoL. Examples of such applications include structural faults such as cracks in metallic structures or die-attach degradation in power semiconductor devices.

Wear predictions: There are many situations where systems undergo expected normal wear and need to be maintained or replaced whenever the wear levels impact functionality. In these cases, the system does not experience a fault condition even under the degraded performance. Therefore, the health of the system is tracked from the very beginning of system deployment and detection and diagnosis are not predecessors for prognostics. As stated earlier, the end point of these trajectories can be used to determine the EoL point so appropriate decisions may be taken. Examples of such applications include battery capacity degradation and valve leakage due to wear.

5.2.2 Prediction Methods

There are several ways to carry out prognostics. In some cases, a detailed physical model of the unit under observation can be used. The model captures the unit's behavior under operational and environmental conditions and provides an expected response that describes the current and (given the proper input) future states. Alternative to a physics-based model, historical data can be utilized to estimate expected time to failure. The key is to either have access to a sufficient amount of existing historical data (e.g., medicine) or to be able to experimentally generate run-to-failure trajectories (e.g., for some mechanical systems). Then, a variety of data-driven or statistical techniques can be applied.

The availability of run-to-failure data allows the straightforward evaluation of prediction performance by comparing the predicted EoL to the actual EoL. However, there are many applications where run-to-failure experiments cannot be afforded or where very little failure history data are available (e.g., aerospace). It becomes somewhat more difficult to assess the performance in such cases due to the absence of knowledge about the future outcomes. Methods are tested on experimental or simulated data and, when fielded, are expected to perform similarly on real systems. However, algorithm functionality does rarely translate without loss of performance from simulation environment or laboratory to the field. Indeed, validation and verification of prognostic methods remain a thorny issue.

5.2.3 Performance Evaluation Methods

Techniques employed for prediction or forecasting in the application areas enumerated above use metrics that are based on accuracy and precision with several slight variations [9]. Mostly, they are customized to better serve a particular domain. In medicine and finance, for example, several statistical measures are used to benefit from the availability of large data sets. In contrast, predictions in medicine are commonly evaluated based on hypothesis testing methodologies. In the finance domain, errors are calculated based on reference prediction models. The precision and accuracy metrics include, for example, mean squared error (MSE), standard deviation (SD), mean absolute deviation (MAD), median absolute deviation (MdAD), mean absolute percentage error (MAPE), and similar variants. Other domains, such as aerospace, electronics, and nuclear are less mature with respect to fielded prognostics applications. There, metrics from other system health techniques, such as diagnostics, have been used with the goal to capture the characteristics of prognostics (with varied success). Metrics used include false-positives (FP), false-negatives (FN), and receiver operator characteristics (ROC) curves [10]. Other metrics include those from the reliability domain such as mean time between failures (MTBF) or the ratio mean time between unit replacements (MTBF/MTBUR). Adaptations include, for example, the augmentation with business metrics such as return on investment (ROI) [11], technical value (TV) [12], net present value (NPV) [13], and life cycle cost (LCC) [14].

It becomes apparent that there are several types of metrics for prognostics based on the purpose of prognostics and the end user. A categorization with these objectives in mind allows a more targeted choice of appropriate metrics. Coble and Hines [15] categorized prognostic algorithms into three categories based on type of models/information used for predictions. Wheeler et al. [16] categorized end users from a health management stakeholder's point of view. The top-level user groups were operations, regulatory, and engineering. We combine and expand on these notions and categorize prognostic metrics based both on their goal as well as end users (see Table 5.1).

A favorable cost-benefit case is the hinge pin of a successfully fielded prognostic solution and cost-benefit metrics allow a quantification of the degree of fulfillment. Similarly, verifiability and certification metrics determine the degree to which a prognostic solution conforms with safety assurance and certification requirements. Both these top-level metrics categories require prognostic estimates to satisfy stringent performance metrics, which are often derived from reliability analysis or condition-based prognostic methods. Computational performance metrics are important for implementation and figure in the trade space of cost-benefit analysis and algorithmic performance. Often, most of the metrics mentioned are connected through a requirement specification process.

TABLE 5.1 Categorization of Prognostic Metrics Based on End Usage

		Operations				Engineering		Regulatory
Metrics	**Assessment Goals**	**Program Manager**	**Plant Manager**	**Operator**	**Maintainer**	**Designer**	**Researcher**	**Policy Maker**
Certification metrics	Assess conformance to safety assurance and certification requirements	X						X
Cost-benefit metrics	Assess the economic viability for specific applications before it can be approved or funded	X						X
Reliability-based metrics	Assess probabilities of failures based on statistical evidence from multiple systems			X		X		X
Algorithm performance metrics	Assess performance of prediction algorithms in predicting EoL	X	X	X	X		X	X
Computational performance metrics	Assess computational requirements					X	X	

5.3 METRICS FOR PROGNOSTIC APPLICATIONS

In this section, the various prognostic metrics are presented by categories in more detail. Particular emphasis is given to performance metrics.

5.3.1 Certification Metrics

Regulatory bodies (such as the FAA) are concerned with whether a fielded system might negatively impact (directly or indirectly) overall system safety. Associated metrics can be expressed as the logical conjunction of an exhaustive set of safety-related use cases.

5.3.2 Cost-Benefit Metrics

Establishing cost-benefits of prognostics is an important step in integrating the health management practices into fielded applications. Thus, metrics that measure economic viability of prognostics have started gaining in importance. Some of the most common cost-benefit metrics include:

5.3.2.1 MTBF-to-MTBUR Ratio Method

This reliability statistics-based expresses the efficiency of a maintenance operation by measuring the ratio between the lengths of time a component is expected to last and the length of time for which it was used before it was replaced [17].

5.3.2.2 Life Cycle Cost

LCC is fundamentally the sum of acquisition cost and cost of operations. To assess the value of prognostics, LCC is compared with and without prognostics [18].

5.3.2.3 Return on Investment

In an ROI calculation, the difference between return and investment (the gain) is divided by the investment. It is one of the most commonly used metrics (not just in the context of prognostics) that assesses the benefits of deploying a PHM system.

5.3.2.4 Technical Value

The benefits achieved through accurate detection, fault isolation, and prediction of critical failure modes are weighed against the costs associated with false alarms, inaccurate diagnoses/prognoses, and resource requirements of implementing and operating specific techniques [19,20].

5.3.2.5 Total Value

Given the coverage a PHM system provides for multiple fault modes in a system, total value quantifies the usefulness of a PHM technology in a particular application. Total value is defined as the summation of the benefits prognostics provides over all the failure modes that it can diagnose or give a prognosis, minus the implementation cost, operation and maintenance cost, and consequential cost of incorrect assessments. This metric connects the algorithm performance in a PHM system to the management and operational performance.

5.3.3 Metrics for Computational Performance

Computational performance is one of the most closely related factors to actual implementation of a system. It provides a mechanism to negotiate between computational and time resources that are demanded and the required fidelity and accuracy of the PHM system. Most of the metrics that can be used to quantify computational performance come from theoretical computer science and computer engineering. These are not specific to prognostics and are just mentioned here for completeness. Computational performance metrics include computational complexity, CPU time, memory size, and data rate. Depending on the length of the prediction horizon in a prognostic application, data processing capabilities are of greater significance from the design and implementation point of view. All the above metrics help specify the hardware requirements or otherwise specify constraints within which a software must work and still satisfy algorithmic performance requirements.

5.3.4 Metrics for Reliability Analysis

Referring again to Table 5.1, reliability analysis metrics are chiefly used by operators, designers, and policy makers. Reliability analysis stems from statistical evidences aggregated from historical data. Failure rates and distributions are extracted from history data or experimental data, which are then used to make failure predictions for a system under test. While prognostics is the science of prediction based on condition and usage of the monitored UUT, the reliability analysis predicts failures based on expected outcome from the observed statistic over a population of the UUT. It has been the traditional way to use these metrics to assess the costs and the risks of using a system. Used correctly, reliability metrics are connected to system improvement due to prognostic performance.

Broadly classifying, there are two types of reliability metrics as described below [21]:

5.3.4.1 Constant Rate Reliability Metrics

These are the most popular reliability metrics in the electronics industry as they represent a good approximation of the flat region of the reliability bathtub curve. Mean life metrics usually assume an exponential distribution, which makes them equivalent to constant rate metrics. These rates are typically measured from field data and are simple and intuitive to explain. Some common examples of these metrics are mean time between failure (MTBF), mean time to failure (MTTF), part return/repair rate, part replacement rate, mean time between service call (MTBSC), and mean time between maintenance action (MTBMA).

5.3.4.2 Probability of Success Metrics

When systems do not show constant failure rates, specifying mean times does not suffice. In such cases a better way is to specify probability of success or, in other words, the probability that a system performs a required function under stated condition for a stated period. Another way to specify probability of success is to measure the percentage of population that survives a specific duration. Therefore, these metrics are usually time dependent, i.e., the probability of success will depend on the length of the mission. These may be specified as the percentiles of the distributions. A common example used in mechanical systems domain is *Lx Life*, which specifies the number of hours after which at least x% of the population would have failed. Other metrics commonly used are failure free operating time, maintenance free operating time, mean mission duration, etc. More discussion on reliability-based metrics may be found in [22].

5.3.5 Metrics for Prognostics Algorithm Performance

Before using performance metrics, an a priori analysis should be conducted to identify the relevant factors in a given application and address them appropriately. A good set of metrics should accommodate all or most of these factors. In this context, the challenges surrounding prognostics (as compared with, say, diagnostics) should be discussed. It should also be noted that the metrics detailed in this chapter are continuously evolving as the field matures further.

5.3.5.1 Challenges

Prognostics requires special consideration in a number of areas. These include the acausality phenomenon, need for run-to-failure data, online performance evaluation, and expression of estimation uncertainty.

Acausality: An acausal system is defined as a system with outputs and internal states that depend on future input values. Prognostics has acausal properties. It requires input from future events (for instance knowledge about operational conditions and load profiles) to make accurate predictions. To accurately assess the performance (both accuracy or precision), one must also know the true EoL to compare with the predicted EoL estimates. In some cases, future operating conditions are well known. This is the case for example for stationary applications with constant operating conditions. However, in nonstationary applications and where knowledge about future events is not available, estimates may be derived based on past usage history, the expected mission profile, and predictions for future operating and environmental conditions that are not controllable (e.g., weather conditions). This, however, adds uncertainty to the overall process and complicates prognostic performance evaluation.

Run-to-failure data from real applications: Assessing the correctness of prognostics benefits greatly from allowing the system to fail such that the prediction can be confirmed. For many systems, this is not feasible because it may be too expensive or because it negatively impacts system safety. However, if a corrective action (such as maintenance or repair) is taken, one has just removed the ability to assess how early the prediction was. This is sometimes referred to as the "paradox of prognostics."

Online performance evaluation: The aforementioned considerations lead to an argument in favor of controlled run-to-failure (RtF) experiments for the algorithm development phase. While this may allow offline performance evaluation, some issues remain: First, it is difficult to extend the results of offline conditions to a real-time scenario; second, an RtF experiment needs often times frequent disassemblies to gather ground truth data. This assembly-disassembly process creates variations in the system performance, and the EoL point shifts from what it may have been in the beginning of the experiment. Since actual EoL is observed only at the end there is no guarantee that a prediction made based on the initial part of data will be very accurate. Whereas, this does not necessarily mean that prognostic algorithm is poorly trained, it is difficult to prove otherwise. Therefore, one must be careful while interpreting the performance assessment results. Third, even controlled subscale RtF experiments can be very expensive and time-consuming, in particular, if one seeks to conduct a statistically significant number of experiments for an exhaustive set of components and fault modes.

Uncertainty in prognostics: The quantification of prediction confidence is indispensable in prognostics. Consider a remaining life estimate of, say, 5 hours. If one knew that the confidence bounds for a given risk acceptance level are also 5 hours, then reactive action has to be taken immediately. If, however, the confidence bounds are at ± 1 hour, then a completely different set of action can be taken. Without such information any prognostic estimate is of limited use and cannot be incorporated in mission critical applications [6]. Uncertainties arise from various sources in a PHM system [23–25]. Some of these sources include:

- Model uncertainties (errors in the representation and parameters of both the system model and fault propagation model).
- Measurement uncertainties (these arise from sensor noise, ability of sensor to detect and disambiguate between various fault modes, loss of information due to data preprocessing, approximations, and simplifications).
- Operating environment uncertainties.
- Future load profile uncertainties (arising from unforeseen future and variability in usage history data).
- Input data uncertainties (estimate of initial state of the system, variability in material properties, manufacturing variability), etc.

Assessing the levels and characteristics of uncertainties arising from each of these sources is often times not a trivial task. It is even more difficult to determine how these uncertainties combine at different stages of the prognostic process and propagate through the possibly complex and nonlinear system. On top of that, statistical properties may not follow any known parametric distributions, thus complicating analytical solutions.

Owing to all of these challenges, uncertainty representation and management has become an active area of research in the field of PHM [25–29]. Methods for prognostic performance evaluation must then be able to incorporate various expressions of uncertainties.

Performance metrics for prognostics can be classified into accuracy, precision, and robustness. We use the working definition for accuracy as those that assess the degree of closeness of predictions to the actual failure time. Precision is defined as the spread of predictions performed at the same time. Robustness is defined as the sensitivity of the predictions with changes of algorithm parameter variations or external disturbances. There are a large number of prognostic performance metrics that have been used. However, as discussed earlier, most of these metrics do not take into consideration the particular challenges of prognostics. Hence, we feature here only a subset of general metrics especially suitable for prognostics. For a comprehensive list of performance metrics, the reader is referred to [9].

5.3.6 Error-Based Metrics

Many metrics are based on the assessment of the error, i.e., the deviation of the actual output from the target. One example of such an error metric is the average scale independent error. This metric provides an exponential weight of the errors in RUL predictions and averages over several UUTs [20,30]:

$$A(i) = \frac{1}{L}\sum_{l=1}^{L} \exp\left\{-\frac{\left|\Delta^l(i)\right|}{D_0}\right\} \tag{5.1}$$

where Δ is the error and D_0 is a normalizing constant whose value depends on the magnitudes in the application. The range of $A(i)$ varies between 0 and 1, where 1 represents perfect score. Other error-based metrics include root mean squared error (RMSPE) and mean absolute percentage error (MAPE).

5.3.6.1 FP, FN, and ROC

FP and FN are at heart also error-based metrics, but they deserve special consideration. A common way to assess performance is to treat predictions as dichotomous forecasts by means of categorizing them into false-positives (FP), false-negatives (FN), true-positives (TP), and true-negatives (TN) [10]. FP assesses unacceptable early predictions and FN assesses unacceptable late predictions at specified time instances. A user must set acceptable ranges (t_{FN} and t_{FP}) for prediction. Early predictions result in excessive lead time, which may lead to unnecessary corrections. Also note that, a prediction that is late more than a critical threshold time units (t_c) is equivalent to not making any prediction and having the failure occurring. Mathematically, FP is defined as

$$\text{FP}\left(r_*^l(i)\right) = \begin{cases} 1 & \text{if } \Delta^l(i) > t_{FP} \\ 0 & \text{otherwise} \end{cases} \tag{5.2}$$

where t_{FP} is the user-defined acceptable early prediction and FN is defined as

$$\text{FN}\left(r_*^l(i)\right) = \begin{cases} 1 & \text{if } -\Delta^l(i) > t_{FN} \\ 0 & \text{otherwise} \end{cases} \tag{5.3}$$

where t_{FN} is the user-defined acceptable late prediction.

FP and FN both can vary between values 0 and 1, where 1 represents perfect score. FP and FN can then be compiled into the ROC curve. The ROC allows to assess the tradeoff between FP and FN [31,32] in a comprehensive fashion by plotting (1 – FNs) over the FPs (see Figure 5.1). The ideal curve would have zero FPs and zero FNs, but such a curve cannot realistically be achieved for real-world problems. Use of time-dependent ROC has been suggested that depicts ROC obtained for forecasts made for different time horizons. Also, each point on the ROC curve may be associated with a point wise fixed width confidence bounds to indicate confidence in predictions. Tuning the prognostic algorithm such that a

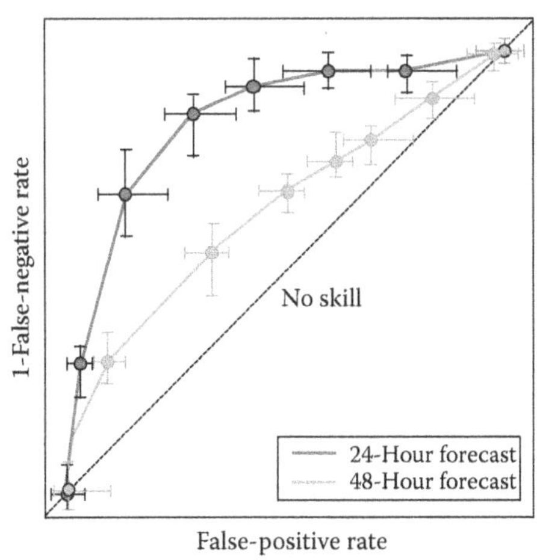

FIGURE 5.1 ROC curve.

ROC can be generated may prove difficult in practice (e.g., due to lack of data or lack of tuning "parameters").

5.3.6.2 Spread-Based Metrics

Spread-based metrics measure the dispersion/spread of the error. The most basic spread-based metric is the sample deviation, which considers the error dispersion with respect to the error sample mean [20,33].

$$S(i)=\sqrt{\frac{\sum_{l=1}^{n}\left(\Delta^{l}(i)-M\right)^{2}}{n-1}}, \tag{5.4}$$

where M is the sample mean of the error. This metric is restricted to the assumption of normal distribution of the error. It is, therefore, recommended to carry out a visual inspection of the error plots. SD can vary between 0 and ∞, where the perfect score is 0. Other spread-based metrics include MAD and MdAD from the sample median.

5.3.6.3 Anomaly Correlation Coefficient

This metric is used to measure correspondence or phase difference between prediction and observations, subtracting out the historical mean at each point and is frequently used to verify output from numerical prediction models [7]. Anomaly correlation coefficient (ACC) is not sensitive to error or bias, so a good anomaly correlation does not guarantee accurate predictions. In the PHM context, ACC computed over a few time-steps after t_P can be used to modify long-term predictions. However, the method requires computing a baseline from history data, which may be difficult to come by.

Mathematically, ACC can be represented as follows:

$$\mathrm{ACC}=\frac{\sum\left(\pi^{l}(i|j)-z_{\#}(i)\right)\left(z_{*}(i)-z_{\#}(i)\right)}{\sqrt{\sum\left(\pi^{l}(i|j)-z_{\#}(i)\right)^{2}\sum\left(z_{*}(i)-z_{\#}(i)\right)^{2}}}, \tag{5.5}$$

where $z_*(i)$ is a prediction variable (e.g., $f^l_{*n}(i)$ or $h^l_*(i)$) and $z_\#(i)$ is the corresponding history data value. ACC can vary between −1 and 1, where 1 represents a perfect score.

Thus, ACC averages the absolute percentage errors in the predictions of multiple UUTs at the same prediction horizon. The percentage is computed based on the mean value of the prediction and ground truth. This prevents the percentage error from being too large for the cases where the ground truth is close to 0. This metric is computed at a particular time and does not capture performance variation with time.

5.3.6.4 Prognostic Horizon

The prognostic horizon (PH) can be formally defined as the difference between the time index i when the predictions first meet the specified performance criteria (based on data accumulated until time index i) and the time index for EoL. PH can be considered as a robustness metric. The basic notion behind the metric is that a longer PH implies more time to act based on a prediction that has some credibility. The performance requirement is specified in terms of an allowable error bound (α) around the true EoL where the choice of α depends on the estimate of time required to take a corrective action. PHs are typically determined offline during the validation phase for an algorithm-application pairing. PH performance is then used as a guideline for algorithm deployment where actual EoL is not known in advance.

$$\mathrm{PH} = \mathrm{EoL} - i \tag{5.6}$$

where $i = \min\left\{ j \mid \left(j \in p \right) \wedge \left((r_* - \alpha \cdot \mathrm{EoL}) \leq r(j) \leq (r_* + \alpha \cdot \mathrm{EoL}) \right) \right\}$.

- p: is a set of all time indices a prediction is made
- r_*: is the True RUL
- $r(j)$: is the predicted RUL at time index j
- α: is the accuracy modifier
- EoL: is the gorund truth End of Life point

PH output is a score that is characterized by both the length of remaining life of a system and the time scales in the problem at hand. As shown in Figure 5.2, the desired level of accuracy with respect to the EoL ground truth is specified as $\pm\alpha$-bounds. A remaining life estimate within those bounds has sufficient utility to a user (it is not too far off from the target to be actionable). The PH for an algorithm is declared as soon the corresponding predictions enter the α-bounds. RUL values can be superimposed for various algorithms, thus providing an easy aid in their comparison. As evident from Figure 5.2, algorithm A1 has a longer PH than algorithm A2.

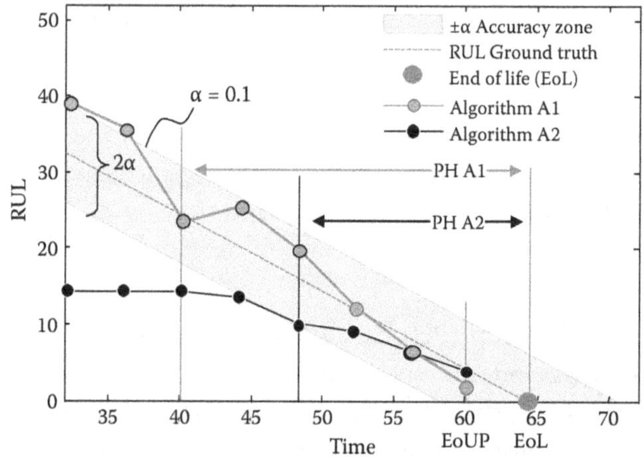

FIGURE 5.2 Prognostic horizon.

5.3.6.5 α-λ Performance

α-λ metric quantifies the prediction quality by determining whether the prediction falls within specified limits at particular times with respect to a performance measure. The evaluation times may be specified either as a ratio to the total remaining life from the point the first prediction is made or it may be specified as a given absolute time interval before EoL is reached. α-λ Performance could be expressed as α-λ accuracy, α-λ precision, or α-λ robustness metric. In the discussion below, we delineate α-λ performance without loss of generality as an accuracy performance measure.

Here we define α-λ accuracy as the prediction accuracy to be within α % of the actual RUL at specific time instance expressed as a fraction of time between the point when an algorithm starts predicting and the actual failure (Figure 5.3). Consider an example case where this metric determines whether a prediction falls within 20% accuracy (i.e., $\alpha = 0.2$) halfway to failure from the time the first prediction is made, (i.e., $\lambda = 0.5$). The α-λ accuracy metric is defined as

$$\alpha\text{-}\lambda \text{ Accuracy} = \begin{cases} 1 & \text{if } (1-\alpha)\cdot r_*(i) \le r^l(i_\lambda) \le (1+\alpha)\cdot r_*(i) \\ 0 & \text{otherwise} \end{cases} \quad (5.7)$$

where α is the accuracy modifier and λ is the time window modifier such that $t_\lambda = t_P + \lambda(t_{EoL} - t_P)$.

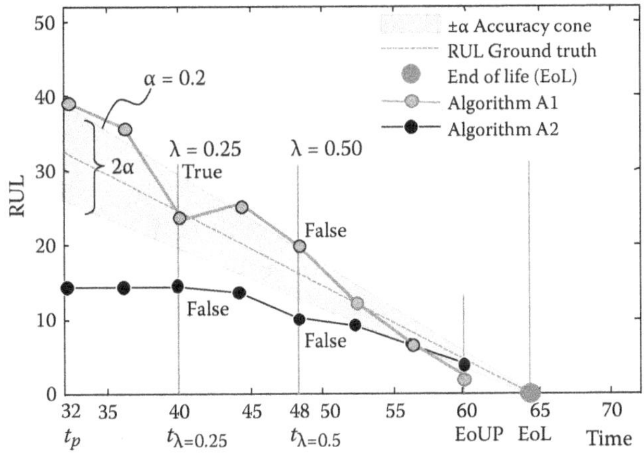

FIGURE 5.3 α-λ Performance.

For illustrating the usage of this performance measure, several prediction algorithms employed in [34–37] are compared in Figure 5.4. Here this metric evaluates whether predictions made by various algorithms lie within 10% error when evaluated at halfway to the EoL.

5.3.6.6 *Relative Accuracy*

Relative accuracy (RA) is defined as a measure of error in RUL prediction relative to the actual RUL.

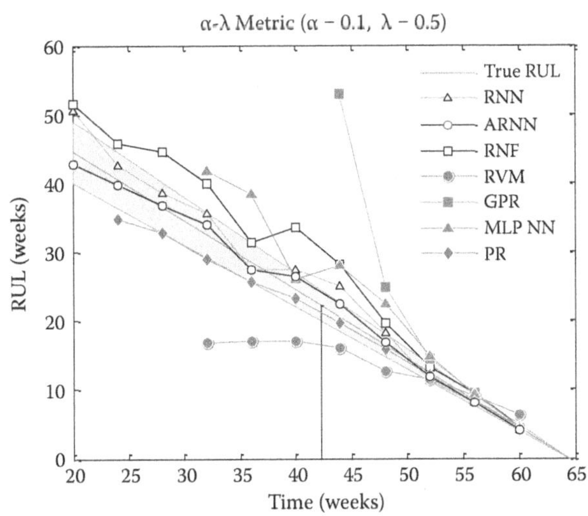

FIGURE 5.4 α-λ Performance comparison.

$$\mathrm{RA}_{\lambda} = 1 - \frac{\left| r_{*}\left(i_{\lambda}\right) - \left\langle r^{l}\left(i_{\lambda}\right)\right\rangle \right|}{r_{*}\left(i_{\lambda}\right)}, \tag{5.8}$$

where λ is the time window modifier such that $t_{\lambda} = t_P + \lambda(t_{EoL} - t_P)$, l is the index for lth UUT, $r_*(i_{\lambda})$ is the ground truth RUL at time index i_{λ}, and $\langle r(i_{\lambda})\rangle$ is an appropriate point estimate of the predicted RUL distribution at time index i_{λ}.

5.3.6.7 Cumulative Relative Accuracy

RA conveys information at a specific time. To estimate the general behavior of the algorithm, RA can be evaluated at multiple time instances to provide an aggregate accuracy level or the cumulative relative accuracy (CRA).

$$\mathrm{CRA}_{\lambda} = \frac{1}{\left|\ell_{\lambda}\right|} \sum_{i \in \ell_{\lambda}} w\left(r(i)\right) \mathrm{RA}_{\lambda}, \tag{5.9}$$

where $w(r(i))$ is a weight factor as a function of RUL at all time indices and l_{λ} is the set of all time indexes when a prediction was made.

It may be desirable to give more weight to RA evaluated at times closer to EoL since good performance close to EoL is important for condition-based decision making. Therefore, one would expect that λ is chosen in a meaningful fashion, e.g., the time required to apply a corrective action. RA evaluated at $\lambda = 0.5$ indicates the time when a system is expected to have consumed half of its remaining life. Alternatively, RA could be evaluated at time instances where the damage magnitude has reached 50% of the failure threshold. This metric is also useful in comparing different algorithms for a given λ to get an idea on how well a particular algorithm does at significant times.

5.3.6.8 Convergence

Convergence expresses the rate at which any metric (M), such as accuracy or precision, improves with time. The error of different algorithm metric evaluation is connected into a curve. Convergence is then defined as the distance between the origin and the centroid of the area under the curve for a given metric.

$$C_M = \sqrt{(x_c - t_P)^2 + y_c^2}, \tag{5.10}$$

where C_M is the Euclidean distance between the center of mass (x_c, y_c) and $(t_P, 0)$, $M(i)$ is a nonnegative prediction accuracy or precision metric with a time varying value, (x_c, y_c) is the center of mass of the area under the curve $M(i)$ between t_P and t_{EoUP}, defined as following

$$x_c = \frac{\frac{1}{2}\sum_{i=P}^{EoUP}(t_{i+1}^2 - t_i^2)M(i)}{\sum_{i=P}^{EoUP}(t_{i+1} - t_i)M(i)},$$

$$y_c = \frac{\frac{1}{2}\sum_{i=P}^{EoUP}(t_{i+1} - t_i)M(i)^2}{\sum_{i=P}^{EoUP}(t_{i+1} - t_i)M(i)}, \tag{5.11}$$

where EoUP (end of useful predictions) is the time index for the last useful prediction made. Alternatively, one may use EoP, but EoUP makes sure that performance is evaluated only based on those predictions that are useful from a practical view point since any prediction made after EoUP does not leave enough time to carry out any corrective measure [37].

As stated earlier, convergence banks on the implicit assumption that algorithm performance should improve with time. For illustration of the concept, consider three cases that converge at different rates in Figure 5.5. Lower distance implies a faster convergence.

5.3.6.9 Robustness

A robustness metric has the task of quantifying the sensitivity of an algorithm with respect to its parameters, such as those found in expressing prior distribution, initial conditions, and training data size. Confidence bounds of a robust algorithm are not expected to change much with variations of algorithm parameters. Mathematically, the robustness metric R_b can be defined as

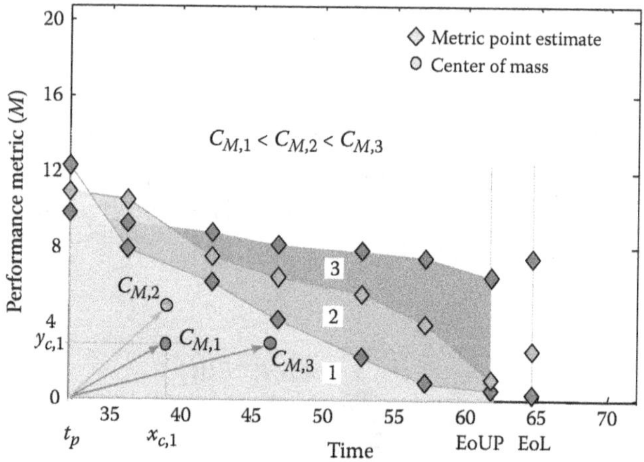

FIGURE 5.5 Convergence metric.

$$R_b = \frac{\int_{Xmean-\eta}^{Xmean+\eta} f(x)dx}{4\eta\delta}, \tag{5.12}$$

where x is the investigated algorithm parameter and $f(x)$ is the confidence bound variation function with respect to x.

The assessment of algorithm robustness is of high value in particular since most of the time an accurate prior is difficult to obtain with limited data source and extensive experiments on the actual engineering system are often prohibitive due to time and cost constraints [38].

5.3.6.10 RUL Online Precision Index

This index quantifies the length of 95% confidence bounds relative to the predicted RUL at any given time instant [39]. The index is normalized between 0 and 1. It can be used as an online performance metric to ensure if I_1 remains close to 1 as system deteriorates, i.e., EoL approaches

$$I_1(i) = e^{-\left(\frac{\sup\{CI(i)\}-\inf\{CI(i)\}}{r(i)}\right)}, \tag{5.13}$$

where $0 < I_1 \leq 1, \forall i \in [1, \langle \text{EoL} \rangle], i \in I^+$.

5.3.7 Incorporating Uncertainty Estimates

Prognostics algorithms typically involve estimating the probability distribution function (PDF) of the EoL and RUL, rather than single point predictions, which enables them to handle uncertainties arising from various sources such as noise, loading conditions, and so on. They also allow for propagation of uncertainties for subsequent predictions [40]. Thus, it is necessary to ensure that prognostic performance metrics include these factors. The most common form of assessing a PDF output is through estimates of mean and variance of the distribution owing to their simplicity and easy interpretation [34]. However, in reality, these distributions are rarely smooth or symmetric and hence mean and variance are not robust enough to evaluate the performance. A combination of mean as the measure of location and quartiles or interquartile range (IQR) as a measure of spread can provide better estimates of the distribution [33].

The metrics shown in the previous sections do not explicitly accommodate for uncertainty estimating capability of the prognostic algorithms. However, a fairly straightforward way to do so is to specify an allowable error bound for a given metric. This error bound could be asymmetric as shown in Figure 5.6. In case of prognostics, typically, a wider error margin to the "left" of the prediction (that is, an early prediction) may be preferred because early predictions have lower cost and safety consequences than late ones.

These concepts can be analytically incorporated into the metrics by calculating the probability mass of a prediction falling within the specified

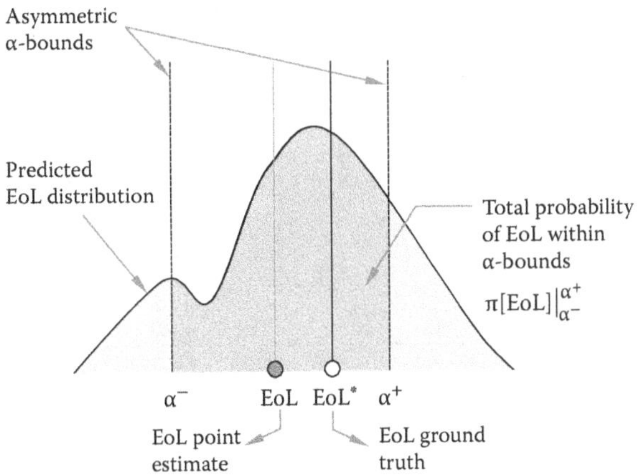

FIGURE 5.6 Concepts for incorporating uncertainties [4] (CC 3.0).

α-bounds. As an illustrative example, consider again the α-λ accuracy. The α-bounds are expressed as a percentage of actual RUL $r(i_\lambda)$ at t_λ.

$$\alpha\text{-}\lambda\ \text{Accuracy} = \begin{cases} 1 & \text{if}\ \ \pi\left[r(i_\lambda)\right]_{-\alpha}^{+\alpha} \geq \beta \\ 0 & \text{otherwise} \end{cases}, \tag{5.14}$$

where $r(i_\lambda)$ is the predicted RUL at time index i_λ, $\pi\left[r(i_\lambda)\right]_{\alpha^-}^{\alpha^+}$ is the probability mass of the prediction PDF within the α-bounds that are given by $\alpha^+ = r_*(i_\lambda) + \alpha \cdot r(i_\lambda)$ and $\alpha^- = r_*(i_\lambda) - \alpha \cdot r(i_\lambda)$.

With α = 0.1 and λ = 0.5, the criteria for matching the metric requirement is determined by assessing the intersection of the uncertainty with the α-cone as shown in Figure 5.7. Note that there may be no prediction assessed at time t_λ for a given λ and the corresponding $i_\lambda \notin p$ because the set of time indexes (p) where a prediction is made is determined by the frequency of prediction step in a prognostic algorithm. In such cases, one can make choose λ′ closest to λ such that $i_{\lambda'} \in p$. To illustrate the application of α-λ accuracy further, refer to Figure 5.8, where the performance of a Recurrent Neural Network algorithm is plotted for every time instant when a prediction was made. Figure 5.8 indicates at any point with either

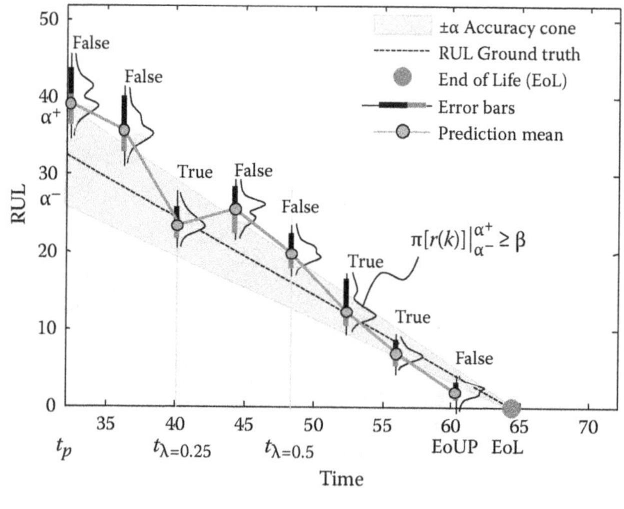

FIGURE 5.7 α-λ Accuracy with the accuracy cone shrinking with time on RUL versus time plot.

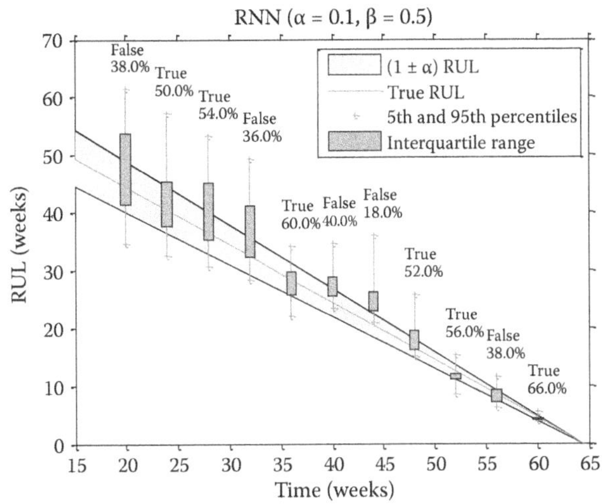

FIGURE 5.8 α-λ Accuracy for a particular algorithm with distribution information [35] (CC 3.0).

"true" or "false" (written above the upper quartiles) whether α-λ accuracy metric is satisfied or not based on β-criterion.

5.3.8 Guidelines for Applying Prognostics Metrics

Given the structure of the prognostics metrics described in this chapter, one can observe a progression in the manner how these metrics characterize the algorithm performance. The first metric, PH, identifies whether an algorithm predicts within a specified error limits around the actual EoL and, if it does, how much time it allows for any corrective action. Therefore, if an algorithm does not allow a sufficient PH it may not be meaningful to continue on computing other metrics. On the other hand, if an algorithm passes the PH test, the next metric, α-λ performance, identifies whether the algorithm performs within desired error margins of the actual RUL at any given time instant that may be of interest to in a particular application. This is a more stringent requirement of staying within a converging cone of the error margins as a system nears EoL. If this criterion is also met, the next step is to quantify the accuracy levels relative to the actual RUL. This is accomplished by the metrics RA and CRA. These metrics assume that prognostic performance improves as more information becomes available with time, and hence, by design, an algorithm will satisfy these metrics criteria if it converges to true RULs. Therefore, the fourth metric, convergence, quantifies how fast the algorithm converges if it does satisfy all previous metrics. These metrics can be considered as a hierarchical test that provides several levels of comparison among different algorithms in addition to the specific information these metrics individually provide regarding algorithm performance. Of course, the use of other metrics such as robustness stands by itself to assess sensitivity of any of these or even other metrics with respect to a key system parameter.

5.3.8.1 Guidelines on Choosing Performance Parameters

The time critical nature of prognostic application resulted in metrics for which the performance evolves with time and needs to be tracked. This required several special parameters that must be specified to define time criticality (λ), confidence level (β), or acceptable error bounds (α). The choice of α depends on the estimate of time required to take a corrective action. Depending on the situation this corrective action may correspond to performing maintenance (manufacturing plants) or bring the system to a safe operating mode (operations in a combat zone). Adjustments to these parameters may translate into significant changes in the cost-benefit-risk equation in a process. Therefore, it is suggested that these parameters be chosen carefully to clearly specify prognostic requirements [5]. Requirements engineering is a discipline that provides guidelines to obtain these requirements in a systematic manner. For instance, in a safety critical military application, first, a failure modes affects and criticality analysis (FMECA) or hazard and operability analysis (HAZOP) must be conducted to identify most critical failures. Then, based on available sensors, measurement quality, and noise levels, etc., desired confidence levels must be derived. For safety critical systems, a more conservative failure threshold may be chosen, while for commercial applications a less conservative but more cost effective threshold is preferred. It must be noted that the choice of metrics and performance specifications is an iterative process that negotiates between user requirements and constraints originating from performance needs, available resources, established maturity level of PHM, and time criticality for that application.

5.3.8.2 Guidelines for Dealing with Uncertainties

A prognostic system models a stochastic process, and hence, the behavior observed from a particular run (single realization of the stochastic process) does not represent the complete behavior of the predicted trajectories. Assuming that all measures practically possible for uncertainty reduction have been taken during the algorithm development phase, such observations should be treated only as isolated realization of the process. A level of confidence or probability of occurrence should be attached to such predictions. Otherwise, multiple trajectories should be aggregated from several runs to achieve statistical significance and more sophisticated stochastic analyses may be carried out. Another aspect dealing with uncertainties is related to prognostic algorithm output. Different algorithms represent uncertainties in different ways. Some specify parametric distribution and others as nonparametric ones. Furthermore, some result in a closed form analytical equation for these distributions and others only result in discretized histograms. It is very important to carefully treat these distributions and not lose critical information by approximating these by known simpler forms such as normal distribution or by computing their statistical moments [4,37]. A common practice has been to compute mean and variance for all types of distributions whereas they may not be very meaningful for nonnormal distributions. Use of more robust estimators such as median, L-estimator, or M-estimator for expressing central tendency and IQR, MAD, or MdAD for expressing the spread is suggested [41].

5.3.8.3 Guidelines to Resolve Ambiguities

In practice, there can be several situations where the definitions discussed above result in ambiguity. Some of such situations are very briefly discussed here with suggested resolutions.

While applying the PH metric, a common situation encountered is when the RUL trajectory jumps out of the $\pm\alpha$ accuracy bounds temporarily. Situations like this result in multiple time indexes where RUL trajectory enters the accuracy zone to satisfy the metric criteria. A simple and conservative approach to deal with this situation is to declare a PH at the latest time instant the predictions enter the accuracy zone and never comes out thereafter. Another option is to use the original PH definition and further evaluate other metrics to determine whether the algorithm satisfies all other requirements. As discussed by Saxena et al. [37], situations such as these can occur due to a variety of reasons.

Inadequate system model: Real systems often exhibit inherent transients at different stages during their life cycles. These transients get reflected as deviations in computed RUL estimates from the true value if the underlying model assumed for the system does not account for these behaviors. In such cases, one must step back and refine the respective models to incorporate such dynamics.

Operational transients: Another source of such behaviors can be due to sudden changes in operational profiles under which a system is operating. Prognostic algorithms may show a time lag in adapting to such changes and hence resulting in temporary deviation from the real values. Therefore, whenever inconsistent behavior of PH metric is observed, one must identify the root cause of it and accordingly interpret the results. The situations discussed here are more common typically toward the end when a system nears EoL. This is because in most cases the fault evolution dynamics are too fast and complex to model or learn from data as the system nears EoL. Therefore, RUL curve deviates from the error band near t_{EoL}. To determine whether such deviations are critical for postprognostic decision making, the concept of t_{EoUP} or EoUP is introduced. This index represents the minimum allowable PH that is required to take a corrective measure. Any predictions made beyond EoUP are of little or no use from a practical viewpoint.

5.4 SUMMARY

This chapter presents several performance metrics for offline evaluation of prognostics algorithms. A brief overview of different methods employed for performance evaluation is also included. Because metrics developed in the context of forecasting differ from prognostics in the systems health management context and to account for the additional considerations, metrics specialized for prognostics (but not necessarily for the application) are needed. These metrics were introduced, and their use was illustrated with recommendations.

ACKNOWLEDGMENTS

The authors would like to express their gratitude to colleagues at the Prognostic Center of Excellence (NASA Ames Research Center) and external partners at Impact Technologies, Georgia Tech, and Clarkson University for participating in research discussions, evaluating metrics in their respective applications, and providing a valuable feedback. This work was funded by NASA Aviation Safety Program-IVHM Project.

REFERENCES

[1] M. Schwabacher. *A Survey of Data Driven Prognostics.* AIAA Infotech@ Aerospace Conference, Arlington, VA, 2005.

[2] M. Schwabacher, K. Goebel. *A Survey of Artificial Intelligence for Prognostics.* AAAI Fall Symposium, Arlington, VA, 2007.

[3] Condition Monitoring and Diagnostics of Machines—Prognostics part 1: General Guidelines, ISO/IEC Directives Part 2 C.F.R., 2004.

[4] A. Saxena, J. Celaya, B. Saha, S. Saha, K. Goebel. Metrics for offline evaluation of prognostic performance. *International Journal of Prognostics and Health Management,* 1(1):21, 2010.

[5] A. Saxena, J. Celaya, S. Saha, B. Saha, I. Roychoudhury, K. Goebel. *Requirements Specification for Prognostics Performance—An Overview.* AIAA Infotech @ Aerospace, Atlanta GA, 2010.

[6] S. Uckun, K. Goebel, P.J.F. Lucas. Standardizing research methods for prognostics. In *International Conference on Prognostics and Health Management (PHM08),* Denver, CO. 2008.

[7] S.J. Engel. Prognosis requirements and V&V: Panel discussion on PHM capabilities: Verification, validation, and certification issues. In *International Conference on Prognostics and Health Management (PHM08),* Denver, CO, 2008.

[8] S.J. Engel, B. Gilmartin, K. Bongort, A. Hess. Prognostics, the real issues involved with predicting life remaining. In *IEEE Aerospace Conference,* Big Sky, MT, 2000.

[9] A. Saxena, J. Celaya, E. Balaban, K. Goebel, B. Saha, S. Saha. Metrics for evaluating performance of prognostics techniques. In *1st International Conference on Prognostics and Health Management (PHM08),* Denver, CO, 2008.

[10] K. Goebel, P.P. Bonissone. Prognostic information fusion for constant load systems. In *7th Annual Conference on Information Fusion,* 2005.

[11] P. Sandborn. A decision support model for determining the applicability of Prognostic Health Management (PHM) approaches to electronic systems. In *Reliability and Maintainability Symposium (RAMS),* Arlington, VA, 2005.

[12] C.S. Byington, M.J. Roemer, P.W. Kalgren. Verification and validation of diagnostic/prognostic algorithms. In *Machinery Failure Prevention Technology Conference (MFPT 59),* Virginia Beach, VA, 2005.

[13] G.J. Kacprzynski, M. Gumina, M.J. Roemer, D.E. Caguiat, T.R. Galie, J.J. McGroarty. A prognostic modeling approach for predicting recurring maintenance for shipboard propulsion system. In *55th Meeting of the Society for Machinery Failure Prevention Technology,* Virginia Beach, VA, 2001.

[14] J.J. Luna. Metrics, models and scenarios for evaluating PHM effects on logistics support. *Annual Conference of the Prognostics and Health Management Society,* San Diego, CA, 2009.

[15] J.B. Coble, J.W. Hines. Prognostic algorithm categorization with PHM challenge application. In *1st International Conference on Prognostics and Health Management (PHM08),* Denver, CO, 2008.

[16] K.R. Wheeler, T. Kurtoglu, S. Poll. A survey of health management user objectives related to diagnostic and prognostic metrics. In *ASME 2009 International Design Engineering Technical Conferences and Computers and Information in Engineering Conference (IDETC/CIE),* San Diego, CA, 2009.

[17] C. Teal, B. Larsen. Technology update II: Wire systems diagnostics and prognostics. In *Digital Avionics Systems Conference*, 2003.
[18] C. Wilkinson, D. Humphrey, B. Vermeire, J. Houston. Prognostic and health management for avionics. In *IEEE Aerospace Conference*, Big Sky, MT, 2004.
[19] J.E. Dzakowic, G.S. Valentine. Advanced techniques for the verification and validation of prognostics and health management capabilities. *In Machinery Failure Prevention Technologies Conference (MFPT 60)*, Virginia Beach, VA, 2006.
[20] G. Vachtsevanos, F.L. Lewis, M. Roemer, A. Hess, B. Wu. Intelligent fault diagnosis and prognosis for engineering systems (1st edition). John Wiley & Sons, Hoboken, New Jersey, 2006.
[21] A.P. Wood. *Reliability-metric varieties and their relationships.* In *Reliability and Maintainability Symposium, 2001*, Philadelphia, 2001.
[22] IEEE-Standard 1413. IEEE Guide for Selecting and Using Reliability Predictions Based on IEEE 1413™ *1413.1TM* (pp. 97): IEEE Standards Coordinating Committee 37, 2002.
[23] A. Coppe, R.T. Haftka, N. Kim, F. Yuan. Reducing uncertainty in damage growth properties by structural health monitoring. In *Annual Conference of the Prognostics and Health Management Society (PHM09)* San Diego, CA, 2009.
[24] D. Hastings, H. McManus. A framework for understanding uncertainty and its mitigation and exploitation in complex systems. In *Engineering Systems Symposium MIT*, Cambridge MA, 2004.
[25] M. Orchard, G. Kacprzynski, K. Goebel, B. Sahaand, G. Vachtsevanos. Advances in uncertainty representation and management for particle filtering applied to prognostics. In *International Conference on Prognostics and Health Management (PHM08)*, Denver, CO, 2008.
[26] R. DeNeufville. Uncertainty management for engineering systems planning and design. In *Engineering Systems Symposium MIT*, Cambridge, MA, 2004.
[27] K.-C. Ng, B. Abramson. Uncertainty management in expert systems. *IEEE Expert Systems*, 5:20, 1990.
[28] S. Sankararaman, Y. Ling, C. Shantz, S. Mahadevan. Uncertainty quantification in fatigue damage prognosis. In *Annual Conference of the Prognostics and Health Management Society (PHM09)*, San Diego, CA, 2009.
[29] L. Tang, G. Kacprzynski, K. Goebel, G. Vachtsevanos. Methodologies for uncertainty management in prognostics. In *IEEE Aerospace Conference*, Big Sky, MT, 2009.
[30] R.J. Hyndman, A.B. Koehler. Another look at measures of forecast accuracy. *International Journal of Forecasting*, Elsevier, 22(4):679–688, 2006.
[31] B. Ebert. Forecast Verification—Issues, Methods and FAQ, from http://www.bom.gov.au/bmrc/wefor/staff/eee/verif/verif_web_page.html, 2007.
[32] T.N. Palmer, A. Alessandri, U. Andersen, et al. Development of a European multimodel system for seasonal-to-interannual prediction (Demeter), *Bulletin of the American Meteorological Society*, 85:853–872, 2004.
[33] D.C. Hoaglin, F. Mosteller, J.W. Tukey, editors. Understanding robust and exploratory data analysis. John Wiley & Sons, 1983.

[34] K. Goebel, B. Saha, A. Saxena, J. Celaya, J.P. Christopherson. Prognostics in battery health management. *IEEE Instrumentation and Measurement Magazine*, 11:33–40, 2008.

[35] J. Liu, A. Saxena, K. Goebel, B. Saha, W. Wang. An adaptive recurrent neural network for remaining useful life prediction of lithium-ion batteries. In *2nd Annual Conference of the Prognostics and Health Management Society (PHM10)*, Portland OR, 2010.

[36] A. Saxena, J. Celaya, B. Saha, S. Saha, K. Goebel. Evaluating algorithmic performance metrics tailored for prognostics. In *IEEE Aerospace Conference*, Big Sky, MT, 2009.

[37] A. Saxena, J. Celaya, B. Saha, S. Saha, K. Goebel. On applying the prognostics performance metrics. In *Annual Conference of the Prognostics and Health Management Society (PHM09)*, San Diego, CA, 2009.

[38] X. Guan, R. Jha, Y. Liu, A. Saxena, J. Celaya, K. Goebel. Comparison of two probabilistic fatigue damage assessment approaches using prognostic performance metrics. *International Journal of Prognostics and Health Management*, 1–13, 2010.

[39] M.E. Orchard, L. Tang, K. Goebel, G. Vachtsevanos. A novel RSPF approach to prediction of high-risk, low-probability failure events. In *Annual Conference of the Prognostics and Health Management Society (PHM09)*, San Diego, CA, 2009.

[40] M.E. Orchard, G. Vachtsevanos. A particle-filtering approach for on-line fault diagnosis and failure prognosis. *Transactions of the Institute of Measurement and Control*, 31(3–4):221–246, 2009.

[41] J.L. Devore. *Probability and Statistics for Engineering and the Sciences* (6th edition). Thomson, 2004.

II

Physics-Based Methods for Systems Health Management

CHAPTER 6

Gaussian Process Damage Prognosis under Random and Flight Profile Fatigue Loading

Aditi Chattopadhyay
Subhasish Mohanty

CONTENTS

6.1 INTRODUCTION

Aircraft structural components are expected to function reliably for a long time under demanding operational conditions involving complex spectrum or flight profile loadings. Traditionally, the design of aircraft structural components against fatigue loading has been based on an assumed load spectrum and using some form of damage accumulation rule. A majority of these damage accumulation methodologies are based on stress life or the S-N diagram [1]. S-N diagrams are generally developed using constant amplitude fatigue test data. However, it is widely accepted that constant amplitude test data insufficiently represents the intrinsic dynamic behavior (such as transient loading effects in the fatigue crack growth rates) that occur with variable amplitude loading (e.g., random gust load that acts on aircraft structures) [2]. Consequently, accurate fatigue predictions are unlikely using S-N diagram approach. Nowadays, the use of flight condition test data are increasingly being used in the design of fatigue critical components [3]. This is because flight condition test data is a more realistic simulation of the load history in service. Depending on design requirements, many standardized load spectrums have been developed for fatigue life prediction. For example, the Load History Transport Wing Standard (TWIST) load spectrum was developed for the design of wing tension skin of a transport aircraft [4], whereas the Fighter Aircraft Loading STAndard For Fatigue (FALSTAFF), load spectrum was developed for the design of a wing root area of a fighter aircraft [5]. Many damage models were developed by considering the effect of spectrum loading [2,6,7]. However, a majority of these models are based on standardized test coupon [8] geometry. Currently, a common practice in damage tolerant design is to use a model that assumes the geometry of these standardized test coupons, but is augmented with a real component flaw size as the initial condition [3]. The real component flaw sizes are estimated either from the individual component fatigue test or from the full-scale fatigue test of the entire aircraft. The mentioned damage models are usually a function of the stress intensity factor [9]. It should be noted that the evaluation of stress intensity factor for a complex geometry is not a trivial task. Due to this reason, the crack growth models of a complex geometry are oversimplified, assuming the crack growth model of the standardized test coupon. However, for complex structures, the damage initiation and growth pattern can be different from the coupon structure. This leads to large error in fatigue damage states and life prediction. To avoid this, the fatigue life of real-life complex geometry can be predicted using finite element (FE)–based crack growth models [10]. Currently, there is no single FE-based damage model available that can be useful for on-board applications. The major drawback in the FE-based approaches is the high computational requirements that make it unsuitable for real-time applications. The computational problems in FE-based approaches as well as the large error in fatigue life prediction using coupon-based damage models can be circumvented by using probabilistic data-driven approaches. The probabilistic model can learn the damage growth dynamics of complex geometry from previous test data. Also, the dynamics can be learned under a complex loading condition. In addition, these models are computationally economical for on-board applications. The Gaussian process (GP) [11–14] approach is one such data-driven approach that can be used for fatigue damage prediction [15,16] of complex geometry and under complex loading patterns. The present chapter discusses the use of GP for fatigue life prediction of an Al-2024 cruciform structure under FALSTAFF and random loading.

6.2 THEORETICAL APPROACH

The theoretical approach behind the proposed data-driven damage model is discussed in this section. However, for suitable implementation of any data-driven approach, it is necessary to first understand the physics of a problem. From a physics-based model, more insight can be made for selecting proper variables for a data-driven approach. Before discussing the proposed data-driven approach, the physics-based model widely used for fatigue life prediction is discussed first. This physics-based model will help to select proper input-output variables of the data-driven approach and is discussed below.

6.2.1 Physics-Based Fatigue Damage Prognosis Model

The fracture mechanics–based Paris law [17,18] is widely used for fatigue life prediction of structural components. Many advanced crack growth models [6,19] have been developed based on the original Paris law. However, all theses models have utility in a limited area or sets of data; none of these models have applicability in a wide range of problems.

According to Paris law, the rate of fatigue crack propagation per fatigue cycle *da*/*dN* is governed by the stress intensity factor range ΔK. This rate at a future damage level $n + 1$ can be expressed as

$$\left(\frac{da}{dN}\right)_{n+1} = C(\Delta K)^m \tag{6.1}$$

Equation 6.1 can be expressed in a simpler functional form expressed as

$$\left(\frac{da}{dN}\right)_{n+1} = \Psi\left\{U_{n+1}^{max}, U_{n+1}^{min}, a_n, C, m\right\} \tag{6.2}$$

where Ψ is a function that describes the damage growth dynamics. The function can either be a physics-based function as described by Equation 6.1 or can be estimated by a data-driven approach that is discussed in the later part of this section. In Equation 6.2 U_{n+1}^{max}, U_{n+1}^{min} are the anticipated maximum and minimum cyclic stresses at the forward damage level, respectively, a_n is the current damage level crack length, and C and m are some experimentally determined parameters. For direct crack length prediction, the equivalent from of Equation 6.2 can be written in functional form as

$$a_{n+1} = \Psi\left\{U_{n+1}^{max}, U_{n+1}^{min}, dN_{n+1}, a_n, C, m\right\} \tag{6.3}$$

where, dN_{n+1} is the number of fatigue cycles elapsed between the current (nth) damage level and the forward step (($n + 1$)th) damage level. It is to be noted that a fatigue damage modeling problem is a type of dynamics modeling problem and requires the loading information to model it. As seen in Equation 6.3, the prediction of future damage state a_{n+1} requires the information of anticipated loading information such as U_{n+1}^{max}, U_{n+1}^{min}. Generally, dynamic problems are modeled using short duration but continuous data. However, a fatigue damage model is generally used for long duration prediction, and hence, it is computationally uneconomical to use continuous data. Accordingly, the load history for a fatigue problem is divided into multiple blocks consisting of single or multiple loading cycles. The individual cycles are represented by maximum and minimum load U^{max}, U^{min} of that particular cycle. The maximum and minimum load of any particular block (say ($n + 1$)th) can be estimated using the following root mean square (RMS) expression.

$$U_{n+1}^{max} = \left[\frac{1}{M} \sum_{i=1}^{M} (U_i^{max})^2 \right]^{1/2} \tag{6.4}$$

$$U_{n+1}^{min} = \left[\frac{1}{M} \sum_{i=1}^{M} (U_i^{min})^2 \right]^{1/2} \tag{6.5}$$

In Equations 6.4 and 6.5, M is the number of cycles in a block.

6.2.2 Bayesian Framework for Damage Prediction

The lack of general applicability of available physics-based damage prediction model is due to the complex dependency of damage propagation on different factors such as (i) material variability (e.g., microstructure), (ii) geometric complexity, (iii) manufacturing process (e.g., heat treatment, cold deformation), (iv) loading, (v) environmental effects (e.g., temperature, humidity). For a physics-based model, it is necessary to find its exact functional form (say given by Equation 6.3) that incorporate this dependency. Finding this functional form is not always straightforward. However, in a Bayesian framework, the functional form of the damage model can easily be estimated by using history data. The schematic of a probabilistic Bayesian framework for future step damage state forecasting is shown in Figure 6.1. The goal of a Bayesian prediction approach is to compute the posterior distribution of future damage state a_{n+1}, i.e., to predict $f(a_{n+1}|D = \{\mathbf{x}_i, a_i\}_{i=0,\ldots n-1,n}, \mathbf{x}_{n+1})$. Referring to Equation 6.3, it can be found that the input vector $\mathbf{x}_{i+1}$ is comprised of a_i (i.e., ith level damage condition), $U_{i+1}^{max}, U_{i+1}^{min}$ (i.e., the i + 1th level maximum and minimum loading), and dN_{i+1} (i.e., the number of fatigue cycles elapsed between the current (ith) and forward ((i + 1)th) damage levels). The parameter equivalent of C and m (refer to Equation 6.3) can be estimated from the history input-output data represented by $D = \{\mathbf{x}_i, a_i\}_{i=0,\ldots n-1,n}$. In a Bayesian framework, let us represent these equivalent parameters as α and β. The corresponding predictive distribution of a target can be found by conditioning the targets $a_0,\ldots,a_{n-1},a_n$, a_{n+1} that are affected by the corresponding random inputs $\mathbf{x}_0,\ldots,\mathbf{x}_{n-1}$, $\mathbf{x}_n$, $\mathbf{x}_{n+1}$. Since $a_0,\ldots,a_{n-1}$ and a_n are conditioned random variables in the observed set of targets, the conditional distribution of a_{n+1} [11–14] can be written as follows:

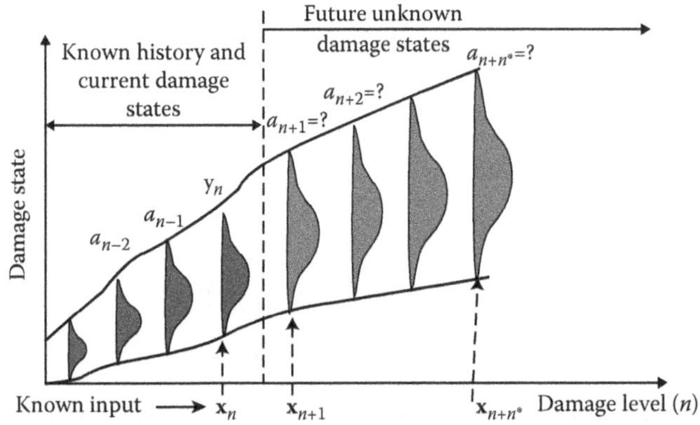

FIGURE 6.1 General Bayesian probabilistic framework for future step damage state forecasting.

$$f\left(a_{n+1} \mid D=\left\{\mathbf{x}_i, a_i\right\}_{i=1,\ldots,n}, \mathbf{x}_{n+1}, \alpha, \beta\right)=\frac{f\left(\mathbf{a}_{n+1} \mid \left\{\mathbf{x}_i\right\}_{i=1,\ldots,n+1}, \alpha, \beta\right)}{f\left(\mathbf{a}_n \mid \left\{\mathbf{x}_i\right\}_{i=1,\ldots,n}, \alpha, \beta\right)}, \tag{6.6}$$

where the numerator and denominator of Equation 6.6 can be evaluated by solving the following integral expression:

$$f\left(\mathbf{a}_n \mid \left\{\mathbf{x}_{i=1,\cdots,n}, \alpha, \beta\right\}\right)=\int\left(\mathbf{a}_n \mid \left\{\mathbf{x}_{i=1,\cdots,n}, a, \vartheta\right\}\right) f(a \mid \alpha) f(\vartheta \mid \beta) dy\, d\vartheta \tag{6.7}$$

where one can assume that $f(a|\alpha)$ represents a priori over the space of possible functions to model the target crack length and α represents a set of parameters that account for modeling the damage prediction function (say Ψ in Equation 6.3). Similarly, assume that $f(\vartheta|\beta)$ is some priori noise function and β is another set of parameters that represents the uncertainty associated with microstructural variability and its dependencies on loads.

6.2.3 Damage Prediction Using GP Regression

GP regression [11–14] is a type of Bayesian prediction approach that is based on the assumption that the target function follows Gaussian distribution. With this assumption, the close form of the integral in Equation 6.6 can be easily found. However, in general, the fatigue crack length follows log-normal distribution [21]. By performing logarithmic scaling, the original crack length data can be made to follow Gaussian distribution and the close form of the integral expression (6.7) can be evaluated. With this, the analytical form of Equation 6.6 can be evaluated as given below:

$$f\left(\mathbf{a}_n \mid \left\{\mathbf{x}_i\right\}_{i=1,\cdots,n}, \mathbf{C}_n\right) = \frac{\theta}{(2\pi)^{n/2} det(\mathbf{C}_n)^{1/2}} exp\left(-\frac{1}{2}(\mathbf{a}_n - \mu)^T \mathbf{C}_n^{-1} (\mathbf{a}_n - \mu)\right), \tag{6.8}$$

where $\mathbf{x}_i$ is the ith fatigue-affecting input parameter, μ is the mean crack length, and $\mathbf{C}_n$ is a $n \times n$ covariance matrix that is a function of parameters α and β. Each element c_{ij} of the covariance matrix $\mathbf{C}_n$ is a measure of the linear dependence between the two random variables $\mathbf{x}_i$ and $\mathbf{x}_j$, and c_{ij}'s are linked to the target crack length a through the parameter θ. The covariance between two random variables $\mathbf{x}_i$ and $\mathbf{x}_j$, with their respective expected values μ_{x_i} and μ_{x_j}, can be found as

$$c_{ij} = \left\langle (\mathbf{x}_i - \mu_{x_i})(\mathbf{x}_j - \mu_{x_j}) \right\rangle. \tag{6.9}$$

6.2.3.1 Covariance Matrix to Kernel Matrix

The expression given in Equation 6.8 is valid if there exists a linear dependence between the two random variables $\mathbf{x}_i$ and $\mathbf{x}_j$. However, in general, for fatigue damage modeling, the input variables $\mathbf{x}_i$ and $\mathbf{x}_j$ often do not have a linear relationship. For instance, if the input space only consists of previous level crack lengths, then the input space variables for ith and jth damage levels can be given as $\mathbf{x}_{i=n-1} = a_{n-1}$ and $\mathbf{x}_{j=n} = a_n$. However, the relationship between a_{n-1} and a_n is not always linear. The nonlinearity is more pronounced during the unstable and transient damage growth regime. In addition to crack length, if other parameters, such as loading information, are introduced in the input space, the relation between input variables $\mathbf{x}_i$ and $\mathbf{x}_j$ becomes more complex. To avoid this nonlinearity problem, the expression given in Equation 6.8 can be reformulated using a high-dimensional transformation of the original input space. The schematic of the high-dimensional transformation Φ is shown in Figure 6.2. The transformation Φ can be performed using an assumed kernel function $k(\mathbf{x}_i,\mathbf{x}_j,\Theta)$ [22]. With the high-dimensional transformation, the covariance matrix $\mathbf{C}_n$ (Equation 6.8) can be replaced by the kernel matrix $\mathbf{K}_n$, and Equation 6.8 can be rewritten [11–14] as

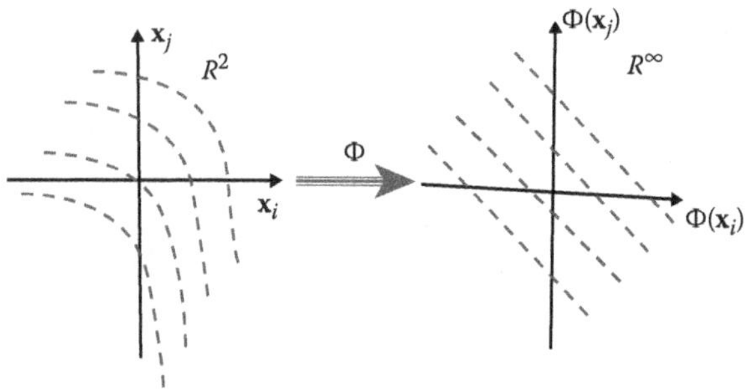

FIGURE 6.2 Schematic of high-dimensional transformation of input space.

$$f\left(\mathbf{a}_n \mid \{\mathbf{x}_i\}_{i=1,\cdots,n}, \mathbf{K}_n\right) = \frac{\theta}{(2\pi)^{n/2} det(\mathbf{K}_n)^{1/2}} exp\left(-\frac{1}{2}(\mathbf{a}_n - \mu)^T \mathbf{K}_n^{-1}(\mathbf{a}_n - \mu)\right). \quad (6.10)$$

6.2.3.2 *Mean and Variance of the Predicted Damage at (n + 1)th Damage Level*

The individual elements k_{ij} of the kernel matrix $\mathbf{K}_n$ can be found from a parameterized kernel function described in the next section. Assuming a zero mean crack length distribution, Equation 6.10 can be written as

$$f\left(a_{n+1} \mid D = \{\mathbf{x}_i, a_i\}_{i=1,\ldots,n}, \mathbf{x}_{n+1}, k_{mn}(x_i, x_j, \Theta)_{i,j=1,2,\ldots n+1}\right)$$

$$= \sqrt{\frac{det(K_n)}{(2\pi) det(K_{n+1})}} exp\left(-\frac{(a_{n+1} - \mu_{n+1})^2}{2\sigma_{n+1}^2}\right), \quad (6.11)$$

where μ_{n+1} is the predicted mean damage (crack length a_{n+1}) at n + 1th fatigue damage level and is given by

$$\mu_{n+1} = \mathbf{k}^T \mathbf{K}_n^{-1} \mathbf{a}_n; \quad k_i = k(\mathbf{x}_{n+1}, \mathbf{x}_i)_{i=1,2,\ldots n}, \quad (6.12)$$

σ_{n+1}^2 is the associated predicted variance at n + 1th fatigue damage level and is given by

$$\sigma_{n+1}^2 = \kappa - \mathbf{k}^T \mathbf{K}_n^{-1} \mathbf{k}; \quad k_i = k(\mathbf{x}_{n+1}, \mathbf{x}_i)_{i=1,2\ldots n}; \quad \kappa = c(\mathbf{x}_{n+1}, \mathbf{x}_{n+1}). \quad (6.13)$$

In Equations 6.12 and 6.13, $\mathbf{K}_n$, $\mathbf{k}$, and κ are the partitioned matrix/vector of the n + 1th fatigue damage level kernel matrix.

6.2.3.3 *Kernel Function Selection*

There are many possible choices for the kernel functions [14]. From a modeling viewpoint, the objective is to select a kernel a priori, which agrees with the assumptions and mathematically represents the structure of the process being modeled. Formally, it is required to specify a function that will generate a positive definite kernel matrix for any set of inputs. In more general terms, the high-dimensional transformation through the assumed kernel function should satisfy Mercer's theorem of functional analysis [23]. For any two sets of input vectors x_i and x_j, the kernel function in Equation 6.10 through 6.13 has the following form

$$k(x_i,x_j,\Theta) = k_a(x_i,x_j,\Theta) + k_{scatter}(x_i,x_j,\Theta), \tag{6.14}$$

where k_a is associated with the interpolation function a and $k_{scatter}$ is associated with the scatter model. In the present case of GP damage modeling, the scatter arises from microstructural variability and related load interactions. For modeling the interpolation function, a neural network–based multilayer perceptron (MLP) kernel [24] is used. The MLP kernel function has the following form:

$$k_a(x_i,x_j,\Theta) = \theta_p Sin^{-1}\left\{\frac{\mathbf{x}_i^T\theta_w\mathbf{x}_j}{\sqrt{(1+\mathbf{x}_m^T\theta_w\mathbf{x}_i)(1+\mathbf{x}_j^T\theta_w\mathbf{x}_j)}}\right\}, \tag{6.15}$$

where θ_p and θ_w represent the process and width hyperparameter, respectively. In Equation 6.14, the second term associated with the scatter model only contributes to the diagonal term of the kernel matrix, and for an input independent scatter model $k_{scatter}$, it can be written in terms of a hyperparameter $\theta_{scatter}$ as of

$$c_{scatter} = (x_m,x_n,\Theta) = \delta_{ij}\theta_{scatter}, \tag{6.16}$$

where δ_{ij} is the Kronecker delta.

6.2.3.4 *Hyperparameter Determination*

So far, only the properties of the prediction model for fixed values of the hyperparameters have been considered. In this section, a discussion on how to obtain the hyperparameters Θ for a fixed training data set $D = \{\mathrm{x}_i,a_i\}_{i=1,\ldots,n}$ is presented. Ideally, one should integrate over all possible hyperparameters to obtain the best possible predictions of the crack length a_{n+1} at $(n + 1)$th damage level. This can be written as

$$f\left(a_{n+1} \mid D,\mathbf{x}_{n+1},K(\cdot)\right)=\int f\left(a_{n+1} \mid D,\mathbf{x}_{n+1},K(\cdot),\Theta\right)f\left(\Theta \mid D,K(\cdot)\right)d\Theta. \quad (6.17)$$

The above integral expression can be evaluated using two popular approaches, which are the maximum evidence [11] and Monte Carlo [25] approaches. The maximum evidence approach is used in this work and is described briefly. Using this approach [11–14], Equation 6.17 can be written in its approximate form as

$$f(a_{n+1}|D,\mathbf{x}_{n+1},K(\cdot)) \cong f(a_{n+1}|D,\mathbf{x}_{n+1},K(\cdot),\Theta_{MAP}). \quad (6.18)$$

The approximation in Equation 6.18 is based on the assumption that the posterior distribution over Θ, i.e., $f(\Theta|D,K(\cdot))$, has a sharp peak around Θ_{MAP}. This approximation is generally good and predictions are often found to be very close to those obtained using the true predictive distribution [26]. To find the peak location of $f(\Theta|D,K(\cdot))$, the posterior distribution needs to be optimized. The posterior distribution can be written as

$$f\left(\Theta \mid D,K(\cdot)\right)=\frac{f\left(\mathbf{a}_n \mid \left\{\mathbf{x}_i\right\}_{i=1,2,\ldots n},K(\cdot),\Theta\right)f(\Theta)}{f\left(\mathbf{a}_n \mid \left\{\mathbf{x}_i\right\}_{i=1,2,\ldots n},K(\cdot)\right)}. \quad (6.19)$$

In Equation 6.19, the denominator is independent of Θ and can be ignored in the optimization process. The remaining two terms, the likelihood $f(\mathbf{a}_n|\{\mathbf{x}_i\}_{i=1,2,\ldots,n},K(\cdot),\Theta)$, and the prior, $f(\Theta)$, need to be considered in the optimization of $f(\Theta|D,K(\cdot))$. Assuming that $a_{i=1,2,\ldots,n}$ follow a Gaussian distribution and using Equation 6.10, the logarithm of the objective function can be written as

$$L \equiv Log\left(f\left(\Theta \mid D,K(\cdot)\right)\right)=-\frac{1}{2}Log(detK_n)-\frac{1}{2}\mathbf{a}_n^T K_n^{-1}\mathbf{a}_n$$
$$-\frac{n}{2}Log(2\pi)+Log f(\Theta). \quad (6.20)$$

The log-likelihood function L in Equation 6.20 is generally multimodal and can be optimized using any multivariate optimization algorithm. In the present work, the conjugate gradient method is used to optimize the log-likelihood function to obtain the optimized hyperparameters. It is common practice [26] to ignore the log prior term in Equation 6.20 due to the lack of knowledge about Θ. The resulting solution may not always be realistic. However, it can be assumed that $Logf(\Theta)$ is implicitly modeled through the optimization of the log-likelihood L.

6.3 NUMERICAL RESULTS AND DISCUSSION

6.3.1 Fatigue Experiment and Data Collection

To validate the prognosis algorithm, three fatigue tests were performed on Al-2024-T351 cruciform specimens using MTS biaxial-torsion load frame. The loaded cruciform specimen is shown in Figure 6.3. The specimens were subjected to two types loading: biaxial random and flight profile FALSTAFF loading. Finite element analysis (FEA) was performed to evaluate the yield stress. Typical tested cruciform specimen and the FEA contour plot are shown in Figure 6.4. From the FEA results, the yield load was determined (7200 lbf). Based on this limiting yield load, both random and FALSTAFF flight profile load patterns were generated using MATLAB® and then coded to the MTS controller.

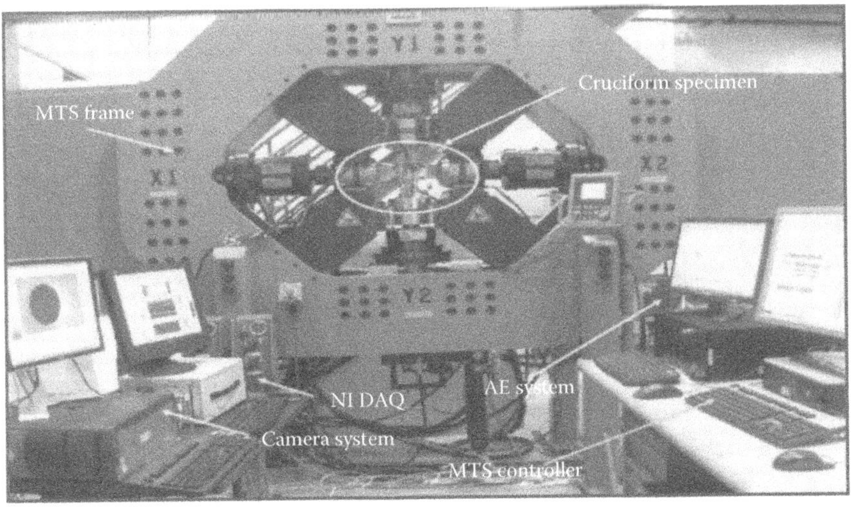

FIGURE 6.3 Biaxial/torsion experimental setup.

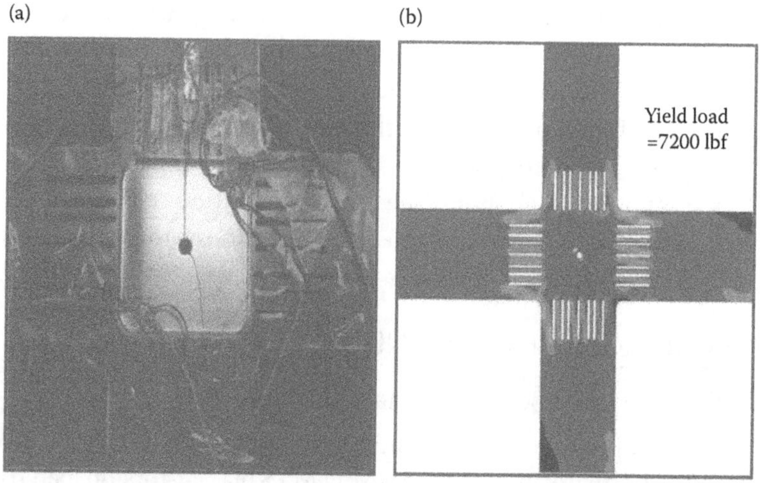

FIGURE 6.4 (a) Al-2024 cruciform specimen. (b) Stress analysis contour plot.

It should be noted that for the random loading case all blocks are non-repetitive. However, for FALSTAFF loading, all blocks are repetitive. Keeping a maximum load limit of 80% of the yield load and a minimum load limit of 6.6% of the yield load, the random loading patterns were generated using MATLAB. For the FALSTAFF loading, the normalized FALSTAFF profile presented by Schijve et al. [5] was modified, using a maximum load limit of 80% of the yield load and a minimum load limit of 6.6% of the yield load. A typical single block (equivalent to 300 cycles) of random load pattern is shown in Figure 6.5, and a typical four block

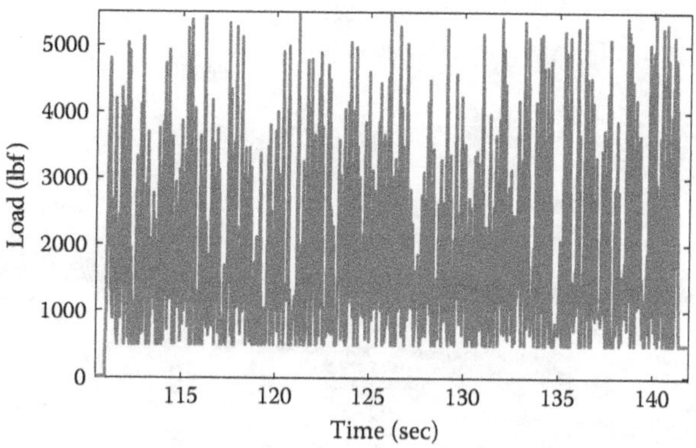

FIGURE 6.5 One block (300 cycles) of random load.

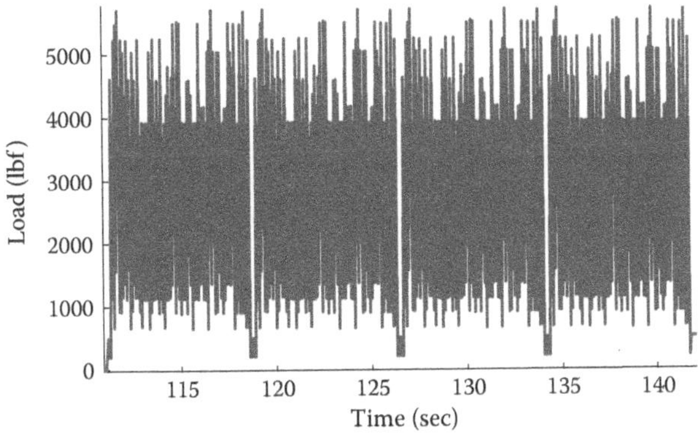

FIGURE 6.6 Four blocks (308 cycles) of modified FALSTAFF load.

(equivalent to 308 cycles) FALSTAFF loading profile is presented in Figure 6.6. To speed up the fatigue process, the extreme low loads in original FALSTAFF profile were augmented with additional loads. The flight profile load patterns obtained with the augmented loading are referred to as modified FALSTAFF loading.

The actuator in the biaxial test frame was operated at a fixed frequency of 10 Hz. Both the *x*- and *y*-axis actuators were subjected to in-phase fatigue loading. A circular hole was made in the center of the specimen to induce crack initiation in the web area of the cruciform. To accelerate damage growth, an EDM notch (45° to the vertical axis) of 1 mm length was made at the bottom right quadrant of the central hole. A 48-channel NI PXI system was used for data collection. A high-resolution camera was used to monitor the crack growth. The data acquisition system and the computer capturing the visual image were synchronized with the biaxial machine controller to collect the time-synchronized data/image at a specified interval of 300 cycles (one block) for the random load and 308 cycles (four blocks) for the modified FALSTAFF. Three specimens (cruciform 7, 8, and 11) were tested; cruciform 7 and 8 under random load and cruciform 11 under modified FALSTAFF loading. Although there was a central hole and an EDM notch, all specimens were first fatigued under constant amplitude loading of 480–4800 lbf to further accelerate the damage process. The constant cycle fatigue tests were performed for 20–30 kilocycles to obtain an initial crack of 1–3 mm. For the demonstration of the prognosis algorithm, crack lengths were estimated at discrete instances (fatigue cycles). Also, at those instances, the *x*- and *y*-axis loads were estimated from the acquired load cell signals.

6.3.2 GP Input–Output Data

The GP input-output relation for biaxial loading condition can be expressed as follows.

$$
\begin{array}{ccccccccc}
a_{k,0} & U^{\max}_{x_{k,1}} & U^{\min}_{x_{k,1}} & U^{\max}_{y_{k,1}} & U^{\min}_{y_{k,1}} & dN_{k,1} & \cdot & \rightarrow & a_{k,1} \\
\cdot & \cdot & \cdot & \cdot & \cdot & \cdot & \cdot & \rightarrow & \cdot \\
a_{k,n-1} & U^{\max}_{x_{k,n}} & U^{\min}_{x_{k,n}} & U^{\max}_{y_{k,n}} & U^{\min}_{y_{k,n}} & dN_{k,n} & \cdot & \rightarrow & a_{k,n} \\
a_{k,n} & U^{\max}_{x_{k,n+1}} & U^{\min}_{x_{k,n+1}} & U^{\max}_{y_{k,n+1}} & U^{\min}_{y_{k,n+1}} & dN_{k,n+1} & \cdot & \rightarrow & a_{k,n+1} = ?
\end{array}
\quad (6.21)
$$

In Equation 6.21, $U^{\max}_{x(\cdot)}, U^{\min}_{x(\cdot)}, U^{\max}_{y(\cdot)}, U^{\min}_{y(\cdot)}$ represent the x-axis maximum and minimum and y-axis maximum and minimum loads at different fatigue cycles, respectively. The quantity, $dN_{(\cdot)}$ represents the number of fatigue cycles elapsed between the current and forward step damage level and $a_{(\cdot)}$ represents the crack length. The parameters that affect fatigue damage are generally time variant. Equation 6.21 represents a generic input-output framework, other parameters such as temperature and humidity can also be included in this framework. For the work reported here, the tests were performed under laboratory condition with minimal changes in temperature and humidity. Once the input-output variables are known, the GP model above can be used to model the multivariate fatigue damage process. For example, at the nth damage level, the corresponding input vector $x_{i=n}$ is given by $x_n = \{a_n, U^{\max}_{x_{n+1}}, U^{\min}_{x_{n+1}}, U^{\max}_{y_{n+1}}, U^{\min}_{y_{n+1}}, dN_{n+1}\}$. In the case of uniaxial constant cyclic loading, since the mean load does not vary and is uniaxial, the corresponding input vector reduced to $x_n = \{a_n, dN_{n+1}\}$. The crack length and loading data used in Equation 6.21 are first normalized against the maximum crack length (experimentally determined to be 72 mm) and yield load (obtained from FEA simulation to be 7200 lbf), respectively. This helps to avoid numerical ill-conditioning of kernel matrix. Also, the normalization of crack length will help in easier integration of present predictive model with an online state estimation model. To note, most of the online structural health monitoring (SHM) algorithms quantify damage in the form of damage indices that vary between 0 and 1. The online state estimation is beyond the scope of this chapter; rather, the damage states considered here are crack lengths and are estimated manually from camera images. In addition to the above-mentioned normalization, both the input-output data are logarithmically scaled. This is to make the crack length data to follow Gaussian distribution. As mentioned earlier, the crack length data in general follows log-normal distribution.

6.3.3 Future Damage State Prediction

The supervised GP-based damage predictions are performed for the following three cases.

1. Case I: prediction under random loading with training data from a different specimen but tested under the same random loading blocks.
2. Case II: prediction under modified FALSTAFF loading with training data from a different specimen and tested under a different loading (such as random loading).
3. Case III: prediction under random loading with training data from a different specimen and tested under a different loading (such as modified FALSTAFF loading).

6.3.3.1 *Case I*

Two cruciform specimens (cruciform 7 and 8) were tested under the same random loading blocks. The fatigue test data acquired from cruciform 8 are used for training the GP model, while the data acquired from cruciform 7 is used as test (or target) case. Figures 6.7 and 6.8 show the prediction results for cruciform 7. At individual damage instances (e.g., $n + 1$th

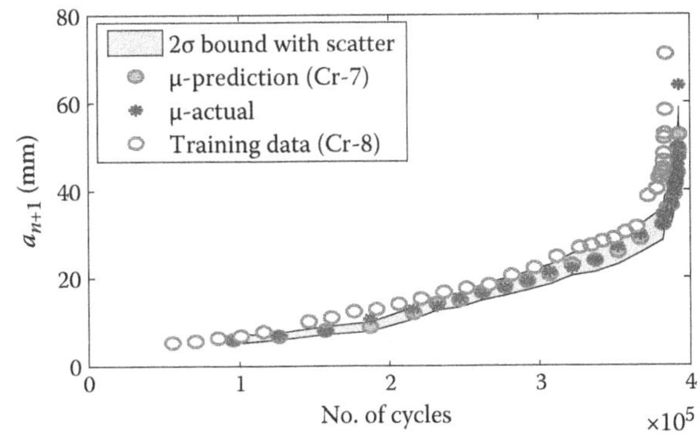

FIGURE 6.7 Prediction under biaxial random loading with training data from random loading test (error bound includes the contribution from noise hyperparameter).

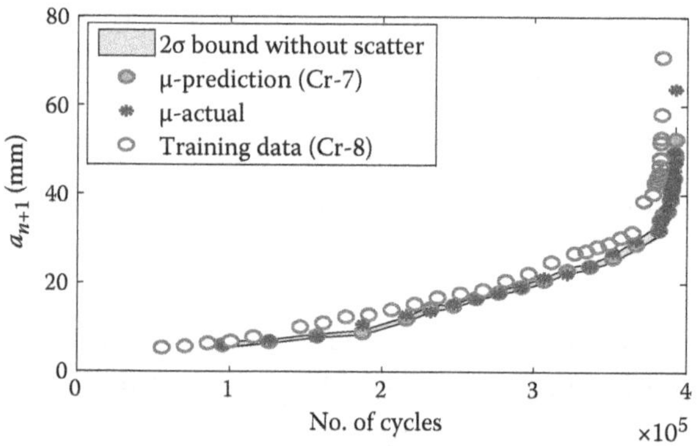

FIGURE 6.8 Prediction under biaxial random loading with training data from random loading test (error bound does not include the contribution from noise hyperparameter).

damage level), the GP input space is defined using six variables such as nth damage level damage state (a_n), $n + 1$th damage level x-axis maximum load ($U_{x_{n+1}}^{max}$), x-axis minimum load ($U_{x_{n+1}}^{min}$), y-axis maximum load ($U_{y_{n+1}}^{max}$), y-axis minimum load ($U_{y_{n+1}}^{min}$), and fatigue cycle interval (dN_{n+1}). The $n + 1$th level damage state, a_{n+1}, is considered a target. It must be noted that the fatigue test data (both camera image and load cell measurements) were collected in an interval of 300 cycles. However, the present training and test cases are selected only at few discrete instances. This is because it is highly time-consuming to manually estimate the crack length from the acquired images. Using the trained GP model, the one-step-ahead predictions are performed for individual input space (or input condition) of cruciform 7. Figure 6.7 shows the training crack growth data as well as the test case prediction (mean prediction). It can be seen that there is a good correlation between mean prediction and mean actual crack length. All the predictions are performed in the scaled domain with interval 0–1. The predicted damage states are scaled back to their original scales (in mm). Figure 6.7 also shows the 2σ (95 %) confidence bound.

It can be seen that the mean predictions fall within the 2σ confidence bound. It is to be noted that the confidence bound has contributions from both modeling error as well as scatter or noise hyperparameter $\theta_{scatter}$ (Equation 6.15). Figure 6.8 shows 2σ confidence bound without considering the noise hyperparameter contribution. Comparison of Figure 6.7 and Figure 6.8 shows that the 2σ confidence bound due to only modeling error contribution is substantially less compared with 2σ confidence bound with both modeling error and noise contribution.

6.3.3.2 Case II

The prediction is performed for cruciform specimen 11, which was fatigue-tested under modified FALSTAFF loading. In this case, test data from cruciform 8, which was fatigued under random loading, was used to train the GP model. Figures 6.9 and 6.10 show the prediction results.

Figure 6.9 shows the training crack length data as well as the mean prediction and its associated 2σ (95%) confidence bound (with both modeling error and noise hyperparameter contribution). Comparison of the mean prediction with actual mean crack length shows a good correlation between actual and prediction. The 2σ confidence bound without noise hyperparameter contribution can be seen in Figure 6.10. Comparing Figures 6.9 and 6.10, it can be seen that there is slightly narrower 2σ bound in the absence of noise. However, compared with case I (where both training and test loading cases are same), the reduction in 2σ bound upon removal of the noise contribution is not significant. This implies that modeling error contribution to the 2σ confidence bound is more significant in case II. This is because the GP framework was trained using a different load case. This scenario is closer to a realistic situation where appropriate training data set for the case being predicted may not be available. From Figure 6.10, it can be observed that although the 2σ bound is wider compared with the case I, it is still reasonable. For a good GP prediction, it is always better if the

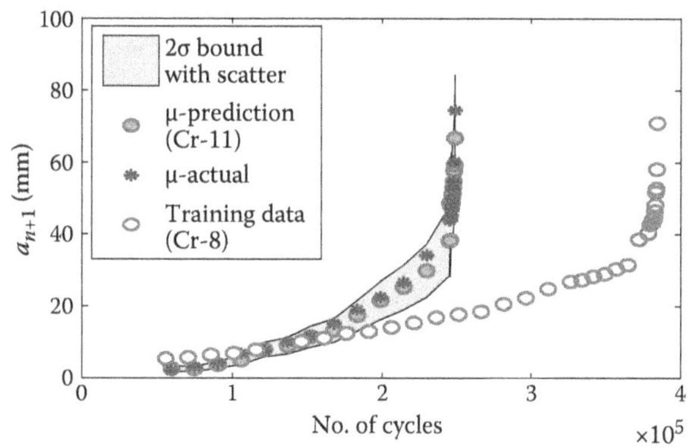

FIGURE 6.9 Prediction under biaxial modified FALSTAFF loading with training data from random loading test (error bound includes the contribution from noise hyperparameter).

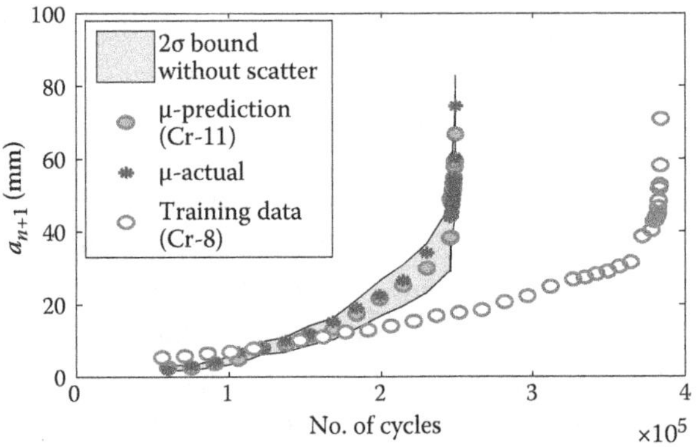

FIGURE 6.10 Prediction under biaxial modified FALSTAFF loading with training data from random loading test (error bound does not includes the contribution from noise hyperparameter).

training data set has a large number of different load patterns compared with the test case load patterns. In the present study, each block of the modified FALSTAFF loading has only 75 different load cycles compared with the random loading case with much larger number of different load ratios.

6.3.3.3 Case III

The prediction is performed for cruciform specimen 8 (tested under random loading) with training data from cruciform specimen 11 (tested under modified FALSTAFF loading). The results are shown in Figures 6.11 and 6.12.

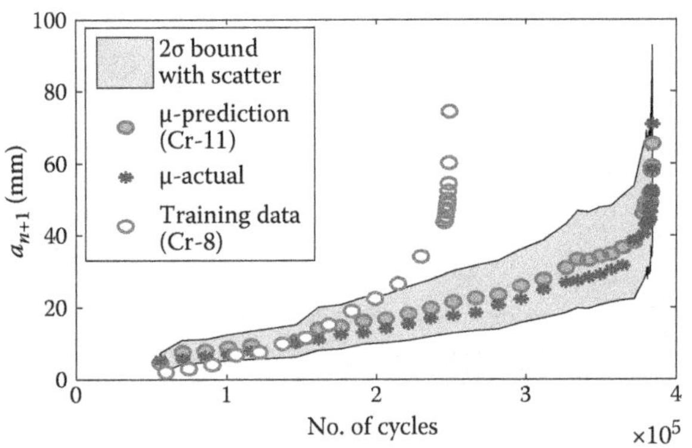

FIGURE 6.11 Prediction under biaxial random loading with training data from modified FALSTAFF loading test (error bound includes the contribution from noise hyperparameter).

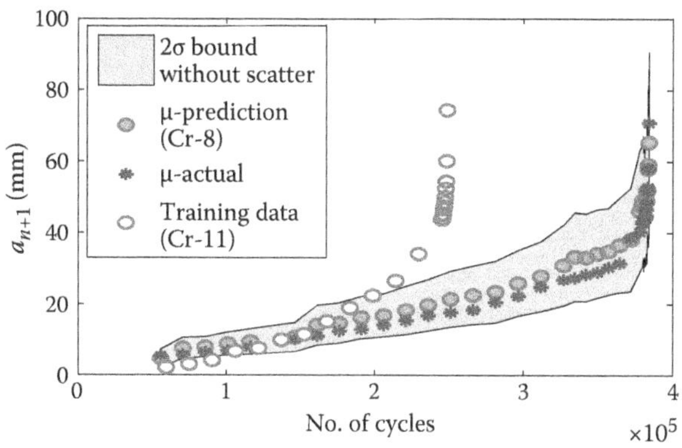

FIGURE 6.12 Prediction under biaxial random loading with training data from modified FALSTAFF loading test (error bound does not include the contribution from noise hyperparameter).

Figure 6.11 shows the training crack length data as well as the mean prediction and its associated 2σ confidence bound (with noise hyperparameter contribution). Comparing the mean prediction with actual mean crack length it can be seen that, unlike cases I and II, there is not much correlation between actual and prediction. In addition, comparing prediction results for random loading case (Figure 6.7) and modified FALSTAFF case (Figure 6.9), the present prediction results (Figure 6.11) shows a much wider 2σ (95%) confidence bound. The 2σ confidence bound without noise hyperparameter contribution is shown in Figure 6.12. Comparison of Figures 6.11 and 6.12 indicate that the contribution of modeling error is quite significant in this case. The reason for large modeling error is due to the use of modified FALSTAFF input-output data as the training set, which contains less information compared with the random loading input-output data set.

6.4 CONCLUSION

A Bayesian GP predictive model has been developed to forecast damage under random and flight profile fatigue loading. The GP model is based on high-dimensional kernel transformation that can perform nonlinear pattern reorganization. The framework is used to predict the future fatigue damage states of aluminum cruciform specimen under biaxial loading. The following are some important observations from the current work.

1. The proposed data-driven approach does not require any geometry dependent stress intensity function generally required in a fracture mechanics–based damage prediction model.

2. The numerical validation under random and flight profile fatigue load shows that prediction under a particular load spectrum can be performed very accurately by training the GP model with the fatigue test data obtained under the same loading condition.

3. Prediction under a different loading spectrum can be performed fairly accurately if the GP model is trained with fatigue test data that contain the majority of the test case load patterns (e.g., the load ratios). In the present case, this is illustrated with an example where the GP model is trained using random load fatigue test data and prediction is made for flight profile load.

REFERENCES

[1] M.A. Miner. Cumulative damage in fatigue. *Journal of Applied Mechanics, ASME*, 12:A-159–164, September, 1945.

[2] I.P. Bond, and I.R. Farrow. Fatigue life prediction under complex loading for xas/914 cfrp incorporating a mechanical fastener. *International Journal of Fatigue*, 22:633–644, 2000.

[3] N. Iyyer, S. Sarkar, R. Merrill, and N. Phan. Aircraft life management using crack initiation and crack growth models-p-3c aircraft experience. *International Journal of Fatigue*, 29:1584–1607, 2007.

[4] D. Jonge, J.B. Schutz, D.H. Lowak, and J. Schijve. A standardized load sequence for flight simulation tests on transport aircraft. In *TR-73029*. National Aerospace Laboratory, NLR, The Netherlands, 1973.

[5] J. Schijve, M. Skorupa, A. Skorupa, T. Machniewicz, and P. Gruszczynski. Fatigue crack growth in the aluminum alloy d16 under constant and variable amplitude loading. *International Journal of Fatigue*, 26(3):1–15, 2004.

[6] M. Sander, and H.A. Richard. Experimental and numerical investigations on the influence of the loading direction on the fatigue crack growth. *International Journal of Fatigue*, 28(3):583–591, 2006.

[7] J. Schijve, A.M. Vlutters, Ichsan, and J.C.P. Kluit. Crack growth in aluminum alloy sheet material under flight-simulation loading. *Int. J. of Fatigue*, 7(3):127–136, 1985.

[8] *ASTM Standard Test Method for Measurement of Fatigue Crack Growth Rates*. American Society of Testing Materials, West Conshohocken (PA), ASTM Standard E 647-00, pp. 1–42, 2000.

[9] S. Suresh. *Fatigue of Materials*. Cambridge University Press, Cambridge, UK, 1991.

[10] N. Sukumar, D.L. Chopp, and B. Moran. Extended finite element method and fast marching method for three-dimensional fatigue crack propagation. *Engg. Fracture Mech*, 70(1):29–48, 2006.
[11] D. MacKay. Bayesian interpolation. *Neural Computation*, 4(1):415–447, 1992.
[12] D. MacKay. Introduction to Gaussian processes. *Neural Networks and Machine Learning*, 168:133–165, 1998.
[13] D. MacKay. *Information Theory, Inference, and Learning Algorithms*. Cambridge University Press, Cambridge, UK, 2003.
[14] C. Rasmussen, and C. Williams. *Gaussian Processes for Machine Learning*. The MIT Press, Cambridge, MA, 2006.
[15] S. Mohanty, S. Das, A. Chattopadhyay, and P. Peralta. Gaussian process time series model for life prognosis of metallic structure. *Journal of Intelligent Material Systems and Structures*, 20(8):887–896, 2008.
[16] S. Mohanty. *Structural Health Monitoring and Condition Based Fatigue Damage Prognosis of Complex Metallic Structures*. PhD thesis, Aerospace Engineering, Arizona State University, 2010.
[17] P.C. Paris, M.P. Gomez, and W.E. Anderson. A rational analytic theory of fatigue. *The Trend in Engineering*, 13:9–14, 1961.
[18] P.C. Paris. *The growth of fatigue cracks due to variations in load*. PhD. Thesis, Leigh University, 1962.
[19] J.C. Newman. Prediction of fatigue crack growth under variable-amplitude and spectrum loading using a closure model. ASTM STP 761: 255–277, 1982.
[20] J.A. Harter. *AFGROW Users' Guide and Technical Manual*. Report No. AFRL-VA-WP-1999-3016, Air force Research Laboratory, 1999.
[21] J.N. Yang, W.H. Hsi, and S.D. Manning. *Stochastic crack propagation with applications to durability and damage tolerance analyses*. Technical Report; Flight Dynamics Laboratory, Wright-Patterson Air Force Base, AFWAL-TR-85-3062, 1985.
[22] C. Cortes, and V. Vapnik. Support vector networks. *Journal of Machine Learning*, 20:273–297, 1995.
[23] R. Courant, and D. Hilbert. *Methods of Mathematical Physics, Vol. 1*. Interscience Publishers Inc., New York, 1953.
[24] C.K.I. Williams. *Advances in Neural Information Processing Systems, Vol. 9*. MIT Press, Cambridge, MA, 1997.
[25] R.M. Neal. *Probabilistic Inference Using Markov Chain Monte Carlo Method*. Technical Report CRG-TR-93-1, Dept. of Computer Science, University of Toronto, 1993.
[26] M.N. Gibbs. *Bayesian Gaussian Processes for Regression and Classification*. PhD thesis, Department of Physics, University of Cambridge, 1997.

CHAPTER 7

Bayesian Analysis for Fatigue Damage Prognostics and Remaining Useful Life Prediction

Xuefei Guan

Yongming Liu

CONTENTS

7.1 INTRODUCTION

Fatigue damage is one the major failure modes caused by usage and aging in many structural systems, such as aircraft, bridges, railroads, rotors, and electronic circuit assemblies [1–6]. Accurate prognostics on fatigue damage propagation are of critical importance for structural health management and risk mitigation. In high-reliability demanding systems, the severity of fatigue damage must be identified, and the remaining useful life (RUL) needs to be quantified for decision-making. Deterministic fatigue life prediction is not appropriate under realistic service conditions due to various uncertainties. These uncertainties arise from difference sources, such as mechanism model choice uncertainty, model parameter uncertainty, material property uncertainty, as well as measurement uncertainty. The model choice uncertainty refers to the situation where more than one parameterized mechanism models are available to explain the fatigue growth phenomenon. Those models may differ in mechanism modeling and in the number of parameters. The comparison and determination between different models are sometimes required. For example, numerous models exist for crack propagation analysis. Models derived based on fracture mechanics include classical Paris' equation [7], Forman's equation [8], Newman's crack closure model [9], and so on. Models derived based on stochastic process include [10–12]. Models based on state space concept [13–15]. Reviews of fatigue crack growth models can be found in [2,16]. Different mechanism models may be valid for specific problems or regimes, and the justification of the using a particular model is case-dependent. For a specific model, model parameter regressions using testing data can introduce the statistical uncertainty, and this refers to the parameter uncertainty. To describe the stochastic process of fatigue damage propagation, measurement data may be used to calibrate the model. Measurement noises may include additional uncertainties into the overall prognostics. The hierarchical uncertainty structure is usually modeled using probabilistic methods. Several studies have been reported to combine the information obtained from inspection with the mechanism model to update the crack growth and fatigue life prediction using Bayesian approach [17–20] and maximum relative entropy inference methods [21]. Those methods focus on one particular model and do not include the model choice uncertainty. The evaluation of Bayes factors using classical Markov chain Monte Carlo (MCMC) simulations is computationally expensive in Bayesian analysis involving multiple models. Additionally, in applications where streaming data require continuous Bayesian updating, the standard simulation methods requires intensive computations and may not be suitable for online state estimates and predictions. Therefore, the systematical inclusion of all major sources of uncertainties and effective and efficient numerical evaluations are two important aspects in fatigue damage prognostics and remaining life predictions.

This study is focused on improving the current methods and developing new algorithms for general probabilistic prognostics and predictions. Fatigue damage problems are of major concern in this study. A hierarchical multimodel Bayesian framework is developed to systematically model various uncertainties. An efficient updating algorithm is developed to reduce the computational cost for applications that require continuous updating in a time-constrained environment. The transdimensional MCMC simulation method is incorporated to perform efficient evaluations for Bayes factors and model parameter distributions. A factorized Metropolis-Hastings (M-H) algorithm is introduced to improve the simulation efficiency.

This chapter is organized as follows. First, the hierarchical multimodel Bayesian framework is developed using probabilistic modeling. Uncertainties are systematically included in the framework. Next, an efficient updating algorithm is formulated based on the principle of minimum discrimination information (MDI) to reduce the computational complexity in updating. Following that, the MCMC methodology in the general state space is introduced. A factorized MCMC algorithm is presented to further enhance the simulation efficiency. After that, a fatigue damage problem evolving multiple crack growth models is presented for methodology illustration and validation. The performance and efficiency of the proposed methods and algorithms are compared with the traditional methods. Finally, several conclusions are drawn based on the current study.

7.2 PROBABILISTIC MODELING FOR HIERARCHICAL UNCERTAINTIES

In engineering applications, parameterized mechanism models are usually developed to describe the underlying mechanism of a target system to infer the future behavior of the system. Uncertainties are introduced in this inference process due to model choice, mechanism modeling, parameter regression, measurement, experimental testing, and numerical evaluations. Those uncertainties are modeled as a two-level hierarchical structure in this study: the first level is the model choice uncertainty and the second level is model associated uncertainties such as parameter, measurement, and mechanism modeling.

7.2.1 Model Choice Uncertainty

The probabilistic inference is usually associated with a specific model, M_k, and it is conditional on the assumption that the model is the correct one that can fully describe the physical phenomenon. However, when the mechanism is not exactly clear, multiple models may be available to simplify the actual complicated mechanism. This is not unusual in fatigue damage analysis, where most of the models are semiempirical [22] and derived based on different crack growth mechanisms. In this case, it may be beneficial to include all useful models to reduce the risk in prognostics and decision-making.

Denote $p(\cdot)$ as the probability density function (PDF) or probability distribution and $P(\cdot)$ the probability. The joint distribution of an event X and model M_k with a n_k-dimensional parameter vector $\theta_k \in \Theta_k$ can be expressed as

$$p(X,\theta_k,M_k) = P(M_k)p(\theta_k|M_k)p(X|\theta_k,M_k), \tag{7.1}$$

where $k \in \mathcal{K}$ is the model index, $P(M_k)$ is the prior probability assigned to model M_k, $p(\theta_k|M_k)$ is the prior PDF of parameter vector θ_k under model M_k, and $p(X|\theta_k,M_k)$ is the conditional PDF of event X given model M_k and the parameter θ_k.

Equation 7.1 can also be interpreted as the product of model probability, parameter prior, and the likelihood. Considering all the possible models in the finite set $k \in \mathcal{K}$, the general state space is $\Theta = \bigcup_{k\in\mathcal{K}}(\{k\} \times \Theta_k)$. The global likelihood for model M_k can be calculated by integrating over its parameter space Θ_k as Equation 7.2.

$$P(X \mid M_k) = \int_{\Theta_k} p(\theta_k \mid M_k) p(X \mid \theta_k, M_k) d\theta_k. \tag{7.2}$$

Evaluating the global likelihood for each of the models according to Equation 7.2 and summing up the entire model set leads to the total probability of event X as

$$P(X) = \sum_{k\in\mathcal{K}} P(M_k)P(X|M_k) = \sum_{k\in\mathcal{K}} P(M_k)\int_{\Theta_k} p(\theta_k|M_k)p(X|\theta_k, M_k)d\theta_k. \tag{7.3}$$

According to the Bayes theorem, the posterior probability of model M_k reads

$$P(M_k|X) = \frac{P(M_k)P(X|M_k)}{P(X)}. \tag{7.4}$$

It is of more interest to consider the model posterior probability ratios of two models, rather than the probability in Equation 7.4 directly. This ratio is referred to as the odds ratio.

$$O_{ij} = \frac{P(M_i|X)}{P(M_j|X)} = \frac{P(M_i)}{P(M_j)} B_{ij}, \tag{7.5}$$

where $P(M_i)/P(M_j)$ is the prior odds ratio and $B_{ij} = P(X|M_i)/P(X|M_j)$ is called the *Bayes factor* [23] in favor of model M_i over model M_j. Given equal prior model probabilities, the odds ratio is reduced to the Bayes factor. Model selection or determination is usually based on the Bayes factors in a Bayesian hypothesis testing context. Next, uncertainties under a specific model are introduced.

7.2.2 Parameter, Mechanism, and Measurement Uncertainties

The parameter uncertainty associated with a specific model M_k is usually introduced by statistical parameter identifications. For example, the model parameter θ_k may be estimated from expert knowledge, empirical judgment, and experimental data regression analysis. Denote the deterministic prediction of model M_k with a parameter θ_k as $\mathcal{M}(\theta_k, M_k; Y)$, where Y is the independent variables. Assume the measurement uncertainty and mechanism modeling uncertainty are described by two variables ε_1 and ε_2, respectively. The probabilistic description of X can be expressed as

$$X = \mathcal{M}(\theta_k, M_k; Y) + \varepsilon_1 + \varepsilon_2. \tag{7.6}$$

For illustration purposes, ε_1 and ε_2 are assumed to be two independent Gaussian variables with standard deviations of σ_{ε_1} and σ_{ε_2}, respectively. The conditional PDF (i.e., $\propto$ likelihood term) of $p(X|\theta_k, M_k)$ is expressed as

$$p(X|\theta_k, M_k) = \frac{1}{\sqrt{2\pi\sigma_\varepsilon^2}} exp\left\{ -\frac{\left[X - \mathcal{M}(\theta_k, M_k; Y)\right]^2}{2\sigma_\varepsilon^2} \right\}, \tag{7.7}$$

where $\sigma_\varepsilon = \sqrt{\sigma_{\varepsilon_1}^2 + \sigma_{\varepsilon_2}^2}$. In case of multiple independent events $X_1, X_2, \ldots, X_n$, the joint conditional PDF is

$$p\left(\bigcap_{d=1}^{n} X_d \,|\, \theta_k, M_k \right) \propto exp\left\{ -\sum_{d=1}^{n} \frac{\left[X_d - \mathcal{M}(\theta_k, M_k; Y_d)\right]^2}{2\sigma_\varepsilon^2} \right\}. \tag{7.8}$$

Substituting Equation 7.7 or Equation 7.8 into Equation 7.2 obtains the global likelihood for model M_k.

The overall Bayesian hierarchical uncertainty management structure is shown in Figure 7.1. For complex models with high dimensional parameters, the direct integral evaluation of Equation 7.2 is difficult, and simulation methods such as MCMC [24] and Gibbs sampling [25] are usually adopted. To evaluate the Bayes factor B_{ij}, two simulation instances are needed. One is used for approximation of $P(X|M_i)$ and one is for $P(X|M_j)$. The required computational effort will increase significantly if multiple model comparisons are desired. For example, to calculate the Bayes

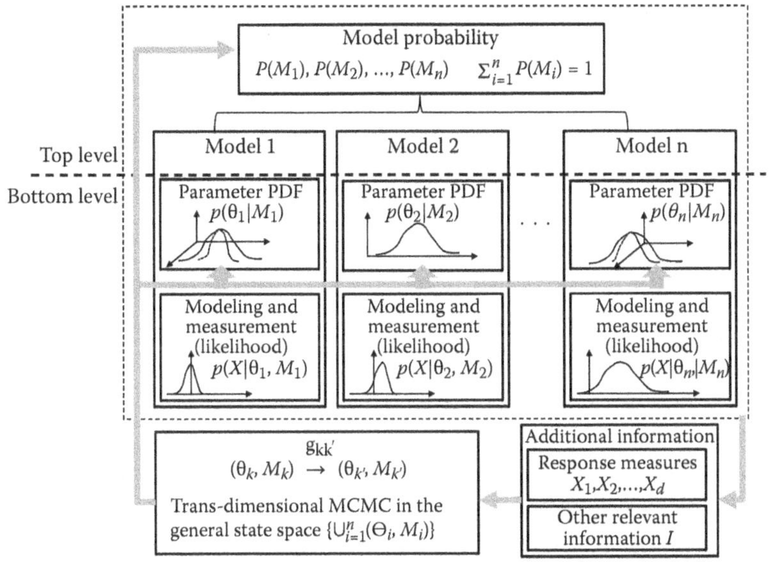

FIGURE 7.1 A hierarchical Bayesian framework for uncertainty management.

factors, B_{i1}, $i \in (2, 3, \ldots, n)$, in favor of each model over model M_1 requires $n - 1$ simulations.

An alternative way to evaluate the Bayes factors is using analytical approximations, such as the Laplace method. One potential drawback for the Laplace method is that it yields inaccurate results if the distribution $p(\theta_k|M_k)p(X|\theta_k,M_k)$ is not asymptotically Gaussian distributed or it has multiple significant modes [23]. Examples are multivariate exponential distributions or mixture Gaussian distributions. Therefore, the applicability of the Laplace method is restricted to a few practical problems.

On the other hand, rearranging Equation 7.1 by treating the model itself as a variable, the Bayesian posterior inference about (θ_k,M_k) can be carried out according to

$$p(\theta_k, M_k|X) = \frac{p(X|\theta_k, M_k)p(\theta_k|M_k)P(M_k)}{P(X)}, \tag{7.9}$$

which is defined across models in the general state space $\Theta = \bigcup_{k \in \mathcal{K}}(\{k\} \times \Theta_k)$.

However, the required measure for move across models with different dimensions is not defined for the standard MCMC simulations. To perform the transdimensional moves in the general state space, Green

[26] and other authors [27,28] proposed the reversible-jump MCMC method by recasting the transition kernel and treating the model itself as a parameter. One advantage of using transdimensional MCMC simulation is that only one simulation instance is needed to obtain all the quantities of interest, including Bayes factors, model probabilities, and parameter distributions.

In practical applications, efficient transitional kernels are difficult to find. Unlike the usual MCMC simulation, neighborhood perturbation or random walking is no longer useful. This is due to the fact that two different models have different parameter structures. Therefore, the design of transition kernels must be carefully made. Otherwise, the rejection rate will be prohibitively high for practical applications [29]. Several implementations of the generic and automatic transdimensional MCMC sampler have been reported in [30,31].

The Bayesian analysis with transdimensional MCMC simulation methods is well suited for problems involving multiple models. In the traditional Bayesian updating, the evaluation of the likelihood term Equation 7.8 requires the computational effort that is linearly proportional to the number of measurement data. For example, given two measurement data points, two model evaluations are needed during the evaluation of the likelihood term. In applications where measurement data arrive continuously and frequently, this updating strategy may not be efficient for time-constrained prognostics. Apart from the overall updating algorithm, improvement of the MCMC simulation in the general state space can also be explored to enhance the computational efficiency. Detailed discussions of those two aspects are presented in the next two sections, respectively.

7.3 AN EFFICIENT ALGORITHM FOR CONTINUOUS BAYESIAN UPDATING

The Bayesian updating is usually performed when new measurement data become available. Following the previous notations, the model parameter θ_k is of our interest and need updating. Given multiple measurement data, the posterior PDF of θ_k is (for convenience, the model notation M_k is omitted in this section)

$$p(\theta_k \mid \bigcap_{d=1}^{n} X_d) \propto p(\theta_k) exp\left\{ -\sum_{d=1}^{n} \frac{\left[X_d - \mathcal{M}(\theta_k; Y_d)\right]^2}{2\sigma_\varepsilon^2} \right\}, \quad (7.10)$$

which is the multiplication of prior PDF of θ_k and the likelihood term in Equation 7.8. In continuous Bayesian updating applications using simulation methods, such as MCMC simulations and other accept/reject-based methods. The computation of the acceptance ratio for one sample proposal involves n model evaluations given n data. For example, upon the arrival of the first data point, to draw one sample from Equation 7.10 requires one model evaluation. If two data points are obtained, drawing one sample needs two model evaluations. Similarly, when the Kth data point is obtained, it takes K model evaluations to draw a sample from Equation 7.10. Hence, the complexity of such an updating algorithm is $\backsim O(n)$ for n measurement data. The overall complexity of the entire updating process is $1 + 2 + \ldots + n \backsim O(n^2)$.

Consider the posterior samples obtained after obtaining the Kth data point. If the information encoded in those samples can be transferred into the next updating in an incremental way such that the information is encoded in the prior PDF analytically for the $(K + 1)$th updating, then only one model evaluation is required. Therefore, the complexity of this algorithm is $\backsim O(1)$ and the overall complexity is $1 + 1 + \ldots +1 \backsim O(n)$.

The performance of the latter algorithm depends entirely on how to transfer the previous information into the current updating computation. One natural intuition of designing such a transfer mechanism is that the information should be conserved as much as possible before and after the transformation. Since the distribution of a parameter is of the primary concern, the MDI can be used as a measure in the transferring process.

7.3.1 Bayesian Updating with MDI Reparameterization

Without loss of generality, assume the updating process for parameter θ start with a parameterized prior PDF $p(\theta_k;\zeta)$ and the first measurement data point X_1. Term ζ is the distribution parameter. For example, a Gaussian distribution is generally characterized by two parameter, the mean value and the standard deviation. Denote the generated MCMC samples as $(\theta_1,\theta_{k,2},\ldots,\theta_{k,t})$. They are asymptotically distributed according to the posterior distribution in Equation 7.10

$$p(\theta;\zeta|X_1) \propto p(\theta;\zeta)exp\left\{-\frac{\left[X_1 - \mathcal{M}(\theta;Y_1)\right]^2}{2\sigma_\varepsilon^2}\right\}, \quad (7.11)$$

where $\mathcal{M}$, Y, and σ_ε are defined as before. For simplicity, the model index k is omitted hereafter in this section. To encode the information contained in those samples into the new prior distribution for the next updating as described in the $O(n)$ algorithm, a reparameterization process for the previous prior $p(\theta;\zeta)$ is required. This means to find an optimal distribution (the new prior PDF for the next updating) given new information. Based on this consideration, the reparameterization can be formulated as the following optimization problem.

Find the optimal distribution $p(\theta;\hat{\zeta})$ *that has the minimal information divergence with respect to* $p(\theta;\zeta)$ *given a set of samples distributed asymptotically according to* $p(\theta;\hat{\zeta})$.

In probability and information theory, the Kullback-Leibler (KL) divergence [32] is a measure of the difference between two distributions $f_1(x)$ and $f_0(x)$. KL divergence is defined as $KL(f_1 \| f_0) = \int_{x \in \mathcal{X}} f_1(x) ln \frac{f_1(x)}{f_0(x)} dx$. If f_0 is the prior PDF of x, the principle of MDI states that given new facts, the optimal posterior PDF f_1 is the one that minimize $KL(f_1 \| f_0)$ subject to constraints [32]. Applying this principle to our optimization problem is to minimize

$$KL[p(\theta;\hat{\zeta}) \,\|\, p(\theta;\zeta)], \tag{7.12}$$

subject to constraints. Minimization of Equation 7.12 using the method of Lagrange multipliers requires explicit setting of constraints. Those constraints include the universal normalization constraint for probability distributions, and constraints formulated using the information associated with the MCMC simulation samples. Since the task is to reparameterize the prior PDF, one of the constructive ways of obtaining those constraints can be derived as follows.

Most common distributions used in engineering are from exponential families, including normal, exponential, gamma, chi-square, beta, Dirichlet, Poisson, and many others [33,34]. A canonical (but not unique) representation of the distribution $p(\theta;\zeta)$ from exponential families is

$$p(\theta;\zeta) = c(\zeta) exp\{\zeta \cdot t(\theta) + ln h(\theta)\}, \tag{7.13}$$

where $t(\theta)$ is a set of real-valued functions and $h(\theta) > 0$ is a real-valued function, and $c(\zeta) > 0$ is a normalization function. Term $\zeta \cdot t(\theta)$ is the inner product of ζ and $t(\theta)$. The commonly used continuous exponential family distributions are summarized in Table 7.1 in the canonical form. They are exponential (Exp), normal (Norm), log-normal (LogNorm), gamma (Gamma), Weibull (We ib), and truncated-normal (TruNorm) distributions. Term $\Phi(\cdot)$ is standard normal (with zero mean and unit standard deviation) cumulative distribution function (CDF).

Consider a general expectation (moment) constraint

$$\mathbb{E}_{p(\theta;\hat{\zeta})} f(\theta) = \int_{\theta\in\Theta} p(\theta;\hat{\zeta}) f(\theta) d\theta = \mathcal{F}, \tag{7.14}$$

and the universal normalizing constraint for any arbitrary continuous PDF, $\int_{\theta\in\Theta} p(\theta;\hat{\zeta}) d\theta = 1$. Lagrangian Λ reads

$$\Lambda = KL[p(\theta;\hat{\zeta}) \,||\, p(\theta;\zeta)] + \beta[\int_{\theta\in\Theta} p(\theta;\hat{\zeta}) d\theta - 1] + \lambda[\int_{\theta\in\Theta} p(\theta;\hat{\zeta}) f(\theta) d\theta - \mathcal{F}], \tag{7.15}$$

TABLE 7.1 Canonical Representations of Commonly Used Exponential Family Distributions

PDF	$t(\theta)$	$c(\zeta)$	ζ	$h(\theta)$
Exp(τ)	θ	τ	$-\tau$	1
Norm(μ,σ^2)	θ, θ^2	$\dfrac{\exp\left\{-\frac{1}{2}\left(\frac{\mu}{\sigma}\right)^2\right\}}{\sqrt{2\pi}\sigma}$	$\dfrac{\mu}{\sigma^2}, -\dfrac{1}{2\sigma^2}$	1
LogNorm(μ,σ^2)	$\ln\theta, (\ln\theta)^2$	$\dfrac{\exp\left\{-\frac{1}{2}\left(\frac{\mu}{\sigma}\right)^2\right\}}{\sqrt{2\pi}\sigma}$	$\dfrac{\mu}{\sigma^2}, -\dfrac{1}{2\sigma^2}$	$\dfrac{1}{\theta}$
Gamma(a,b)	$\theta, \ln\theta$	$\dfrac{b^a}{\Gamma(a)}$	$-b, a-1$	1
Weib(a,b)	$\ln\theta, -\theta^a$	ab^a	$a-1, b^a$	1
TruNorm(μ,σ^2)	θ, θ^2 $\theta\in(\Theta_1,\Theta_2)$	$\dfrac{\exp\left\{-\frac{1}{2}\left(\frac{\mu}{\sigma}\right)^2\right\}}{\sqrt{2\pi}\sigma\left[\Phi\left(\frac{\Theta_2-\mu}{\sigma}\right)-\Phi\left(\frac{\Theta_1-\mu}{\sigma}\right)\right]}$	$\dfrac{\mu}{\sigma^2}, -\dfrac{1}{2\sigma^2}$	1

where β and λ are Lagrange multipliers. Minimizing $KL[p(\theta;\hat{\zeta}) \| p(\theta;\zeta)]$ by $\delta\Lambda/\delta p(\theta;\hat{\zeta})=0$ leads to

$$p(\theta;\hat{\zeta})=\frac{1}{Z}\exp\{\zeta\cdot t(\theta)+lnh(\theta)+\lambda\cdot f(\theta)\}, \tag{7.16}$$

where λ is the Lagrange multipliers associated with $f(\theta)$ and $\mathcal{Z}$ is the normalization constant. A close review of Equation 7.16 indicates that to keep the statistical characteristics of the prior PDF $p(\theta;\zeta)$ and only perform the reparameterization, constraint function $f(\theta)$ should be chosen with the same function as $t(\theta)$ and/or $h(\theta)$, so that the exponential terms $t(\theta)$ and $f(\theta)$ can be collected to form an analytical and normalized PDF $p(\theta;\hat{\zeta})$. In addition, Z can be readily computed according to $c(\zeta)$ in the table. For example, given a gamma prior PDF, $f(\theta)$ take $(\theta,\ln\theta)$ according to $t(\theta)$ for the gamma distribution in Table 7.1. Correspondingly, constraints are $\mathbb{E}_{p(\theta;\hat{\zeta})}\theta$ and $\mathbb{E}_{p(\theta;\hat{\zeta})}ln\theta$ computed from MCMC simulation samples.

7.3.2 The Updating Algorithm

Based on the above mechanism, the updating algorithm with the MDI reparameterization is summarized as follows:

Step 1: Perform Bayesian updating using Equation 7.11 using MCMC simulations and obtain a set of samples.

Step 2: Compute constraints $\mathbb{E}_{p(\theta;\hat{\zeta})}t(\theta)$ using samples generated in Step 1. Perform MDI reparameterization to obtain $p(\theta;\hat{\zeta})$. Continue to Step 1 with $p(\theta;\hat{\zeta})$ as the new prior PDF when new data become available for Bayesian updating.

The procedural flowchart of this updating algorithm and conventional updating algorithm are shown in Figure 7.2.

The same procedure also applies to multiple models cases. The Bayesian updating with MDI reparameterization reduce the computational complexity from $O(n)$ to $O(1)$ at each updating stage, and from $O(n^2)$ to $O(n)$ for the overall complexity during the entire target system service time.

Since MCMC simulations are employed intensively in the updating process, its capability and performance plays an important role. Next, transdimensional MCMC methodology is briefly introduced. A factorized M-H algorithm is proposed to further enhance the computational efficiency.

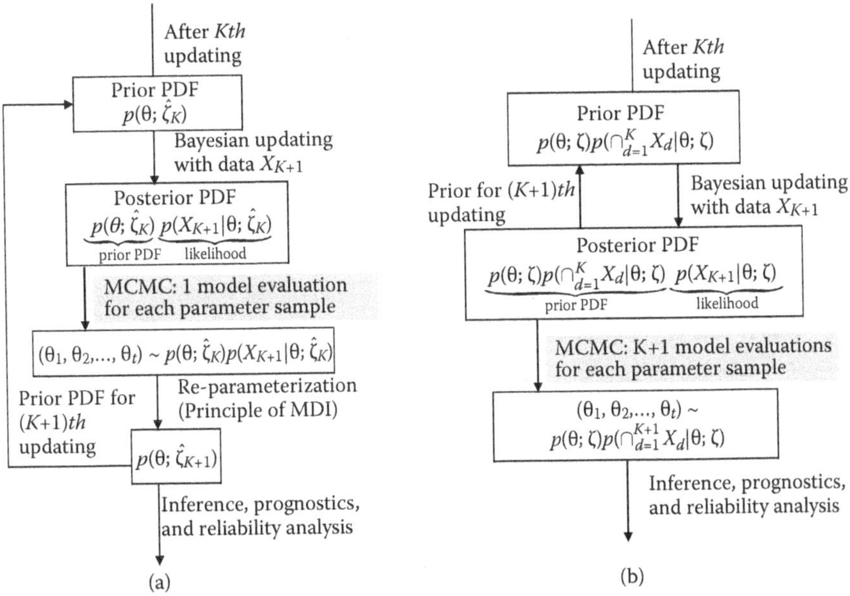

FIGURE 7.2 Procedures for updating two algorithms. (a) Bayesian updating with the MDI reparameterization and (b) conventional Bayesian updating.

7.4 MCMC METHODOLOGY IN THE GENERAL STATE SPACE

The standard MCMC simulation technique was first introduced in [24] as a method to simulate discrete-time homogeneous Markov chain using the random walk (RW) algorithm. The RW algorithm is based on symmetric proposal distributions. In [35], the more general M-H algorithm is formulated to use asymmetric proposal distributions. The M-H algorithm generates a discrete Markov chain in the state space where the current state of the chain depends only on the previous one. Following the convention in MCMC methodology, denote the target distribution as $\pi(x)$. For the distribution of x generated by the M-H algorithm to converge to the target distribution, the Markov chain must have three properties [36]: First, it must be *irreducible*. It refers to that the Markov chain must be able to move to all states in the target distribution with positive probabilities. Second it must be *aperiodic*. This property prevents the chain from oscillating between different states in a regular periodic movement. Finally, the chain must be *positive recurrent*, which means if an initial value of x_0 is sampled from $\pi(x)$, then all the subsequent samples will also distribute according to $\pi(x)$.

The standard M-H algorithm is straightforward (shown in Algorithm 1).

Algorithm 1. The standard M-H algorithm

$t \leftarrow 0$, initialize x_0
repeat
  Draw a proposed sample y from the proposal distribution $q(x_t, y)$
  Calculate the *acceptance probability* $\alpha(x_t, y) = \min\left\{1, \dfrac{\pi(y)}{\pi(x_t)}\dfrac{q(y, x_t)}{q(x_t, y)}\right\}$
  Draw a random number u from uniform distribution $U(0,1)$
  if $\alpha(x_t, y) > u$ **then**
    $x_{t+1} \leftarrow y$
  else
    $x_{t+1} \leftarrow x_t$
  end if
    $t \leftarrow t + 1$
until chain is converged

The acceptance/rejection rule of M-H algorithm is devised such that the chain will eventually converge on the target distribution. This is achieved by maintaining the detailed balance equation in Equation 7.17.

$$\pi(x_t)q(x_t, x_{t+1})\alpha(x_t, x_{t+1}) = \pi(x_{t+1})q(x_{t+1}, x_t)\alpha(x_{t+1}, x_t). \tag{7.17}$$

Inference in terms of model averaging or model selection requires multiple Markov chains in usual MCMC simulations. Simultaneous inference on both model and parameters via one simulation requires the Markov chain to explore the joint space of (M_k, θ_k). Parameter θ_k under different models usually differs in dimension and structure. The acceptance probability in the usual MCMC simulations with the M-H algorithm is not defined for variable dimensional state space. To perform the transdimensional move in the variable dimensional state space, the more general M-H algorithm is required and introduced next.

7.4.1 The M-H Algorithms in the General State Space

The detailed balance equation in the general state space $\Theta = \bigcup_{k \in \mathcal{K}}(\{k\} \times \Theta_k)$ requires

$$\pi(x_i)p(u_i)P(i,j)q(x_i, x_j)\alpha(x_i, x_j) = \pi(x_j)p(u_j)P(j,i)q(x_j, x_i)\alpha(x_j, x_i), \tag{7.18}$$

where x_k defines the state of (θ_k, M_k) and the target distribution $\pi(x_k)$ is $p(\theta_k, M_k|X)$ according to Equation 7.9. Terms u_i and u_j are random quantities used to maintain the equal dimensions on both sides of the equation, i.e., $n_i + \dim(u_i) = n_j + \dim(u_j)$. Term $P(i,j)$ is the probability of proposing to move from model M_i to M_j. To perform the transformation from the coordinate of model M_i to M_j, a bijective transformation g_{ij} is needed so that $g_{ij}(x_i,u_i) = (x_j,u_j)$. The transformation should be reversible, which means the inverse of g_{ij} can also perform the transformation from M_j to M_i.

If g_{ij} and its inverse g_{ji} are differentiable, the change-of-variable formula can be applied to the right hand side of Equation 7.18. The acceptance probability $\alpha(x_i,x_j)$ can be derived as

$$\alpha(x_i,x_j) = \min\left\{1, \frac{\pi(x_j)p(u_j)P(j,i)}{\pi(x_i)p(u_i)P(i,j)}\left|\frac{\partial(x_j,u_j)}{\partial(x_i,u_i)}\right|\right\}. \tag{7.19}$$

Using the (θ_k,M_k) notations, Equation 7.19 can be conveniently written as

$$\alpha((\theta_i,M_i),(\theta_j,M_j)) = \min\left\{1, \frac{p(\theta_j,M_j\,|\,X)P(j,i)p(u_j)}{p(\theta_i,M_i\,|\,X)P(i,j)p(u_i)}\left|\frac{\partial(\theta_j,u_j)}{\partial(\theta_i,u_i)}\right|\right\}. \tag{7.20}$$

Based on Equation 7.20, Algorithm 2 present one possible version of the transdimensional MCMC simulations.

Algorithm 2. The generalized M-H algorithm

$t \leftarrow 0$, initialize $x_t = (\theta_i, M_i)$

repeat

  Propose a jump to model M_j according to $P(i, j)$.

  if $i = j$ **then**

    Generate the proposal sample according to Algorithm 1.

  else

    Draw random quantities u_i and u_j from $p(u_i)$ and $p(u_j)$, respectively.

    Generate the proposal sample under M_j via $(\theta_j, u_j) = g_{ij}(\theta_i,u_i)$.

    Calculate the acceptance probability $\alpha((\theta_i, M_i),(\theta_j, M_j))$ by Equation 7.20.

    Draw a random number u from uniform distribution $U(0,1)$

  end if

if $\alpha((\theta_i, M_i),(\theta_j, M_j)) > u$ **then**
  $x_{t+1} \leftarrow (\theta_j, M_j)$
else
  $x_{t+1} \leftarrow x_t$
endif
  $t \leftarrow t+1$
until chain is converged

The inference on (θ_k, M_k) allows us to perform simultaneous evaluations for model probabilities and parameter distributions in one simulation instead of multiples. For example, given a Markov chain in the general state space Θ, the model probability $P(M_k|X)$ is

$$P(M_k|X) = \frac{P(M_k)P(X|M_k)}{\sum_{k\in\mathcal{K}} P(M_k)P(X|M_k)} = \frac{P(M_k,X)}{\sum_{k\in\mathcal{K}} P(M_k,X)} = \frac{L_k}{L}, \quad (7.21)$$

where L is the total number of samples and L_k is the number of samples under model M_j. In addition, the Bayes factor is

$$B_{ij} = \frac{P(M_i|X)}{P(M_j|X)} = \frac{L_i}{L_j}, \quad (7.22)$$

which is just the ratio of number of samples under model M_i over that under model M_k. For samples under model M_k, they are distributed according to

$$p(\theta_k|M_k,X) = p(X|\theta_k,M_k)p(\theta_k|M_k)\frac{P(M_k)}{P(X)}. \quad (7.23)$$

The last term of Equation 7.23 is a constant under model M_k, so $p(\theta_k|M_k,X)$ is distributed according to $\propto p(X|\theta_k,M_k)p(\theta_k|M_k)$. This indicates that the samples under model M_k from transdimensional MCMC simulations are statistically identical to those we can obtain using the standard MCMC simulations for model M_k.

In the transdimensional MCMC algorithm, the chain explores both across and within models in the general state space. Across model moves involves model coordinate transformations as mentioned above. Local moves within models can rely on the usual RW or M-H algorithms because no model coordinate transformation is needed. To further enhance the computational efficiency for local moves within models, a factorized M-H algorithm is presented next.

7.4.2 A Factorized M-H Algorithm

The standard M-H algorithm (shown in Algorithm 1) is used in usual MCMC simulations for fixed dimensional problems. In the transdimensional cases, Algorithm 2 also employs it for local model moves. When the standard M-H algorithm is used, modifications can be made to improve the computational efficiency. A factorized M-H algorithm is proposed here for this purpose. The basic idea is to factorized the target distribution into several independent terms and evaluate them one by one; therefore, early rejection can be made to avoid computing additional terms. The proof of the factorized M-H algorithm maintaining the detailed balance equation is shown in Appendix 7.1.

Using the previous notation, the parameter PDF under model M_k reads

$$p(\theta_k|M_k,X) \propto p(\theta_k|M_k)p(X|\theta_k,M_k). \tag{7.24}$$

The acceptance probability of the standard M-H algorithm in Algorithm 1 is

$$\alpha(x_t,y) = \min\left\{1, \frac{\pi(y)q(y,x_t)}{\pi(x_t)q(x_t,y)}\right\}. \tag{7.25}$$

Given that $\pi(x)$ can be factorized into several independent components, early rejections can be made for an inappropriately proposed sample. For example, to perform MCMC simulation for the target distribution $\pi(x) = p(\theta_k|M_k,X)$, it can be factorized into the prior term $p(\theta_k|M_k)$ and likelihood term $p(X|\theta_k,M_k)$. A proposed sample can be evaluated for acceptance sequentially based on these two terms using the following procedure:

1. Propose a new sample θ'_k according to $q(\theta'_k,\theta_k)$.
2. Generate a uniformly distributed random variable $r_1 \backsim U(0,1)$.
3. If $r_1 < \min\left\{1, \frac{p(\theta'_k|M_k)q(\theta_k,\theta'_k)}{p(\theta_k|M_k)q(\theta'_k,\theta_k)}\right\}$, continue to the next step. Otherwise, set the current state to θ_k and continue from Step 1.
4. Generate a uniformly distributed random variable $r_2 \backsim U(0,1)$.

5. If $r_2 < \min\left\{1, \frac{p(X|\theta_k', M_k)}{p(X|\theta_k, M_k}\right\}$, continue to the next step. Otherwise, set the current state to θ_k and continue from Step 1.

6. Set the current state as θ_k' and continue from Step 1.

If $p(\theta_k|M_k)$ and $p(X|\theta_k,M_k)$ can be further factorized, the acceptance/rejection procedure can be made based on each of individual terms.

The logic of this procedure is to reject an inappropriately proposed sample earlier to avoid evaluating the entire expression of $p(\theta_k|M_k,X)$. For example, inappropriate proposal can be rejected in step 3, so the evaluation of the likelihood term can be avoided to save computational time. Examples are given in Appendix 7.1 for illustrations.

The previous sections present the two-level hierarchical Bayesian framework for uncertainty quantification and response updating and the transdimensional MCMC simulations for estimations for model probabilities, the Bayes factors, parameter distributions. To improve the overall computational efficiency, the Bayesian updating with MDI reparameterization procedure is presented to reduce the computational complexity from $O(n)$ to $O(1)$ at each updating stage. A factorized M-H algorithm is incorporated to further enhance the simulation efficiency for within model moves in the transdimensional MCMC simulations. Next, a practical example is given to illustrate the proposed methods and algorithms. Experimental data are used to validate the results. In addition, comparisons between the proposed approach and the traditional approach are made to demonstrate the advantages of the proposed approach.

7.5 FATIGUE DAMAGE PROGNOSTICS AND RUL PREDICTION

A fatigue damage problem is used to demonstrate the presented methods and algorithms. Experimental data are used for validation. Traditional approach is employed to generate a baseline solution. The efficiency and accuracy of the proposed approach is compared with the baseline solution.

7.5.1 Component Experimental Data

An aluminum 2024-T3 plate component with a center-through crack is used in this example. The schematic of its geometry and applied cyclic loads are shown in Figure 7.3, where the width is w = 152.4 mm, length

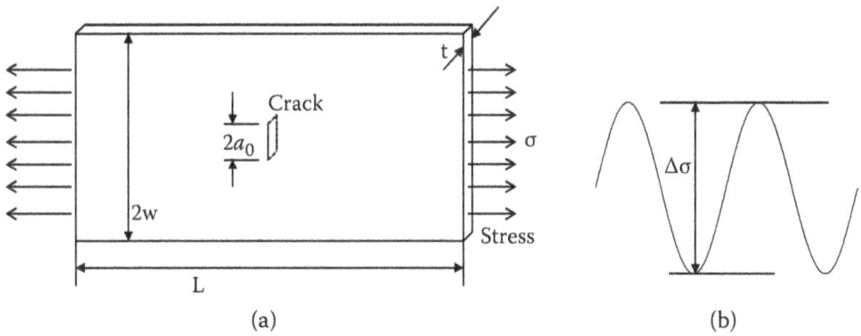

FIGURE 7.3 Schematic illustration of specimen geometries and applied loading. (a) Component geometry and (b) applied cyclic loads. The initial crack length a_0 is 9 mm.

L = 558.8 mm, thickness **t** = 2.54 mm, initial crack size a_0 = 9.0 mm, and the variation of applied stress in one cyclic load is $\Delta\sigma$ = 48.28 MPa with a frequency of 20Hz. The stress ratio is R = 0.2. Experimental data on this component are reported in [37]. The data set consists of 68 crack growth trajectories with identical geometry, material, and loading configurations. Due to uncertainties from material, environment, and specimen manufacturing, those crack growth trajectories are quite different. The last 10 crack growth trajectories are used to identify model parameters statistically, and from the rest of the data set one crack trajectory is arbitrarily chosen to represent the actual target component for fatigue crack growth prognostics and RUL predictions. Four models are included in this example. Four crack size measurements representing the actual inspection data are used for Bayesian updating with transdimensional MCMC simulations. The proposed factorized M-H algorithm is used in the transdimensional MCMC simulations.

7.5.2 Fatigue Crack Growth Models

Four fatigue crack growth models are included for illustration purposes. Model 1 is the classical Paris' model with two parameters, model 2 is Forman's model with two parameters, model 3 is McEvily's model with one parameter, and model 4 is the small time scale model developed by the author [38]. The first three models are valid for constant amplitude loading and the small time scale model can be used in general random loading conditions. More details of those models can be found in the referred articles. For the completeness of this example, only short descriptions of those models are given below.

The Paris' model [38] has two parameters, c and m, as shown in Equation 7.26.

$$\frac{da}{dN} = c(\Delta K)^m, \tag{7.26}$$

where a is the crack length and N is the number of applied cyclic loads. Term $\Delta K = \Delta\sigma\sqrt{\pi a}F(a/w)$ is the stress intensity variation range during one cyclic load and $F(a/w)$ is the geometry correction factor. For the component configuration in this example, $F(a/w) = \sqrt{sec(\pi a/w)}$.

The Forman's model [8] also has two parameters. In addition, the model includes the load ratio (R) and the material toughness (critical stress intensity factor K_c).

$$\frac{da}{dN} = \frac{c(\Delta K)^m}{(1-R)K_c - \Delta K}. \tag{7.27}$$

The McEvily's model [39] includes material toughness and threshold intensity range (ΔK_{th}). Knowledge of ΔK_{th} permits the calculation of permissible crack lengths and applied stresses to avoid fatigue crack growth. Above the threshold value, the crack growth rate increases relatively rapidly with increase ΔK [22].

$$\frac{da}{dN} = c(\Delta K - \Delta K_{th})^2\left(1 + \frac{\Delta K}{K_c - K_{max}}\right). \tag{7.28}$$

The small time scale model [38] is derived from the geometric relationship between crack length increment and crack tip opening displacement (CTOD). The model describes the crack growth behavior within one cyclic load, and the crack only grows when the stress intensity is larger than a reference level. The reference level is calculated according to the reverse and forward plastic zones of the crack tip. The small time scale crack model is shown in Equation 7.29.

$$\dot{a} = H(\dot{\sigma})H(\sigma - \sigma_{ref})\frac{2\phi\gamma}{1-\phi\gamma\sigma^2}\dot{\sigma}a, \tag{7.29}$$

where H(·) is the Heaviside step function. The function takes value of 1 if its variable is larger than 0; otherwise, it takes value of 0. The first Heaviside function ensures the crack only grow during the load amplitude increasing phase and the second one ensures that the crack only grow when the applied stress is larger than the reference level σ_{ref}. Terms $\dot{a}$ and $\dot{\sigma}$ are time derivatives of crack size and applied stress, respectively. Term $\gamma = 4\kappa/(E\sigma_y)$ includes material properties, Young's modulus E, and yielding stress σ_y into the model. Term ϕ describes the crack tip opening angle and is defined as $\phi = \tan((\pi/2)(K_{max} - K_{ref} - \Delta K_{th})/(K_c - \Delta K_{th}))$. Terms K_c, K_{max}, K_{th}, and K_{ref} are critical, maximum, threshold, and reference stress intensities, respectively. The details of this model are discussed in [38]. The small time scale model has one parameter κ.

The material related parameters K_c, ΔK_{th}, σ_y, and E are identified from the standard testing and they are treated as constant values. For this type of material, values of $K_c = 37\ \text{MPa}\sqrt{m}$, $\Delta K_{th} = 2.0\ \text{MPa}\sqrt{m}$, $\sigma_y = 483.1$ MPa, and $E = 73.1$ GPa [40] are used.

Solving those model equations using da accumulation cycle by cycle or more efficient ordinary differential equation (ODE) solvers can obtain the crack growth trajectory. Model parameter prior distributions are estimated from the last 10 crack growth trajectories using the method of maximum likelihood estimate (MLE). The standard deviation of the Gaussian likelihood is estimated to be $\sigma_\varepsilon = 0.35$ mm using MLE. The prior parameter distributions for the four models are presented in Table 7.2. For models 3 and 4, the covariance is reduced to variance since they have only one parameter.

The experimental data used to identify parameters are shown in Figure 7.4a. The 95% bounds of fatigue crack growth predictions based on the prior distributions are shown in Figure 7.4b, where the prediction bounds approximately cover the experimental data envelop boundaries as expected.

TABLE 7.2 Prior Distributions of Model Parameters

Model	1	2	3	4
mean[a]	[−25.865, 2.7993]	[−13.2,1.7231]	−21.226	0.3157
covariance	$\begin{bmatrix} 1.4744 & -0.2511 \\ -0.2511 & 0.0428 \end{bmatrix}$	$\begin{bmatrix} 1.9712 & -0.3337 \\ -0.3337 & 0.0566 \end{bmatrix}$	0.0032	0.0003

[a] Following the convention, ln(c) are estimated in models 1, 2, and 3 instead of c.

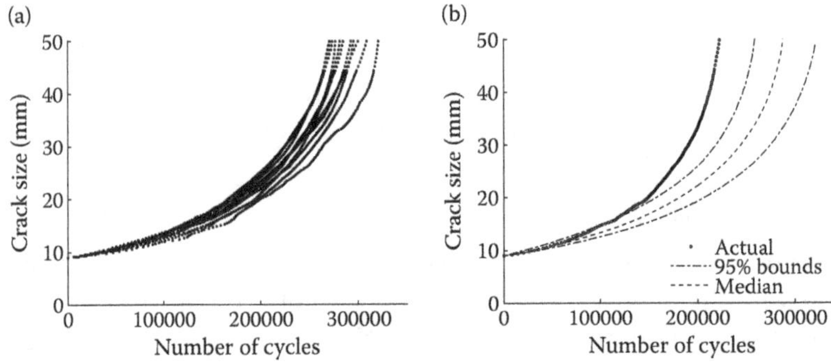

FIGURE 7.4 Experimental data and fatigue crack growth prognosis with prior PDFs. (a) Experimental data for parameter prior PDF estimates and (b) the actual component (solid line) and the fatigue crack growth prognosis with prior PDFs. Results are averaged using the transdimensional MCMC simulation results.

It is also observed in Figure 7.4b that the crack growth trajectory of the actual target component is significantly deviated from the prior prediction bounds. This is not unusual in fatigue crack growth analysis. Reasons causing this include uncertainties from material, usage profile, and environment among others. The prior distributions may reflect the factory manufacturing uncertainties, but prognostics are usually focused on each individual component. For example, a batch of components is manufactured in a factory and uncertainties are characterized by some parameter distributions identified based on the standard testing. Those distributions represent the whole factory manufacturing uncertainties. However, each specimen in the batch may be used in different systems under different usage conditions. The measurement data associated with a specific component can be used for updating and reduce prognosis uncertainties. The resulting updated distributions and prognostics will become more relevant to the component under its usage profile.

7.5.3 Bayesian Updating for Crack Growth Prognostics and RUL Predictions

Four measurement data (crack size and the corresponding number of applied cyclic loads) from the target component crack growth curve are arbitrarily chosen to represent the actual inspection data. They are 12.8 mm, 79,514; 15.8 mm, 115,922; 19.8 mm, 148,636; and 23.8 mm, 167,726. At each updating, 1,000,000 samples are drawn using transdimensional MCMC simulations. Samples under models 1, 2, 3, and 4 are

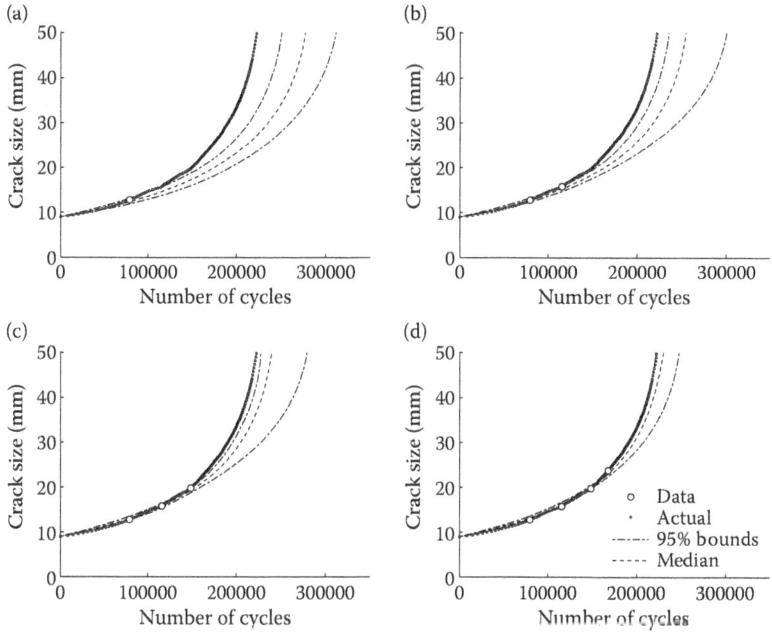

FIGURE 7.5 Fatigue crack growth prognosis results obtained by performing the Bayesian updating with transdimensional MCMC simulations. Results are averaged across the four models. (a) Updated with one data point, (b) updated with two data points, (c) updated with three data points, and (d) updated with four data points.

then used to compute the crack growth curves using Equations 7.26 through 7.29, respectively. Median and bounds results can readily be evaluated according to the computation results. Details are shown in Figure 7.5. It is observed that, as more measurement data are used for updating, the prediction results become more consistent with the crack growth trajectory of the actual target component. Additionally, uncertainties of 95% bounds reduce as the number of updating increases.

Estimates on RUL are performed based on the simulation samples. Integrating model equations (*dN* expression) from the current crack length to the predefined critical length (49.8 mm in this example) gives the RUL results. Figure 7.6 shows the RUL distributions. RUL estimations can be used to determine the maintenance schedule. For example, in some systems, the performance does not deteriorate significantly before the crack length reaches the predefined critical length and the system can still function properly [41]. A maintenance task can be scheduled according to

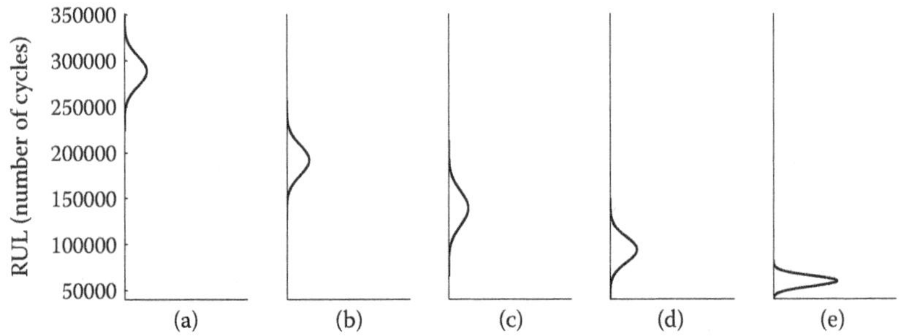

FIGURE 7.6 RUL estimates. Results are averaged over the four models. (a) Prior, (b) updated with one data point, (c) updated with two data points, (d) updated with three data points, and (e) updated with four data points.

the RUL estimates to reduce the cost. For example, a replacement of the component is needed after approximately 40,000 cycles according to the results shown in Figure 7.6e.

7.5.4 Model Probabilities, Bayes Factors, and Parameter Statistics

The transdimensional MCMC simulation can produce many useful statistical results. First, it estimates the model probabilities as the ratio of the number of samples associated with one model to the total number of samples. The results are shown in Table 7.3. Second, samples under one specific model are also distributed according to the model posterior distribution, and they are statistically identical with those that one can obtain using standard fixed dimensional MCMC simulations.

The Bayes factors are easily evaluated using model probabilities. For example, after updating with one measurement data point, the Bayes factor in favor of model 1 over model 2 is 0.239/0.216 = 1.11. However, the Bayes factors need careful interpretations when model determination

TABLE 7.3 Model Probability Estimates and the Bayes Factors

Data	**0**	**1**	**2**	**3**	**4**
Model 1	0.251	0.239	0.232	0.232	0.214
Model 2	0.249	0.216	0.209	0.223	0.222
Model 3	0.250	0.272	0.266	0.251	0.257
Model 4	0.250	0.274	0.294	0.290	0.308

Note: Models 1, 2, 3, and 4 are Paris', Forman's, McEvily's, and the small time scale models, respectively. The first row is the number of data points used in the Bayesian updating. Value 0 corresponds to prior estimates.

(based on Bayesian hypothesis testing) is desired because the interpretation may depend on the context [23]. The weighing evidence against the null hypothesis in terms of the Bayes factors is discussed in [42]. According to the results shown in Table 7.3, the four models perform equally well, and the small time scale model (model 4) slightly outperforms others. However, there is no evidence for leaving out other models based on Jeffreys' criteria [42].

The prior and posterior distributions are shown in Figure 7.7. For illustration purposes, the posterior distributions after the second updating are shown. The significant changes of mean values of model parameters indicate the prior distributions are difficult to capture the actual crack growth trajectory of the target component. As more measurement data are integrated into the updating process, the resulting distributions become more relevant to the target component and the measurement data diminish the effect of the prior distributions.

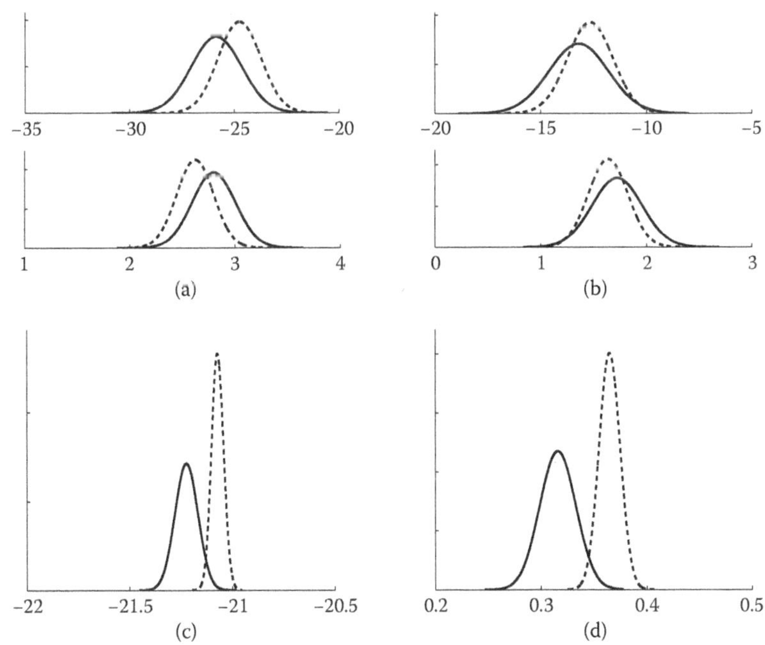

FIGURE 7.7 Prior and the posterior distributions of model parameters. Solid lines are prior distributions and dashed lines are posterior distributions after updating with two measurement data points. (a) Paris' model parameters, (b) Forman's model parameters, (c) McEvily's model parameter, and (d) the small time scale model parameter.

7.5.5 Bayesian Model Averaging

Model averaging is usually used when there is no strong evidence of rejecting the alternative model in the Bayesian hypothesis testing. According to Jeffreys' criteria, the Bayes factor in favor of one model to the alternative model that is larger than 10 is considered as strong evidence. Therefore, there is no strong evidence of supporting one particular model in this example. Results in Figures 7.4 and 7.5 are averaged over the four models. The difference between the predictions under four models and the averaged results are shown in Figure 7.8. It should be noted that one model may give conservative results in one case but may give risky results in another case. This is because none of the four models are exact in modeling the underlying crack growth mechanisms. This is one situation where multiple models should be used. The Bayesian model averaging can take the advantage of multiple useful models and weigh those models by measurement data.

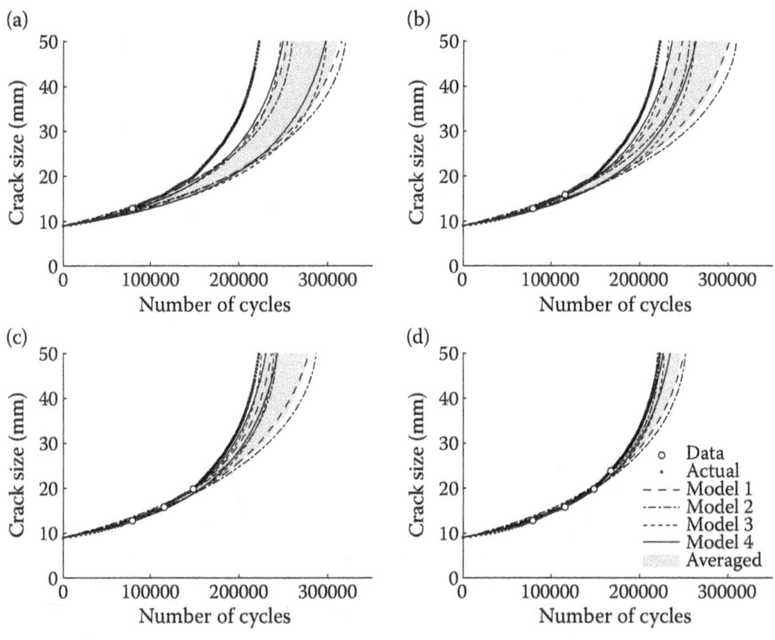

FIGURE 7.8 Fatigue crack growth prognostics (95% bounds) under four models and the Bayesian model averaging results (in shaded areas). Models 1, 2, 3, and 4 are Paris', Forman's, McEvily's, and the small time scale models, respectively. (a) Updated with one data point, (b) updated with two data points, (c) updated with three data points, and (d) updated with four data points.

7.5.6 Comparisons of the Overall Performance and Efficiency

The results in the example are obtained using the Bayesian updating algorithm with MDI reparameterization and the transdimensional MCMC with the factorized M-H algorithm. The standard Bayesian updating and the transdimensional MCMC with the usual M-H algorithm are also performed for this example, serving as the baseline results for comparisons. The performance is quantified in terms of relative error (RE) of proposed approach results with respect to the baseline results. REs of crack growth prognostics and RUL predictions are based on the median estimates. Detailed results of performance comparisons are shown in Table 7.4. The efficiency is measured in the overall computational time. The computational time is normalized against the baseline cost. Detailed results are shown in Table 7.5.

The results produced by the proposed methods and algorithms are close to the baseline results in Table 7.4, and overall maximum RE is less than 5%.

From Table 7.5, it is observed that using the proposed approach (Bayesian updating via MDI reparameterization and the factorized M-H algorithm) the overall cost is approximately constant. This agrees with the theoretical analysis of the computational complexity ($O(1)$). The standard Bayesian updating with the usual M-H algorithm has a linear computational complexity ($O(n)$). As more data are used for updating, the computational ratio is reduced gradually. On the other hand, the overall relative error introduced by the proposed approach is lower than 5%. The advantage

TABLE 7.4 Relative Errors (%) of the Proposed Results with Respect to the Baseline Results

Data		0	1	2	3	4
Model probabilities	Model 1	0	0.17	0.22	0.43	0.59
	Model 2	0	0.26	0.29	1.12	4.64
	Model 3	0	0.10	0.10	0.39	1.42
	Model 4	0	0.04	0.29	0.19	1.64
Crack growth prognostics (median)		0	0.07	3.62	1.83	0.76
RUL predictions (median)		0	0.02	0.80	3.12	4.77

Note: Models 1, 2, 3, and 4 are Paris', Forman's, McEvily's, and the small time scale models, respectively. The relative errors of crack growth prognostics and RUL predictions are reported based on the median estimates (averaged over the four models).

TABLE 7.5 Comparison of the Overall Computational Efficiency

Data	0	1	2	3	4
Proposed (T)	1.00	1.32	1.30	1.30	1.30
Standard (T_0)	(1.00)	1.56	2.08	2.60	3.26
Ratio[a] (T/T_0)	1.00	0.85	0.63	0.50	0.40

Note: Efficiency is measured in terms of computational time. All the values are normalized against the computational time of the value in parentheses. The first row is the number of data points used in the Bayesian updating. Value 0 corresponds to prior estimates.

[a] Results are based on MATLAB® 2008a.

of performing the given task in a bounded time ($O(1)$) provides the user a viable solution to problems where the time constrained prognostics is required. The cost in achieving this is a slight relative error introduced by the MDI reparameterization process. It should be noted that the level of the resulting relative error also depends on the underlying problems and the models.

Appendix 7.2 presents the detailed analysis of the proposed Bayesian updating algorithm with MDI reparameterization using the same example, focusing on the performance and efficiency of the algorithm itself. Appendix 7.1 presents the detailed analysis of the proposed factorized M-H algorithms. The proof that the factorized M-H algorithm maintains the detailed balance equation is included, and two examples are given in detail to demonstrate the improvement of simulation efficiency.

7.5.7 Discussion

This example is based on the experimental data of the standard testing specimens under constant cyclic loading conditions. The stress intensity factor has an analytical solution and the loading spectrum has no uncertainty. For practical problems, several issues must be addressed for prognostics. First, the geometry of the target component could be very complex and no analytical solution for the stress intensity factor is available. Second, the nonstationary loading spectrum has uncertainty.

Several methods are viable for fatigue crack prognostics under complex geometry configurations. The methods include using the finite element model to calculate the stress intensity factor so the classical fracture mechanics fatigue crack growth model can be used as usual. The usual finite element methods (FEM) have some known limitations. The extended finite element methods (XFEM) can be adopted for fatigue crack analysis [43–45]. Simplified surrogate models fitted to the crack growth rate can also be used to reduce the computational cost [46,47]. If sufficient experimental data about the target system or similar systems are available, data-driven methods may be used to avoid the computation of the stress intensity factors. The data-driven methods may not consider the underlying physics and therefore cannot be generic. The classical fracture mechanics model with XFEM is generic, but it is computationally expensive, particularly for complex geometry configurations. Choosing the appropriate methods depends on the actual problems and computational and time constraints.

To perform fatigue crack growth prognostics and RUL prediction under nonstationary loading spectrum can be challenging. First, the nonstationary loading spectrum should be properly modeled using probabilistic methods so that the uncertainty can be expressed explicitly. Second, the future loading spectrum must be extrapolated accurately. Since the loading spectrum can be considered as time series data, modeling methods such as using autoregressive (AR) models and its variants [48–50], stochastic rainflow reconstruction methods [51,52], and Markov methods [53,54].

Despite various studies on fatigue crack growth analysis with complex geometry configurations under nonstationary load conditions, many more studies are required to conduct efficient and effective prognostics in realistic applications. A generic variable-order Bayesian AR model for loading spectrum prediction is currently under investigation and will be the future focus of the authors.

7.6 CONCLUSION

In this study, a Bayesian framework for fatigue damage prognostics and RUL predictions is presented. This framework consists of a two-level hierarchical structure for uncertainty management and the corresponding computational implementations. The hierarchical structure can include multiple competitive models for probabilistic inference. The MCMC methodology in the general state space is presented as a generic computation method for Bayesian analysis involving multiple models. An efficient algorithm based on the principle of MDI is developed to reduce the computational complexity for continuous Bayesian updating. A factorized M-H algorithm is developed to improve the computational efficiency for simulations. The overall methods and algorithms are demonstrated using a fatigue damage problem. Experimental data are used for validation. The advantages of the proposed methods and algorithms are demonstrated by comparisons with the traditional approach. The presented Bayesian updating algorithm with MDI reparameterization and the factorized M-H algorithms are generic to apply to other fields for Bayesian analysis. Based on the current study, several conclusions are drawn.

1. The presented two-level hierarchical structure for uncertainty management can include several major uncertainties, such as model choice uncertainty, model parameter uncertainty, measurement uncertainty, and so on. Additional uncertainty can be easily included using probabilistic modeling. Measurement data can be used to perform the Bayesian updating. As more measurement data are integrated into the updating process, uncertainties in prognostics and predictions can be reduced.

2. The efficient updating algorithm based on the principle of MDI can reduce the computational complexity from $O(n)$ to $O(1)$ at each updating stage and reduce the overall computational complexity from $O(n^2)$ to $O(n)$ in the entire system service time. It provides a viable approach for time-constrained Bayesian applications where data arrive frequently for updating.

3. The transdimensional MCMC method provides a universal computation methods for Bayesian analysis involving multiple models. All the quantities of interest can be estimated from one transdimensional MCMC simulation.

4. The factorized M-H algorithm can enhance the computational efficiency for the usual MCMC simulations by rejecting the proposal sample earlier. The improvement of computational efficiency is demonstrated and verified by examples.

5. The fatigue problem demonstrates the overall methods and algorithms presented in this study. The results indicate that the proposed work can improve the computational efficiency significantly ($O(1)$) and keep the maximum RE less than 5%.

6. As a natural results of transdimensional MCMC simulations, Bayesian model averaging is included to highlight the necessity of including all useful models when the underlying mechanism cannot be precisely modeled. In such cases, model averaging can produce more conservative results for decision-making.

APPENDICES

APPENDIX 7.1 PROOF OF DETAILED BALANCE EQUATION OF THE FACTORIZED M-H ALGORITHM IN SECTION 7.4.2 AND DEMONSTRATION EXAMPLES

Lemma 7.1: The factorized M-H algorithm in Section 7.4.2 maintains the detailed balance equation in MCMC simulations.

Suppose the target distribution $p(\theta_k)$ can be factorized into several independent components as $p(\theta_k) = p(\vartheta_1)p(\vartheta_2)\cdot\cdot p(\vartheta_k)$ and the proposal density $q(\theta'_k,\theta_k)$ can be factorized as $q(\vartheta'_1,\vartheta_1)q(\vartheta'_2,\vartheta_2)\cdot q(\vartheta'_k,\vartheta_k)$ accordingly. ■

Proof 1: The joint probability distribution of (θ_k,θ'_k),

$$
\begin{aligned}
& p(\theta_k)p(\theta'_k \mid \theta_k) \\
&= p(\theta_k)q(\theta_k,\theta_k)\alpha(\theta_k,\theta'_k) \\
&= p(\vartheta_1)q(\vartheta'_1,\vartheta_1)\alpha(\vartheta_1,\vartheta'_1)\times\ldots\times p(\vartheta_k)q(\vartheta_k,\vartheta_k)\alpha(\vartheta_k,\vartheta'_k) \\
&= p(\vartheta_1)q(\vartheta'_1,\vartheta_1)min\left\{1,\frac{p(\vartheta'_1)q(\vartheta_1,\vartheta'_1)}{p(\vartheta_1)q(\vartheta_1,\vartheta_1)}\right\}\times\ldots \\
&\quad \times p(\vartheta_k)q(\vartheta'_k,\vartheta_k)min\left\{1,\frac{p(\vartheta'_k)q(\vartheta_k,\vartheta_k)}{p(\vartheta_k)q(\vartheta'_k,\vartheta_k)}\right\} \\
&= min\left\{p(\vartheta_1)q(\vartheta'_1,\vartheta_1),p(\vartheta'_1)q(\vartheta_1,\vartheta'_1)\right\}\times\ldots \\
&\quad min\left\{p(\vartheta_k)q(\vartheta'_k,\vartheta_k),p(\vartheta'_k)q(\vartheta_k,\vartheta'_k)\right\} \\
&= p(\vartheta'_1)q(\vartheta_1,\vartheta'_1)\alpha(\vartheta'_1,\vartheta_1)\times\ldots\times p(\vartheta'_k)q(\vartheta_k,\vartheta'_k)\alpha(\vartheta'_k,\vartheta_k) \\
&= p(\theta'_k)p(\theta_k|\theta'_k)
\end{aligned}
\tag{7.30}
$$

This proves that the factorized M-H algorithm maintains the detailed balance equation in MCMC simulations and ensures the generated Markov chain will eventually converge to the target distribution. Two examples are presented here to review the computational efficiency and simulation performance. The usual MCMC simulations with the standard M-H algorithm are performed to provide the baseline results.

Example 1: Draw Samples from a Three-Dimensional Distribution

$$p(x) \propto \exp\left[-\frac{1}{2}\left(\frac{x_1-\mu_1}{\sigma_1}\right)^2-\frac{1}{2}\left(\frac{x_2-\mu_2}{\sigma_2}\right)^2-\frac{1}{2}\left(\frac{x_3-\mu_3}{\sigma_3}\right)^2\right], \quad (7.31)$$

where $(\mu_1,\sigma_1) = (0,1)$, $(\mu_2,\sigma_2) = (1,2)$, and $(\mu_3,\sigma_3) = (0.5,2)$.

In practice, drawing samples from such a distribution can be easily done without using MCMC simulations. The example is adopted here for illustration purposes only. 100,000 samples are drawn from $p(x)$ using the factorized M-H algorithm. The MCMC simulation with the standard M-H algorithm is also performed to provide a baseline solution. Both simulations use Gaussian proposal distributions with scaling parameters equal to the corresponding variable standard deviations. For example, the new sample x_i' is proposed as $x_i' = x_i + \text{Norm}(0,1) \cdot \sigma_i$ for $i = 1,2,3$. Table 7.6 shows simulation results. The last column gives the normalized computational time. In this example, the factorized MCMC sampler improves 20% efficiency in terms of the computational time. The mean and standard deviation estimates $(\hat{\mu}_{1,2,3},\hat{\sigma}_{1,2,3})$ indicate both samplers perform well. Sample histograms and exact distributions agree closely as shown in Figure 7.9.

The basic idea of gaining efficiency in the factorized MCMC algorithm is that early rejection can be made to avoid computing other term in the target distribution. For example, in Example 1, an inappropriate proposal x_1' for x_1 may lead to a rejection immediately after evaluating the acceptance probability of $\alpha_1(x_1',x_1)$. Therefore, additional proposal for x_2 and x_3 and the acceptance probability computations can be saved. The efficiency can be improved much more

TABLE 7.6 Comparisons between Standard and Factorized M-H Algorithms of Example 1

Algorithms	$\hat{\mu}_1,\hat{\sigma}_1$	$\hat{\mu}_2,\hat{\sigma}_2$	$\hat{\mu}_3,\hat{\sigma}_3$	**Cost**[a]
standard M-H	0.0182, 1.0116	0.9790, 1.9897	0.5159, 1.9878	(1.00)
factorized M-H	−0.0039, 1.0095	0.9825, 2.0240,	0.4968, 2.0159	0.78

Note: Mean and standard deviation estimates are reported. Cost is measured in terms of computational time and normalized against the time spent using standard M-H algorithm (in parentheses).

[a] Results are based on MATLAB 2008a.

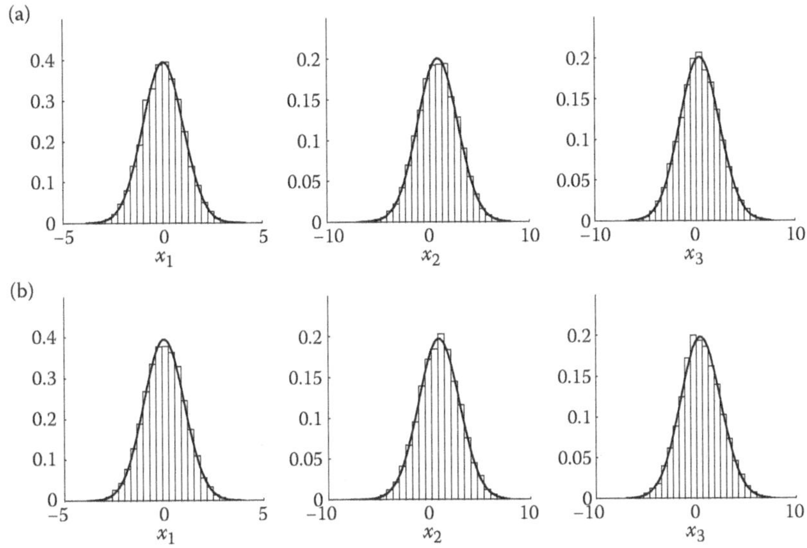

FIGURE 7.9 Sample histograms of MCMC simulations. Solid lines are exact distributions. (a) Standard M-H and (b) factorized M-H.

if evaluations of the prior PDF and the likelihood term are both computational intensive, such as finite element model or structural dynamics model. How to group components to improve the overall efficiency depends on actual problems. A common factorizing practice is to separate the prior PDF and the likelihood term. This is demonstrated in Example 2.

Example 2: The Fatigue Damage Problem Revisit

Here we consider the fatigue damage problem in Section 7.5. To focus on the factorized M-H algorithm, Paris' model is used here. All the information needed is identical to that in Section 7.5. In the factorized M-H algorithm, the prior term and the likelihood term are separated. The proposed sample is sequentially evaluated for acceptance. The computational efficiency is measured in terms of the computation time and the results are shown in Table 7.7. The simulation performance comparison is made based on the sample statistics. In each simulation, 200,000 samples are generated. Estimates on the mean vector and covariance matrix are reported in Table 7.8. The MCMC simulations using the standard M-H algorithm are performed for comparisons.

TABLE 7.7 Comparisons of Efficiency between the Standard and Factorized M-H Algorithms of Example 2

Data	0	1	2	3	4
Standard M-H (T_0)	(1.00)	5.76	10.52	14.97	19.42
Factorized M-H (T_1)	1.00	3.70	6.19	9.37	11.19
Ratio[a] (T_1/T_0)	1.00	0.642	0.588	0.626	0.576

Note: Efficiency is measured in terms of computational time. All the values are normalized against the value in parentheses. The value in the first row indicates the number of data points used in the Bayesian updating. The value 0 means prior distributions.

[a] Results are based on MATLAB 2008a.

It is observed in Table 7.7 that using the factorized M-H algorithm can improve the computational efficiency >30% at least. The performance in terms of estimated sample statistics (mean vector and covariance matrix) is very close for the two algorithms. It should be noted that, in general, the acceptance rate of the factorized M-H algorithm is smaller than that of the standard M-H algorithm. This is

TABLE 7.8 Comparisons of Simulation Performance between Standard and Factorized M-H Algorithms of Example 2

Data	Standard M-H	Factorized M-H
0	$\begin{bmatrix} -25.8587 \\ 2.7983 \end{bmatrix}, \begin{bmatrix} 1.4870 & -0.2534 \\ -0.2534 & 0.0433 \end{bmatrix}$	$\begin{bmatrix} -25.8587 \\ 2.7983 \end{bmatrix}, \begin{bmatrix} 1.4870 & -0.2534 \\ -0.2534 & 0.0433 \end{bmatrix}$
1	$\begin{bmatrix} -25.3424 \\ 2.7146 \end{bmatrix}, \begin{bmatrix} 1.1862 & -0.2044 \\ -0.2044 & 0.0353 \end{bmatrix}$	$\begin{bmatrix} -25.3503 \\ 2.7160 \end{bmatrix}, \begin{bmatrix} 1.1710 & -0.2019 \\ -0.2019 & 0.0349 \end{bmatrix}$
2	$\begin{bmatrix} -24.7486 \\ 2.6214 \end{bmatrix}, \begin{bmatrix} 1.0423 & -0.1817 \\ -0.1817 & 0.0317 \end{bmatrix}$	$\begin{bmatrix} -24.7470 \\ 2.6211 \end{bmatrix}, \begin{bmatrix} 1.0343 & -0.1803 \\ -0.1803 & 0.0315 \end{bmatrix}$
3	$\begin{bmatrix} -24.8969 \\ 2.6566 \end{bmatrix}, \begin{bmatrix} 1.0366 & -0.1806 \\ -0.1806 & 0.0315 \end{bmatrix}$	$\begin{bmatrix} -24.9052 \\ 2.6581 \end{bmatrix}, \begin{bmatrix} 1.0322 & -0.1799 \\ -0.1799 & 0.0313 \end{bmatrix}$
4	$\begin{bmatrix} -26.1870 \\ 2.8879 \end{bmatrix},t \begin{bmatrix} 0.9285 & -0.1610 \\ -0.1610 & 0.0279 \end{bmatrix}$	$\begin{bmatrix} -26.1714 \\ 2.8852 \end{bmatrix}, \begin{bmatrix} 0.9402 & -0.1631 \\ -0.1631 & 0.0283 \end{bmatrix}$

Note: Estimates on mean vector and covariance matrix are reported. The value in the first column indicates the number of data points used in the Bayesian updating. The value 0 means prior distributions.

because a proposal that is rejected earlier in the factorized M-H algorithm could in fact be accepted in the standard M-H algorithm. For example, in Example 2, a proposal may be rejected when it is evaluated for acceptance against the prior term, but the proposal could be accepted when it is evaluated against the entire prior and likelihood term. In MCMC simulations where the proposal distribution is not well designed, the difference in the acceptance rate between those two algorithms is small.

APPENDIX 7.2 DEMONSTRATION OF THE PERFORMANCE AND EFFICIENCY OF THE PROPOSED BAYESIAN UPDATING ALGORITHM WITH MDI REPARAMETERIZATION IN SECTION 7.3

To review the performance of the proposed Bayesian updating algorithm with MDI reparameterization, the fatigue problem is used here as an concrete example. To focus on the algorithmic performance, we only consider one model here. The results considering the four models are summarized in Tables 7.5 and 7.4. A complete inclusion of four models is straightforward using the same procedure presented below. All the information related to the fatigue problem is identical to the problem description in Section 7.5. The computational efficiency in terms of computational time and the prognostics performance are compared with standard Bayesian updating. At each updating, 200,000 samples are generated using the MCMC simulation with the standard M-H algorithm. Fatigue crack growth curves are then calculated using the simulation samples. The computational times are shown in Table 7.9. All the results are normalized against the value in the parentheses.

TABLE 7.9 Comparisons on Computational Efficiency between the Standard Bayesian Updating Algorithm and the Proposed Bayesian Updating Algorithm with MDI Reparameterization

Data	**1**	**2**	**3**	**4**
Standard (T_0)	(1.00)	1.68	2.45	3.17
with MDI (T_1)	1.00	1.02	1.06	1.04
Ratio[a] (T_1/T_0)	1.00	0.607	0.433	0.328

Note: The value in the first row indicates the number of data points used in the Bayesian updating.

[a] Results are based on MATLAB 2008a.

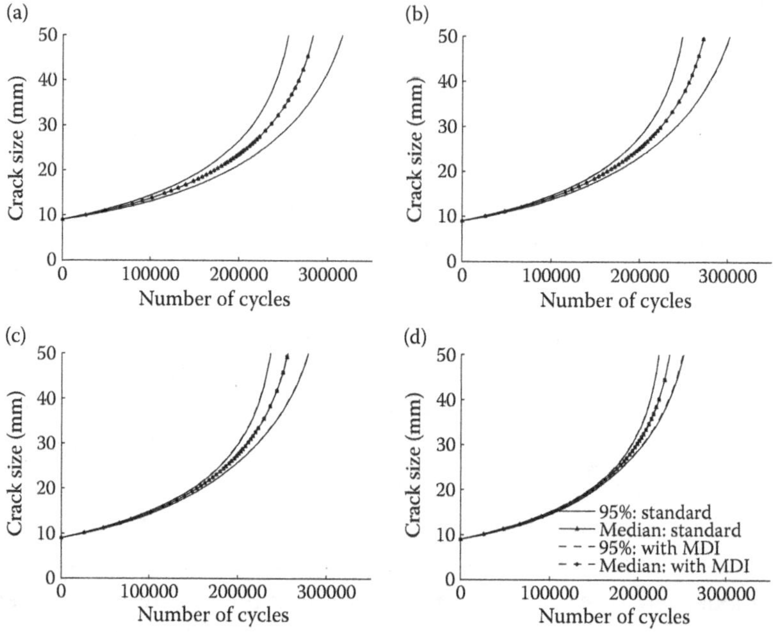

FIGURE 7.10 Comparisons on fatigue crack growth prognostics between the standard Bayesian updating algorithm and the proposed Bayesian updating algorithm with MDI reparameterization. (a) Updating with one data point, (b) updating with two data points, (c) updating with three data points, and (d) updating with four data points.

The computational time of the proposed Bayesian updating algorithm is approximately constant for each updating ($O(1)$). The computational time of the standard Bayesian updating algorithm is approximately linear ($O(n)$). The improvement of the efficiency is very close to the analysis presented in Section 7.3. The performance of prognostics is compared, and the results are shown in Figure 7.10. The prognostics results of the crack growth are almost identical. In fact, the maximum relative error over the four updatings is 3.4%. Comparing to the computational cost reduction (in a factor of $3.17/1.04 \gg 3$), the relative error is not significant. The RUL predictions along with updating are shown in Figure 7.11, where the estimates associated with the Bayesian updating algorithm with MDI reparameterization are almost identical to those with the standard Bayesian updating algorithm.

It should be noted that the comparison of the computational efficiency is based on MCMC simulations with the standard M-H algorithm.

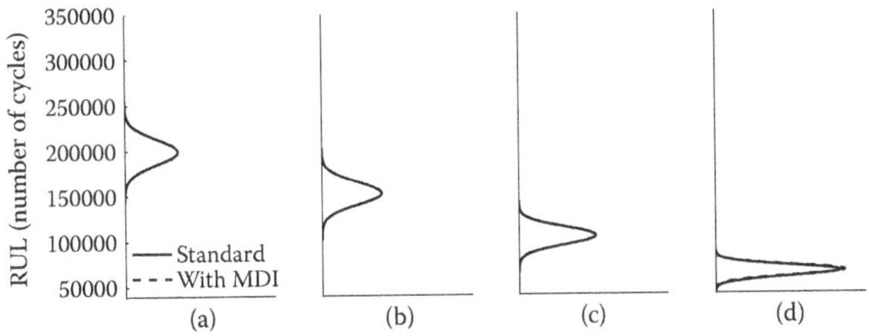

FIGURE 7.11 Comparison of the RUL estimates between the standard Bayesian updating algorithm and the proposed Bayesian updating algorithm with MDI reparameterization. (a) Updating with one data point, (b) updating with two data points, (c) updating with three data points, and (d) updating with four data points.

Finally, if we employ the factorized M-H algorithm in MCMC simulations, the improvement can be further enhanced. For example, using the proposed Bayesian updating algorithm and the factorized M-H algorithm for MCMC simulations, a computational cost reduction factor of $(3.17/1.04) \times (19.42/11.19) \approx 5.3$ is expected. In general, the proposed

---

Nomenclature

$p(\cdot)$	Probability density function (PDF); Probability distribution
$P(\cdot)$	Probability
M_k	Model indexed by $k \in \mathcal{K}$
θ_k	The model parameter under model M_k
Θ_k	Parameter state space for θ_k
X	Measurement data
$M(\theta_k, M_k; Y)$	Deterministic prediction of model M_k
Y	Model independent variable
O_{ij}	Odds ratio
B_{ij}	Bayes factor in favor of model M_i over model M_j
$p(\theta;\zeta)$	The distribution or PDF of variable θ, parameterized by ζ
$KL(f_1 \| f_0)$	Kullback-Leibler divergence of f_1 with respect to f_0
$\mathbb{E}_{p(\theta)} f(\theta)$	Mathematical expectation of $f(\theta)$ under distribution $p(\theta)$
Λ	Lagrangian
λ,β	Lagrange multipliers
$\pi(x)$	The conventional notation of the target distribution
$\alpha(x_t,y)$	The acceptance probability of the proposed sample y

u	Random quantities
g_{ij}	The bijective function
$P(i,j)$	Probability of a proposed move from model M_i to model M_j
ΔK	Stress intensity variation range
K_c	Critical stress intensity or material toughness
ΔK_{th}	Threshold intensity variation range
K_{max}	The maximum stress intensity
σ	Applied stress in fatigue crack growth models
σ_{ref}	Reference stress level in the small time scale model
$H(\cdot)$	Heaviside function
σ_y	Material yielding stress
E	Material Young's modulus

methods and algorithms reduce the complexity to $O(1)$, comparing with the conventional $O(n)$.

REFERENCES

[1] J. Schijve. Fatigue of structures and materials in the 20th century and the state of the art* 1. *International Journal of Fatigue*, 25(8):679–702, 2003.

[2] W. Cui. A state-of-the-art review on fatigue life prediction methods for metal structures. *Journal of Marine Science and Technology*, 7(1):43–56, 2002.

[3] A. Fatemi, L. Yang. Cumulative fatigue damage and life prediction theories: a survey of the state of the art for homogeneous materials. *International Journal of Fatigue*, 20(1):9–34, 1998.

[4] A.D. Dimarogonas. Vibration of cracked structures: a state of the art review. *Engineering Fracture Mechanics*, 55(5):831–857, 1996.

[5] H.D. Merchant, M.G. Minor, Y.L. Liu. Mechanical fatigue of thin copper foil. *Journal of Electronic Materials*, 28(9):998–1007, 1999.

[6] D.T. Read. Tension-tension fatigue of copper thin films. *International Journal of Fatigue*, 20(3):203–209, 1998.

[7] P. Paris, F. Erdogan. A critical analysis of crack propagation laws. *Journal of Basic Engineering*, 85(4):528–534, 1963.

[8] R.G. Forman, V.E. Kearney, R.M. Engle. Numerical analysis of crack propagation in cyclic-loaded structures. *Journal of Basic Engineering*, 89(3):459–464, 1967.

[9] J.C. Newman. A crack closure model for predicting fatigue crack growth under aircraft spectrum loading. *Methods and Models for Predicting Fatigue Crack Growth Under Randomloading-Stp 748*, p. 53, 1981.

[10] K. Ortiz, A.S. Kiremidjian. Stochastic modeling of fatigue crack growth. *Engineering Fracture Mechanics*, 29(3):317–334, 1988.

[11] J. Chiquet, N. Limnios, M. Eid. Piecewise deterministic Markov processes applied to fatigue crack growth modelling. *Journal of Statistical Planning and Inference*, 139(5):1657–1667, 2009.

[12] S. Mohanty, S. Das, A. Chattopadhyay, P. Peralta. Gaussian process time series model for life prognosis of metallic structures. *Journal of Intelligent Material Systems and Structures*, 20(8):887, 2009.

[13] A. Ray, S. Tangirala. Stochastic modeling of fatigue crack dynamics for on-line failure prognostics. *Control Systems Technology, IEEE Transactions On*, 4(4):443–451, 2002.

[14] R. Patankar, A. Ray, A. Lakhtakia. A state-space model of fatigue crack growth. *International Journal of Fracture*, 90(3):235–249, 1998.

[15] A. Ray, R. Patankar. Fatigue crack growth under variable-amplitude loading: Part I—Model formulation in state-space setting* 1. *Applied Mathematical Modelling*, 25(11):979–994, 2001.

[16] S.M. Beden, S. Abdullah, A.K. Ariffin. Review of fatigue crack propagation models for metallic components. *European Journal of Scientific Research*, 28(3):364–397, 2009.

[17] D.N.V. As, T. Parkvej. Stochastic modeling of fatigue crack growth and inspection. *Probabilistic methods for structural design*, p. 59, 1997.

[18] R. Zhang, S. Mahadevan. Model uncertainty and Bayesian updating in reliability-based inspection. *Structural Safety*, 22(2):145–160, 2000.

[19] Z. Zhao, A. Haldar. Bridge fatigue damage evaluation and updating using non-destructive inspections. *Engineering Fracture Mechanics*, 53(5):775–788, 1996.

[20] J.P. Klein, P.K. Goel. *Survival Analysis: State of the Art*. Springer, 1992.

[21] X. Guan, R. Jha, Y. Liu. Probabilistic fatigue damage prognosis using maximum entropy approach. *Journal of Intelligent Manufacturing*, pp. 1–9. 2009. 10.1007/s10845-009-0341-3.

[22] M. Janssen, J. Zuidema, R.J.H. Wanhill. *Fracture Mechanics*. London Spon Press, 2004.

[23] R.E. Kass, A.E. Raftery. Bayes factors. *Journal of the American Statistical Association*, 90(430):773–795, 1995.

[24] N. Metropolis, A.W. Rosenbluth, M.N. Rosenbluth, A.H. Teller, E. Teller, et al. Equation of state calculations by fast computing machines. *Journal of Chemical Physics*, 21(6):1087, 1953.

[25] W.R. Gilks, P. Wild. Adaptive rejection sampling for Gibbs sampling. *Applied Statistics*, 41(2):337–348, 1992.

[26] P.J. Green. Reversible jump Markov chain Monte Carlo computation and Bayesian model determination. *Biometrika*, 82(4):711, 1995.

[27] S. Richardson, P.J. Green. On Bayesian analysis of mixtures with an unknown number of components (with discussion). *Journal of the Royal Statistical Society: Series B (Statistical Methodology)*, 59(4):731–792, 1997.

[28] C.P. Robert, T. Ryden, D.M. Titterington. Bayesian inference in hidden Markov models through the reversible jump Markov chain Monte Carlo method. *Journal of the Royal Statistical Society: Series B (Statistical Methodology)*, 62(1):57–75, 2000.

[29] Y. Fan, G.W. Peters, S.A. Sisson. Automating and evaluating reversible jump MCMC proposal distributions. *Statistics and Computing*, 19(4):409–421, 2009.

[30] P.J. Green, N.L. Hjort, S. Richardson. *Highly structured stochastic systems.* Oxford University Press, 2003.

[31] D. Hastie. *Towards automatic reversible jump Markov chain Monte Carlo.* PhD thesis, Citeseer, 2005.

[32] S. Kullback, R.A. Leibler. On information and sufficiency. *The Annals of Mathematical Statistics*, pp. 79–86, 1951.

[33] R.D. Gupta, D. Kundu. Generalized exponential distributions. *Australian & New Zealand Journal of Statistics*, 41(2):173–188, 1999.

[34] R.Y. Rubinstein, D.P. Kroese. *The cross-entropy method: a unified approach to combinatorial optimization, Monte-Carlo simulation, and machine learning.* Springer-Verlag, New York, 2004.

[35] W.K. Hastings. Monte Carlo sampling methods using Markov chains and their applications. *Biometrika*, 57(1):97, 1970.

[36] G.O. Roberts. Markov chain concepts related to sampling algorithms. *Markov chain Monte Carlo in practice*, 57, 1996.

[37] D.A. Virkler, B.M. Hillberry, P.K. Goel, and Purdue Univ Lafayette in School of Mechanical Engineering. The statistical nature of fatigue crack propagation. Technical report, 1977.

[38] Z. Lu, Y. Liu. Small time scale fatigue crack growth analysis. *International Journal of Fatigue*, 32(8):1306–1321, 2010.

[39] A.J. McEvily. On the quantitative analysis of fatigue crack propagation. *Fatigue Mechanisms: Advances in Quantitative Measurement of Physical Damage*, p. 283, 1983.

[40] J.C. Newman Jr., E.P. Phillips, M.H. Swain. Fatigue-life prediction methodology using small-crack theory. *International Journal of Fatigue*, 21(2):109–119, 1999.

[41] W.A. Logsdon, P.K. Liaw, A. Saxena, V.E. Hulina. Residual life prediction and retirement for cause criteria for sstg upper casings—I. Mechanical and fracture mechanics material properties development. *Engineering Fracture Mechanics*, 25(3):259–288, 1986.

[42] S.H. Jeffreys. *Theory of Probability*. Oxford University Press, USA, 1998.

[43] G. Zi, T. Belytschko. New crack-tip elements for XFEM and applications to cohesive cracks. *International Journal for Numerical Methods in Engineering*, 57(15):2221–2240, 2003.

[44] X.Y. Liu, Q.Z. Xiao, B.L. Karihaloo. XFEM for direct evaluation of mixed mode SIFs in homogeneous and bi-materials. *International Journal for Numerical Methods in Engineering*, 59(8):1103–1118, 2004.

[45] P.M.A. Areias and T. Belytschko. Non-linear analysis of shells with arbitrary evolving cracks using XFEM. *International Journal for Numerical Methods in Engineering*, 62(3):384–415, 2005.

[46] Y. Liu, S. Mahadevan. Probabilistic fatigue life prediction using an equivalent initial flaw size distribution. *International Journal of Fatigue*, 31(3):476–487, 2009.

[47] S. Sankararaman, Y. Ling, S. Mahadevan. Statistical inference of equivalent initial flaw size with complicated structural geometry and multi-axial variable amplitude loading. *International Journal of Fatigue*, 32(10):1689–1700, 2010.

[48] D.A. Reed, R.H. Scanlan. Autoregressive representation of longitudinal, lateral, and vertical turbulence spectra. *Journal of Wind Engineering and Industrial Aerodynamics*, 17(2):199–214, 1984.
[49] C. Leser. *On Stationary and Nonstationary Fatigue Load Modeling Using Autoregressive Moving Average (ARMA) Models*. PhD thesis, Virginia Polytechnic Institute and State University, 1993.
[50] D.F. Sociex, M.A. Pompetzki. Modeling variability in service loading spectra. *Probabilistic Aspects of Life Prediction*, 1(2):46, 2004.
[51] A.K. Khosrovaneh, N.E. Dowling. Fatigue loading history reconstruction based on the rainflow technique. *International Journal of Fatigue*, 12(2):99–106, 1990.
[52] J. Klemenc, M. Fajdiga. A neural network approach to the simulation of load histories by considering the influence of a sequence of rainflow load cycles. *International Journal of Fatigue*, 24(11):1109–1125, 2002.
[53] G. Jiao, T. Moan. Probabilistic analysis of fatigue due to Gaussian load processes. *Probabilistic Engineering Mechanics*, 5(2):76–83, 1990.
[54] M. Frendahl, I. Rychlik. Rainflow analysis: Markov method. *International Journal of Fatigue*, 15(4):265–272, 1993.

CHAPTER 8

Physics-Based Methods of Failure Analysis and Diagnostics in Human Space Flight

Vadim N. Smelyanskiy, Dmitry G. Luchinsky, Vasyl V. Hafiychuk, Viatcheslav V. Osipov, Igor Kulikov, and Ann Patterson-Hine

CONTENTS

8.1 INTRODUCTION

The integrated health management (IHM) for the future aerospace systems requires to interface models of multiple subsystems in an efficient and accurate information environment at the earlier stages of system design. The complexity of modern aeronautic and aircraft systems dictates employment of hybrid models and high-level reasoners for analyzing continuous and discrete information flow involving multiple modes of operation in uncertain environments, unknown state variables, and heterogeneous software and hardware components.

To provide the information link between key design/performance parameters and high-level reasoners our research relies on development of multiphysics performance models, distributed sensors networks, and fault diagnostic and prognostic (FD&P) technologies [1] in close collaboration with system designers. The main challenges are related to the in-flight assessment of the structural stability, engine performance, and trajectory control. The main goal is to develop an intelligent IHM that enhances components and system reliability and provides a postflight feedback for optimization of the next generation of aerospace systems. Our efforts are concentrated on several directions of the research including an innovative approach to the diagnostics/prognostics based on the real time dynamical inference (DI) technologies extended to encompass hybrid systems with hidden state trajectories. The major investments are into the multiphysics performance modeling that provides an access to the main performance parameters of, e.g., solid and liquid rocket motors and composite materials of the nozzle and case.

Some of the recent results of our research are discussed in this chapter. We begin by introducing the problem of dynamical inference of stochastic nonlinear models and reviewing earlier results. Next, we present our analytical approach to the solution of this problem based on the path integral formulation. The resulting algorithm does not require an extensive global search for the model parameters, provides optimal compensation for the effects of dynamical noise, and is robust for a broad range of dynamical models. In the following section, the strengths of the algorithm are illustrated by inferring parameters of the stochastic Lorenz system and comparing the results with those of earlier research. Next, we discuss a number of recent applications of the method to the IHM of aerospace system.

First, we apply the DI approach to a solution of classical three tank problem with unknown continuous and binary parameters. The problem is considered in the context of ground support system for filling fuel tanks of liquid rocket motors. It is shown that the DI algorithm is well suited for solution of a hybrid version of this benchmark problem in the presence of additional periodic and stochastic perturbation of unknown strength.

Second, the approach is applied to an analysis of the nozzle fault in a solid rocket motor (SRM) modeled as a set of one-dimensional partial differential equations coupled to the dynamics of the propellant regression. This example is focused on the inference of discrete and continuous parameters of the nozzle blocking fault and on reducing the probability of "misses" of an on-board FD&P for SRM.

Finally, Bayesian inferential framework is formulated for development of the IHM system for in-flight structural health monitoring (SHM) of composite materials. We consider the signal (elastic waves) generated by piezoelectric actuator mounted on composite structure and scattering by the localized damaged area. The possibility of damage inference is discussed in the context of development of the SHM.

The conclusions are drawn and the future work is discussed in the conclusions.

8.2 DYNAMICAL INFERENCE OF STOCHASTIC NONLINEAR MODELS

Complex phenomena in nature and technology can often be modeled successfully by stochastic nonlinear dynamical systems, thereby facilitating the diagnosis, prognosis, and control of faults. Examples range from molecular motors [2] to epidemiology [3] and from solid rocket motors [4] to coupled matter radiation systems in astrophysics [5].

The problem of inferring parameters of such models from time-series data has attracted much attention over the past decade. Although no general method exists for inferring the parameters of stochastic nonlinear dynamical models from measurements, various schemes have been proposed [6–13]. An important numerical technique, suggested in [7,14], is based on estimating drift and diffusion coefficients at a number of points in the phase space of the dynamical system. A particle filter approach [15] and the Markov chain Monte-Carlo (MCMC) approach [16] were applied successfully to reconstruct the model parameters alone.

Arguably the most general approach to the solution of this problem can be obtained within Bayesian inferential framework [8,9,17]. The Bayesian method was used for parameter estimation in maps in the presence of dynamical [18] and weak measurement [9] noise and was applied to inference of continuous systems in [10].

A common drawback of earlier works is their reliance on numerical methods for the optimization of cost functions. This disadvantage becomes increasingly more pronounced when inferring on-board complex aerospace systems. Another deficiency is that most of the earlier works deal with discrete maps, and the corresponding results are not immediately applicable to continuous systems. This problem was addressed in [10]; however, an *ad hoc* likelihood function was used there.

We now formulate the problem and present its solution using path integral approach within Bayesian inferential framework. The time-series data of the control variables is usually observed at sequential time instants and are related to the unknown "true" system L-dimensional states via the measurement model. In these settings the following formulation of the problem of dynamical inference is commonly adopted. Given M-dimensional time-series data $\mathcal{Y} = \{\mathbf{y}_n \equiv \mathbf{y}(t_n)\}$ $(t_n = nh)$, how can one infer the time variation of the unknown model parameters and the unknown dynamical trajectory $\mathcal{M} = \left\{\mathbf{c}(t), \mathbf{b}(t), \hat{\mathbf{D}}, \hat{\mathbf{M}}, \{\mathbf{x}_n\}\right\}$? It is assumed that the underlying dynamics can be described by a set of L-dimensional $(L \geq M)$ stochastic differential equations of form

$$\dot{\mathbf{x}}(t) = \mathbf{f}(\mathbf{x} \mid \mathbf{c}) + \sqrt{\hat{\mathbf{D}}}\xi(t), \tag{8.1}$$

and the observations $\mathcal{Y}$ are related to the actual unknown dynamical variables $\mathcal{X} = \{\mathbf{x}_n \equiv \mathbf{x}(t_n)\}$ via the following measurement equation

$$\mathbf{y}(t) = \mathbf{g}(\mathbf{x} \mid \mathbf{b}) + \sqrt{\hat{\mathbf{M}}}\eta(t). \tag{8.2}$$

Here $\hat{X}$ is an $M \times L$ measurement matrix, $\xi(t)$ and $\eta(t)$ are L- and M-dimensional Gaussian white noises, and $\hat{\mathbf{D}}$ and $\hat{\mathbf{M}}$ are $L \times L$ and $M \times M$ dimensional dynamical and measurement diffusion matrices respectively.

The solution of this problem is given by the so-called posterior probability density function (PDF) $\rho_{\text{post}}(\mathcal{M}|\mathcal{Y})$ of the unknown parameters $\mathcal{M}$ conditioned on observations. In Bayesian model inference, the posterior PDF is related to the prior $p_{\text{pr}}(\mathcal{M})$ PDF via Bayes' theorem [19]:

$$\rho_{\text{post}}(\mathcal{M}\,|\,\mathcal{Y}) = \frac{\ell(\mathcal{Y}\,|\,\mathcal{M})\rho_{\text{prior}}(\mathcal{M})}{\int \ell(\mathcal{Y}\,|\,\mathcal{M})\rho_{\text{prior}}(\mathcal{M})d\mathcal{M}}. \tag{8.3}$$

Here, the likelihood function $\ell(\mathcal{Y}|\mathcal{M})$ is the probability density to observe $\{\mathbf{y}_n(t)\}$ given choice $\mathcal{M}$ of the dynamical model. The prior acts as a regularizer, concentrating the parameter search to those regions of the model space favored by our expertise and any available auxiliary information. To apply Equation 8.3 iteratively using a sequence of data blocks $\mathcal{Y},\mathcal{Y}',\ldots$; the posterior computed from block $\mathcal{Y}$ can be used as the prior for the next block $\mathcal{Y}'$, etc. For a sufficiently large number of observations, $p_{\text{ps}}(\mathcal{M}|\mathcal{Y},\mathcal{Y}',\ldots)$ becomes sharply peaked around the most probable model $\mathcal{M}^*$.

If the sampling is dense enough the problem can be conveniently solved using Euler midpoint discretization of Equations 8.1 and 8.2 in the form

$$\left.\begin{aligned} \mathbf{x}_{n+1} &= \mathbf{x}_n + h\mathbf{f}(\mathbf{x}_n^*\,|\,\mathbf{c}) + \sqrt{h}\,\hat{\mathbf{D}}\xi_n, \\ \mathbf{y}_n &= \mathbf{g}(\mathbf{x}_n\,|\,\mathbf{b}) + \hat{\mathbf{M}}\eta_n, \end{aligned}\right\}, \tag{8.4}$$

where $\mathbf{x}_n^* = (\mathbf{x}_{n+1} + \mathbf{x}_n)/2$. It was shown earlier (see, e.g., [13,20]) that for independent sources of white Gaussian noise in 8.4 the probability to observe $\mathbf{y}_n$ at each time step can be factorized and written in the form

$$\begin{aligned} \rho(\mathbf{y}_{n+1}\,|\,\mathbf{x}_n,\mathbf{c}) = \int & \frac{d\mathbf{x}_{n+1}}{\sqrt{(2\pi)^M\,|\hat{\mathbf{M}}|}} e^{-\frac{1}{2}[\mathbf{y}_{n+1}-\mathbf{g}(\mathbf{x}_{n+1}|\mathbf{b})]^T\hat{\mathbf{M}}^{-1}[\mathbf{y}_{n+1}-\mathbf{g}(\mathbf{x}_{n+1}|\mathbf{b})]} \\ & \times \frac{1}{\sqrt{(2\pi h)^L\,|\hat{\mathbf{D}}|}} e^{-\frac{h}{2}[\dot{\mathbf{x}}_n-\mathbf{f}(\mathbf{x}_n^*|\mathbf{c})]^T\hat{\mathbf{D}}^{-1}[\dot{\mathbf{x}}_n-\mathbf{f}(\mathbf{x}_n^*|\mathbf{c})]-\frac{h}{2}\nabla\cdot\mathbf{f}(\mathbf{x}_n|\mathbf{c})}. \end{aligned} \tag{8.5}$$

Summation over all the discretization points $n = 0\ldots N-1$ yields the following result for the minus log-likelihood function $S = S_{dyn} + S_{meas} = -\ln \ell(\mathcal{Y}|\mathcal{M})$

$$S=\frac{N}{2}\ln|\hat{\mathbf{D}}|+\frac{h}{2}\sum_{n=0}^{N-1}\left\{\nabla\cdot\mathbf{f}(\mathbf{x}_n\,|\,\mathbf{c})+[\dot{\mathbf{x}}_n-\mathbf{f}(\mathbf{x}_n^*\,|\,\mathbf{c})]^{\mathrm{T}}\hat{\mathbf{D}}^{-1}[\dot{\mathbf{x}}_n-\mathbf{f}(\mathbf{x}_n^*\,|\,\mathbf{c})]\right\}$$

$$+\frac{N}{2}\ln|\hat{\mathbf{M}}|+\frac{1}{2}\sum_{n=1}^{N}[\mathbf{y}_n-\mathbf{g}(\mathbf{y}_n,\mathbf{x}_n\,|\,\mathbf{b})]^T\hat{\mathbf{M}}^{-1}[\mathbf{y}_n-\mathbf{g}(\mathbf{y}_n,\mathbf{x}_n\,|\,\mathbf{b})]$$

$$+(L+M)N\ln(2\pi h), \tag{8.6}$$

where $\dot{\mathbf{x}}_n=\frac{\mathbf{x}_{n+1}-\mathbf{x}_n}{h}$. Here S_{dyn} and S_{meas} are the dynamical (first two terms) and measurement (next two terms) parts of the minus log-likelihood function. We note that S_{dyn} is the minus log-probability density in the space of dynamical paths and, in the limit of $N\to\infty, h\to 0, T=Nh=\text{const}$, it coincides with the path-integral presentation obtained earlier in [21,22].

To find the general solution of the problems 8.1 and 8.2 one can iterate optimization of S in the space of dynamical paths $\{\mathbf{x}_n\}$ and in the space of parameters $\{\mathbf{c},\mathbf{b},\hat{\mathbf{D}},\hat{\mathbf{M}}\}$ (see [20]). From the view point of practical applications for an on-board IVHM system it is important to avoid heavy numerical computations and to restrict the solution to an analytical one whenever it is possible. It was shown in our earlier work that analytical solution can be obtained if prior PDFs for $\mathbf{c}$ and $\mathbf{b}$ are chosen in the form of Gaussian distributions, and prior PDFs for $\hat{\mathbf{D}}$ and $\hat{\mathbf{M}}$ are uniform.

Another key ingredient of analytical approach is successful factorization of the vector field. A general form of factorization that has been proven to be very effective in many interdisciplinary applications, including specifically aerospace applications [25,26], can be written as follows [13,20,23,24]

$$\mathbf{f}(\mathbf{x}\,|\,\mathbf{c})=\hat{\mathbf{F}}(\mathbf{x})\mathbf{c},\quad \mathbf{g}(\mathbf{y},\mathbf{x}\,|\,\mathbf{b})=\hat{\mathbf{G}}(\mathbf{y},\mathbf{x})\mathbf{b}, \tag{8.7}$$

where $\hat{\mathbf{F}}(\mathbf{x})$ and $\hat{\mathbf{G}}(\mathbf{y},\mathbf{x})$ are $((F\times L)\times L)$ and $((G\times M)\times M)$ matrices consisting of F (G) of diagonal blocks multiplied by $\phi_i(\mathbf{x})$ ($\psi_j(\mathbf{y},\mathbf{x})$) base functions of factorization. The F- and G-dimensional sets of base functions $\{\phi_i\}$ and $\{\psi_i\}$ are arbitrary and known.

For the chosen above prior PDFs and factorized vector fields the analytical solution can be used to infer model parameters (cf. with [20,24])

$$\langle\hat{\mathbf{D}}\rangle=\frac{h}{N}\sum_{n=0}^{N-1}\left[\dot{\mathbf{x}}_n-\hat{\mathbf{F}}_n\mathbf{c}\right]\left[\dot{\mathbf{x}}_n-\hat{\mathbf{F}}_n\mathbf{c}\right]^T,\ \langle\mathbf{c}\rangle=\hat{\Xi}_{\mathcal{X}}^{-1}(\hat{\mathbf{D}})\mathbf{w}_{\mathcal{X}}(\hat{\mathbf{D}}), \tag{8.8}$$

$$\mathbf{w}_{\mathcal{X}}(\hat{\mathbf{D}}) = h\sum_{n=0}^{N-1}\left[\hat{\mathbf{F}}_n^T\mathbf{D}^{-1}\dot{\mathbf{x}}_n - \frac{\mathbf{v}(\mathbf{x}_n)}{2}\right],\quad \hat{\Xi}_{\mathcal{X}}(\hat{\mathbf{D}}) = h\sum_{n=0}^{N-1}\hat{\mathbf{F}}_n^T\hat{\mathbf{D}}^{-1}\hat{\mathbf{F}}_n, \tag{8.9}$$

where $\hat{\mathbf{F}}_n \equiv \hat{\mathbf{F}}(\mathbf{x}_n)$, and the components of the vector $\mathbf{v}(\mathbf{x})$ are

$$v_m(\mathbf{x}) = \sum_{l=1}^{L}\frac{\partial F_{lm}(\mathbf{x})}{\partial x_l},\quad m = 1,\ldots,F. \tag{8.10}$$

The parameters of the measurement model can be estimated using the conditions $\frac{\partial S_{meas}}{\partial \mathbf{b}} = 0$ and $\frac{\partial S_{meas}}{\partial M_{nm}} = 0$, recovering the least square results in the form

$$\left\langle\hat{\mathbf{M}}\right\rangle = \frac{1}{N}\sum_{n=1}^{N}\left[\mathbf{y}_n - \hat{\mathbf{G}}_n\mathbf{b}\right]\left[\mathbf{y}_n - \hat{\mathbf{G}}_n\mathbf{b}\right]^T, \tag{8.11}$$

$$\left\langle\mathbf{b}\right\rangle = \hat{\Theta}_{\mathcal{X},\mathcal{Y}}^{-1}(\hat{\mathbf{M}})\mathbf{z}_{\mathcal{X},\mathcal{Y}}(\hat{\mathbf{M}}), \tag{8.12}$$

$$\mathbf{z}_{\mathcal{X},\mathcal{Y}}(\hat{\mathbf{M}}) = \sum_{n=1}^{N}\left[\hat{\mathbf{G}}_n^T\hat{\mathbf{M}}^{-1}\mathbf{y}_n\right], \tag{8.13}$$

$$\hat{\Theta}_{\mathcal{X},\mathcal{Y}}^{-1}(\hat{\mathbf{M}}) = h\sum_{n=0}^{N-1}\hat{\mathbf{G}}_n^T\hat{\mathbf{M}}^{-1}\hat{\mathbf{G}}_n, \tag{8.14}$$

where $\hat{\mathbf{G}}_n \equiv \hat{\mathbf{G}}(\mathbf{y}_n, \mathbf{x}_n)$.

Equations 8.8 through 8.14, coupled with the optimization procedure in the paths space, represent the general Bayesian framework for learning a nonlinear stochastic dynamical system from measurements that are corrupted by noise. Using this approach we can develop a method of fast online tracking of the time-varying parameters of nonstationary systems, as described below.

8.3 THE LORENZ SYSTEM

We start with the archetypical chaotic nonlinear system of Lorenz,

$$\left.\begin{aligned} \dot{x}_1 &= \sigma(x_2 - x_1) + \xi_1(t), \\ \dot{x}_2 &= r x_1 - x_2 - x_1 x_3 + \xi_2(t), \\ \dot{x}_3 &= x_1 x_2 - b x_3 + \xi_3(t), \end{aligned}\right\} \tag{8.15}$$

augmented by zero-mean Gaussian noise processes $\xi_l(t)$ with covariance $\langle \xi_l(t)\xi_{l'}(t')\rangle = D_{ll'}\delta(t - t')$. We emphasize that the algorithm works in general for nondiagonal diffusion matrix when noise terms in Equation 8.15 are correlated. Synthetic data (with no measurement noise) were generated by simulating 8.15 using the standard parameter set $\sigma = 10$, $r = 28$, $b = \frac{8}{3}$, and for various levels of dynamical noise intensities as explained below. An example of the phase portrait of the Lorenz system with dynamical noise is shown in Figure 8.1 along with the noiseless case to visually convey the difficulty of the inference problem. We now demonstrate that the DI formalism allows one both to estimate parameters of the known dynamical equations and to perform model discovery when the vector-field of the Lorenz model is assumed to be unknown in the presence of strong dynamical noise.

8.3.1 Parameter Estimation with Strong Dynamical Noise

In parameter estimation, the functional form of the nonlinear force field of 8.15 is assumed known, and the unknown values of the associated

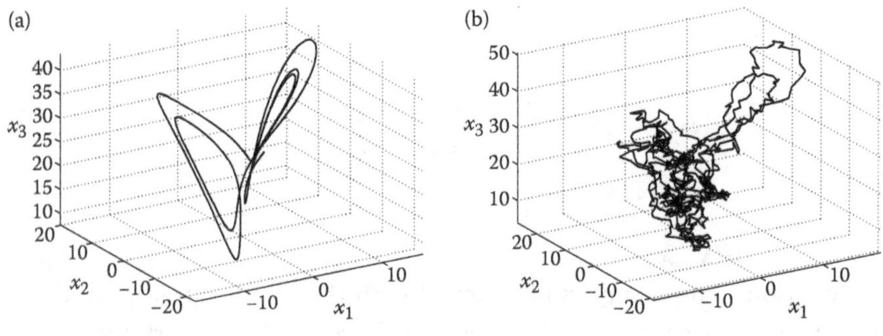

FIGURE 8.1 The phase portrait of the chaotic nonlinear Lorenz system 8.15 with the standard parameters (see text): (a) deterministic system; (b) stochastic system with strong dynamical noise, simulated with a diagonal diffusion matrix having elements $D_{11} = 1500$, $D_{22} = 1600$, and $D_{33} = 1700$.

coefficients are then estimated from data. This is the approach reported in [10], where the diffusion matrix is taken in the form $\hat{\mathbf{D}} = \tau^2 \hat{\mathbf{I}}$, and the unknown parameters $\{\sigma, r, b, \tau^2\}$ are estimated via extensive numerical optimization of a cost function by simulated annealing and back-propagation techniques. We now demonstrate that our algorithm can estimate the parameters of the system 8.15 analytically very efficiently and with high accuracy.

Note first that for the diagonal diffusion matrix the algorithm is reduced to the one-dimensional analytical solution of the problem for each equation in the form (cf. with Equations 8.8 and 8.9) $\mathbf{c}_i = \hat{\mathbf{H}}_i^{-1} \mathbf{w}_i$, where $i = 1\ldots3$

$$w_{il} = \sum_{n=0}^{N-1} \left(c_{il}\phi_{il} - \frac{\tau^2}{2} \frac{\partial \phi_{il}}{\partial x_i} \right) \text{and } \hat{\mathbf{H}}_i = \sum_{n=0}^{N-1} \begin{pmatrix} \phi_{i1}\phi_{i1} & \cdots & \phi_{i1}\phi_{iL} \\ \vdots & \ddots & \vdots \\ \phi_{iL}\phi_{i1} & \cdots & \phi_{iL}\phi_{iL} \end{pmatrix}.$$

Noise intensity is found according to 8.8. Note that each equation now has different basis functions ϕ_{il}: $\phi_{11} = x_1$ and $\phi_{12} = x_2$ for the first equation; $\phi_{21} = x_1$, $\phi_{22} = x_2$, and $\phi_{23} = x_1 x_3$ for the second equation; and $\phi_{31} = x_1 x_2$, $\phi_{32} = x_3$ for the last equation. Thus there are a total of 8 unknown parameters to be estimated: a seven-dimensional coefficient vector c and the noise intensity τ^2.

The convergence of our scheme is so rapid that it is feasible to use the algorithm in real time on "streaming" data. To make a fair comparison we use the same number of data points as in [10]. As an indication of the inference accuracy, we quote in Table 8.1 results for data simulated with the standard Lorenz parameter set, a diffusion matrix $\hat{\mathbf{D}} = \tau^2 \hat{\mathbf{I}}$, and relatively strong dynamical noise. The accuracy of the algorithm can be estimated by building distribution of the inferred model parameters obtained

TABLE 8.1 Inference Results for the Parameters of System 8.15 with Relatively Strong Dynamical Noise

Parameter	Value	Estimate
σ	10.00	9.9039
r	28.00	28.3004
b	2.667	2.8410
τ	40.00	39.9108

Note: A synthetic data set consists of 4000 points sampled with $h = 0.002$.

using N different blocks of data. An example of error estimation for the Lorenz system will be given in the next section (see also [20]). The algorithm works well [20] even when the noise intensity is increased by two orders of magnitude and the amplitude of random fluctuations becomes much larger then the amplitude of dynamical trajectories as shown for example in the Figure 8.1.

8.3.2 Model Reconstruction with Strong Dynamical Noise

We now assume that the analytical form of the nonlinear force field of the system that generated trajectory shown in Figure 8.1b is not known a priori. In this setting, it is more appropriate to refer to the inference problem as model reconstruction. In practical terms, the main difference between parameter estimation and model reconstruction is in the number of unknown parameters involved, which is typically an order of magnitude larger in the latter case. To solve this problem within the formalism of dynamical inference one may adopt a parametric model of the form

$$\dot{x}_l = \sum_{l'=1}^{3} a_{ll'}\, x_{l'}(t) + \sum_{l',l''=1}^{3} b_{ll'l''}\, x_{l'}(t) x_{l''}(t) + \xi_l(t), \tag{8.16}$$

$l,l',l'' = 1,2,3$. Including the elements of the (symmetric) diffusion matrix $\hat{\mathbf{D}}$, we now have a total of 33 unknown parameters comprising the set $\mathcal{M} = \{\{a_{ll'}\},\{b_{ll'l''}\},\{D_{ll'}\}\}$. Despite the restriction to linear, bilinear, and quadratic polynomial basis functions, 8.16 represents a very broad class of dynamical models. Assuming no measurement noise for simplicity, the application of our algorithm entails the use of Equations 8.8 and 8.9.

The accuracy of the reconstruction depends on a number of factors. We have observed that it is generally possible to achieve arbitrarily accurate inference results with a (sufficiently small) fixed sampling interval by increasing the total duration of observation; this is true even in the case of a full (i.e., nondiagonal) diffusion matrix. Indeed, we were able to achieve highly accurate parameter estimates for sampling intervals ranging from 10^{-6} to 0.01 and noise intensities ranging from 0 to 10^2. An example of inference results is summarized Table 8.2 for model 8.16 with a full diffusion matrix, showing extremely high accuracy.

Finally, note the importance of the Jacobian prefactor included in our likelihood function. Analysis shows [20] that the omission of the prefactor in the likelihood function results in a systematic underestimation of this

TABLE 8.2 Inference Results for a Representative Set of Parameters of Model 8.16, Obtained Using 200 Blocks of 600,000 Data Points Each, Sampled at $h = 0.005$

Parameter	Value	Estimate	Error
a_{11}	-10.0000	-9.9984	0.0161
a_{21}	28.0000	28.0139	0.0496
a_{31}	0.0	-0.0052	-0.5180
a_{21}	10.0000	9.9982	0.0178
a_{22}	-1.0000	-1.0051	0.5120
a_{23}	0.0	0.0031	0.3072
a_{33}	-2.6667	-2.6661	0.0196
b_{111}	0.0	0.0002	0.0179
b_{211}	0.0	0.0002	0.0238
b_{311}	0.0	-0.0004	-0.0401
b_{113}	0.0	-0.0001	-0.0111
b_{213}	-1.0000	-1.0004	0.0446
D_{11}	0.2867	0.2865	0.0587
$D_{13} = D_{31}$	0.1069	0.1061	0.7657

Note: True and inferred parameter values are shown along with the corresponding error, which is below 1% for all parameters.

parameter, whereas the inclusion of this term leads to an accurate inference as it optimally compensates for the effects of dynamical noise.

The computational efficiency of the DI algorithm also allows to lift the practical limitation on the total number of data points used for inference in previous work and to extend the dimensionality of the model space. As a consequence it can be efficiently applied to model reconstruction, when the functional form of a nonlinear vector field is unknown. We now consider a selected set of examples of dynamical inference in aerospace applications.

8.4 THE THREE TANK PROBLEM

In this section we describe an application of the DI algorithm to a system formed by a sequence of three interconnected tanks that has been declared as a benchmark for fault detection and diagnosis in dynamical system [27–29]. The system schematic as shown in Figure 8.2 will be considered in the context of ground support for liquid fuel filling. The key feature of the analysis presented below is that the underlying dynamics is mixed, i.e., exhibits both continuous flow and discrete jumps. Accordingly a hybrid probabilistic approach is required for its inference. It is shown that the DI method introduced above can be used to develop hybrid probabilistic algorithm accommodating abrupt changes in the model parameters.

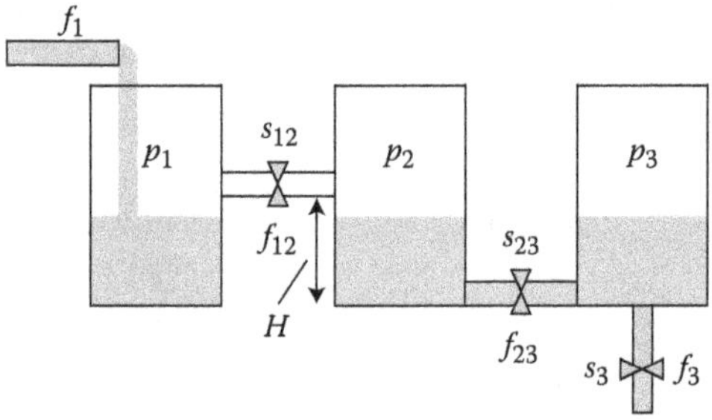

FIGURE 8.2 Schematic of three-tank bench mark system.

Accordingly the model equations are slightly modified to have the form

$$\begin{aligned}
\dot{h}_1 &= \varepsilon_1 \sin(\omega t) - c_{12}\left(\xi\left[h_1 - H_{12}\right] - \xi\left[h_2 - H_{12}\right]\right) \\
&\quad - L_{12}\sqrt{f_0\left[h_1 - H_{12}\right]} - c_{13}\left(\xi\left[h_1 - H_{13}\right] - \xi\left[h_3 - H_{13}\right]\right) \\
&\quad - L_{13}\sqrt{f_0\left[h_1 - H_{13}\right]} + f_1\left[h_1\right] + \sigma_{11}\xi_1(t), \\
\dot{h}_2 &= -\varepsilon_2 \sin(\omega t) + c_{12}\left(\xi\left[h_1 - H_{12}\right] - \xi\left[h_2 - H_{12}\right]\right) \\
&\quad + c_{23}\left(\xi\left[h_3 - H_{23}\right] - \xi\left[h_2 - H_{23}\right]\right) - L_{23}\sqrt{f_0\left[h_2 - H_{23}\right]} + \sigma_{22}\xi_1(t), \\
\dot{h}_3 &= \varepsilon_3 \sin(\omega t) + c_{23}\left(\xi\left[h_3 - H_{23}\right] - \xi\left[h_2 - H_{23}\right]\right) + \sigma_{33}\xi_1(t),
\end{aligned} \tag{8.17}$$

where

$$\begin{aligned}
&\xi(t) = t \cdot \theta(t); \quad f_0(t) = (1 + \tanh(a \cdot t))/2; \\
&f_1(t) = 0.2 \cdot f_0(0.2 - t) + \left(f_0(0.8 - x) - f_0(0.2 - t)\right) \\
&\quad + 0.2 \cdot \left(f_0(1 - t) - f_0(0.8 - t)\right)
\end{aligned} \tag{8.18}$$

and $\theta(t)$ is the unit step function. Here H_{ij} are liquid levels in each tank with respect to the connecting pipes, $f_1(t)$ is the known input flow, c_{ij} are parameters of coupling between the tanks, terms $\propto \varepsilon_i$ correspond to the vibrations of the tanks, terms $\propto L_{ij}$ correspond to the leaks, finally $\xi_i(t)$ are white Gaussian noises with amplitudes of random vibrations σ_{ij}.

The known input flow $f_1(t)$ and measured signals $h_1(t)$, $h_2(t)$, and $h_3(t)$ are shown in Figure 8.3. The values of the known relative heights in these simulations were $H_{12} = 0.5$, $H_{23} = 0.2$, $H_{13} = 0.6$. We emphasize that the time instants of the jumps in the flow signals $h_2(t)$ and $h_3(t)$ is not known and have to be inferred along with other model parameters. The coefficients corresponding to the jumps are discrete parameters of our hybrid model. Note also that all three measured flow signals are perturbed by the periodic vibrations with known frequencies, but unknown amplitudes, and by random forces. To apply algorithm introduced in 2 we have to specify the set of unknown variables

$$c = \{\varepsilon_1, \varepsilon_2, \varepsilon_3, c_{12}, c_{13}, c_{23}, L_{12}, L_{13}, L_{23}, D_{11}, D_{22}, D_{33}\}, \tag{8.19}$$

and F-dimensional ($F = 9$) set of the base functions

$$\begin{aligned}\{\phi_i(x,t)\} = \Big\{&\sin(\omega t), \big(\xi[h_1 - H_{12}] - \xi[h_2 - H_{12}]\big), \sqrt{\xi[h_1 - H_{12}]},\\ &\big(\xi[h_1 - H_{13}] - \xi[h_3 - H_{13}]\big), \sqrt{\xi[h_1 - H_{13}]},\\ &\big(\xi[h_1 - H_{12}] - \xi[h_2 - H_{12}]\big), \big(\xi[h_3 - H_{23}] - \xi[h_2 - H_{23}]\big),\\ &\sqrt{\xi[h_2 - H_{23}]}, \big(\xi[h_3 - H_{23}] - \xi[h_2 - H_{23}]\big)\Big\}.\end{aligned} \tag{8.20}$$

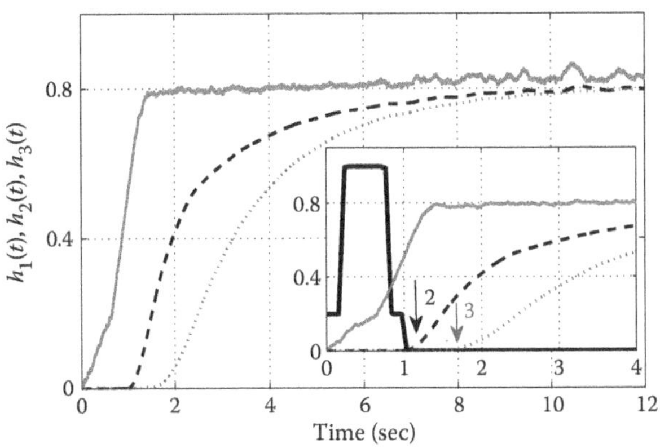

FIGURE 8.3 Measured flow signals for the three tank benchmark system are shown by solid line for $h_1(t)$, dashed line for $h_2(t)$, and by dotted line for $h_3(t)$. The measured signals are compared with the input flow (thick line) in the inset. The location of jumps for flows $h_2(t)$ and $h_3(t)$ is indicated by the arrows.

It is assumed for simplicity that the measurement noise is small, which is justified in the case of application to the ground fuel filling system. Under these assumptions the inference of the unknown parameters of model 8.17 from the measured time-series of the flow variables is performed using Equations 8.8 and 8.9. To accommodate for abrupt changes of the model parameters one has to check at each step of iterations if the base functions corresponding to the discrete parameters are identically zero and exclude related zero columns and rows from matrix $\hat{\Xi}_{\mathcal{X}}(\hat{\mathbf{D}})$ to avoid singularities in Equation 8.8.

An example of convergence of one of the model parameters is shown in Figure 8.4(b). The circles show dynamics of the mean inferred value of the parameter c_{12}. The bars show the dynamics of the corresponding standard deviation. The horizontal line indicates the actual value of the model parameter. It can be seen from the figure that the inferred value of c_{12} approaches its true value as the total time of inference is increasing; simultaneously its standard deviation is decreasing indicating the convergence of the algorithm. Note also that initial value of coupling between tank 1 and 2 is inferred as zero. This value of the coupling parameter c_{12} corresponds to the fact that at the initial time the second tank is closed and there is no flow between the two tanks. Next we can see that a jump to the correct value of $c_{12} = -2$ occurs when the second tank begins to fill. Additional jumps occurs when the third tank is also open (at around 2 sec).

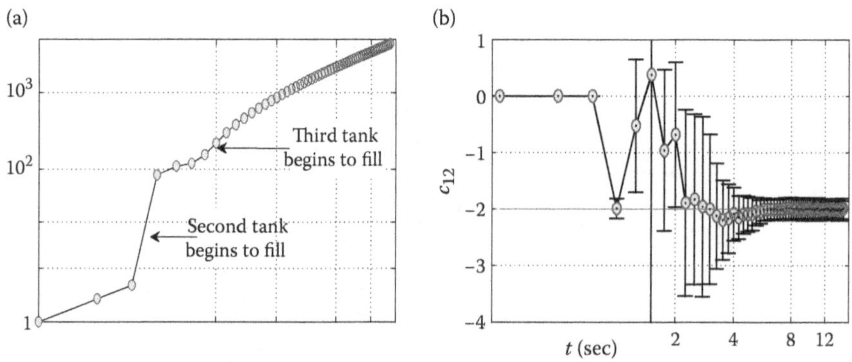

FIGURE 8.4 (a) Number of operations as a function of the total time interval of inference. (b) Convergence of the dynamical inference for the c_{12} parameter. Time is shown in logarithmic scale. The horizontal line shows the true value of the parameter c_{12}.

At this moment we can also observe a stepwise increase of the standard deviation indicating that the dimensionality of the system was increased in a stepwise manner. The number of operations required for convergence as a function of time is shown in Figure 8.4a and the results of inference are summarized in Table 8.3.

It can be seen from the table that discrete jumps in the coupling c_{ij} and leak L_{ij} parameters are also detected and their continuous values are inferred. However, the convergence of the parameters is highly nonuniform.

To clarify this issue let us make a few remarks. It was mentioned earlier that each block of data can be measured independently and used at the next step of inference provided that the results at previous steps are taken into account in the form of a prior distribution. In this case Equations 8.8 and 8.9 can be written in the form (see [20])

$$\mathbf{c}_k = \hat{\Xi}_{k-1}^{-1}\mathbf{c}_{k-1} + h\sum_{n\in N_k}\left[\hat{\mathbf{F}}_n^T\hat{\mathbf{D}}^{-1}\dot{\mathbf{x}}_n - \frac{\mathbf{v}_n}{2}\right], \tag{8.21}$$

$$\hat{\Xi}_k = \hat{\Xi}_{k-1} + h\sum_{n\in N_k}\hat{\mathbf{F}}_n^T\hat{\mathbf{D}}^{-1}\hat{\mathbf{F}}_n. \tag{8.22}$$

TABLE 8.3 Inference Results for a Set of Parameters of Model 8.17, Obtained Using 300 Blocks of 6000 Data Points Each, Sampled at $h = 0.002$

Parameter	Value	Estimate	STD
ε_1	0.1	0.0993	0.01
c_{12}	2.0	1.988	0.18
c_{13}	2.0	1.999	0.01
L_{12}	0.07	0.047	0.19
L_{13}	0.04	0.068	0.23
L_{23}	0.08	0.08	0.001
c_{23}	1.0	1.004	0.006
D_{11}	9.0×10^{-4}	8.9×10^{-4}	1.4^{-5}
D_{22}	4.0×10^{-6}	4.0×10^{-6}	1.3×10^{-7}

Note: True and inferred parameter values are shown along with the corresponding standard deviation (STD).

Applying this equations recursively one can show that the covariance matrix $\hat{\Xi}_k$ of the *posterior* distribution is a sum over all the blocks and has the structure of a Kronecker product

$$\hat{\Xi}_k = h\hat{\Phi} \otimes \hat{\mathbf{D}}^{-1}, \tag{8.23}$$

where

$$\hat{\Phi} = \sum_{n \in N_1,\ldots,N_k} \begin{pmatrix} \phi_{1,n}\phi_{1,n} & \cdots & \phi_{1,n}\phi_{B,n} \\ \vdots & \ddots & \vdots \\ \phi_{B,n}\phi_{1,n} & \cdots & \phi_{B,n}\phi_{B,n} \end{pmatrix}, \quad \phi_{i,n} \equiv \phi_i(x_n).$$

Accordingly, all nonzero elements of this matrix grow in time as $T = hN$. On the other hand the second term in 8.22 remains finite for a fixed number of points in one block N_k. Therefore, $\hat{\Xi}_{k-1}$ approaches $\hat{\Xi}_k$ for large T and c_k becomes

$$c_k \approx c_{k-1} + \hat{\mathbf{D}} \otimes \hat{\Phi}_k^{-1} \sum_{n \in N_k} \left[\hat{\mathbf{F}}_n^T \hat{\mathbf{D}}^{-1} \dot{\mathbf{x}}_n - \frac{\mathbf{v}_n}{2} \right]. \tag{8.24}$$

To analyze the leading order contribution to the residuals of $\{c_k\}$ given by the second term in 8.24 we can assume that noise matrix is known and constant, the contribution from the sum can always be finite and small for small and finite N_k. The only factor that represents the accumulating effect of convergence is $\hat{\Phi}_k^{-1}$. It is clear now that convergence of the residuals is inversely proportional to the sum of eigenvalues of $\hat{\Phi}_k$. Therefore the presence of large eigenvalues of $\hat{\Phi}_k^{-1}$ slows down the convergence of all coefficients $\{c_k\}$.

It is also clear that the choice of the base functions has a strong effect on convergence of the dynamics of the system. In particular, for polynomial base functions and small noise intensities the smallest elements of $\hat{\Phi}$ correspond to the highest powers of polynomials and usually correspond to large eigenvalues of $\hat{\Phi}_k^{-1}$.

Therefore, to achieve the best results in dynamical inference on-board, it is important to identify the key nonstationary parameters for each fault mode and to learn most of the stationary parameters in a preliminary analysis of the system. Next, by incorporating real-time inference into the inferential learning framework and excluding all but the most important nonstationary parameters from the tracking procedure one can improve the time resolution of the method by orders of magnitude.

We next consider an application of the dynamical inference in the context of developing on-board FD&P system for solid rocket motors.

8.5 IN-FLIGHT DECISION SUPPORT FOR SRMS

Safe exploration of the space requires development of new algorithms with the overarching goal of extracting information from available data with no "misses" and no "false alarms." The existing algorithms rely heavily on the linearization of the equations of motion and setting conservative margins for the nominal values of the control parameters. Yet a result of the prognostic reliability of such algorithms is limited because of limited number of sensors available on-board, limited thrust vector control authority, and short time window between the detectable onset of a fault and possibly catastrophic failure. It is very desirable to embed standard algorithms into a Bayesian framework to facilitate FD&P systems.

Here we demonstrate an application of the dynamical inference algorithm to two fault scenarios: (i) stepwise nozzle blocking with progressive steady burnout and (ii) nozzle blocking with time varying fault parameter and neutral thrust curve modeling a possible "miss" situation, where sudden increase of the pressure follows a prolonged period of small deviations of the pressure from the nominal value. To simplify the discussion we restrict our analysis to the case of a subscale motor. An extension of this approach to a large segmented motors can be found in [30].

8.5.1 Internal Ballistics of SRMs

The internal ballistics of the SRMs in the presence of the fault can be described by the following set of stochastic partial differential equations representing conservation laws for mass momentum, and energy of the gas (see, e.g., [4,30–32])

$$\begin{cases} \partial_t(A\rho) = -\partial_x(A\rho u) + \rho_p r_b l + d_1\xi_1(t), \\ \partial_t(A\rho u) = -\partial_x(A\rho u^2) - A\partial_x p + \rho_p r_b l u_S + d_2\xi_3(t), \\ \partial_t(A\rho e_t) = -\partial_x(A\rho h_t u) + H\rho_p r_b l + d_3\xi_2(t). \end{cases} \tag{8.25}$$

Here A is the port area, $e_t = c_V T + u^2/2$ and $h_t = c_P T + u^2/2$ are the total energy and total enthalpy of the gas flowing with velocity u, temperature T, pressure p, and density ρ. The propellant properties are described by the heat of combustion H, density ρ_p, and the burning rate given by the following expression $r_b = r_c(p/p_c)^n$. To model performance of the functionally graded propellants we introduce the port perimeter $l(t,x)$ as a function of the burned distance $R(x)$ given by design, e.g., for a progressive burning $l = 2\pi R$ and for a neutral thrust curve $l = const$. The burning surface is given by the integral $A_b(t) = \int_0^L l(t,x)dx$ and, therefore, is determined by the propellant grading function. The increase of the volume of the combustion chamber during the time interval h is $\Delta V = A_b(t)\Delta R$, where ΔR is given by the burning rate

$$\frac{dR}{dt} = r_b\left(p_0\right)p^n + d_4\xi_3(t). \tag{8.26}$$

To model various uncontrollable sources of noise (such as cracks and case vibrations) that may become essential in off-nominal conditions a random component in the propellant density is introduced that results in the random forcing in Equations 8.25 and 8.26 with amplitude d_i. Various fault modes in SRMs can now be modeled within the set of Equation 8.25 by choosing the time scale and direction of the geometrical alternations of the grain and case. For example, the bore choking fault can be modeled by introducing fault induced changes to the local port area $A(x)$; the crack dynamics can be modeled by introducing crack induced changes to an effective port perimeter; the nozzle blocking and the case breach faults can be introduced as changes to the nozzle throat area A_t (see, e.g., [33] and discussion below).

To be able to employ dynamical inference algorithm on-board one has to reduce model 8.25 to a set of ordinary differential equations allowing for an analytical estimation of the model parameters and their fault induced deviations. To this end, we introduce a low-dimensional performance model of SRM in the "filling volume" approximation [4,32]. For a subscale motor the ratio of the gas velocity to the stagnation speed of sound c_0 is small ($u^2/c_0^2 \ll 1$) everywhere along the propellant grain. Under these conditions one can integrate Equation 8.25 along the grain axis to take into account the sonic condition at the nozzle throat in the form $(A\rho u)\big|_L = \rho_t u_t A_t$, and introduce dimensionless variables (normalized by arbitrary reference values of pressure and density p_s and ρ_s) in the form

$p \to p_0/p_s$; $\rho \to \rho_0/\rho_s$ to arrive (see, e.g., [33]) at the following form of the low-dimensional performance model (LDPM) of the SRM operation in nominal and off-nominal regimes

$$\frac{dp}{dt} = -\frac{c_0\gamma\Gamma A_t(t)}{V(t)} p\sqrt{p/\rho} + \frac{A_b(t)r_b}{V(t)}\left(\gamma\rho_p - p\right)p^n + \frac{\tilde{d}_2(\gamma-1)}{V(t)}\xi_2(t),$$

$$\frac{d\rho}{dt} = -\frac{c_0\Gamma A_t(t)}{V(t)}\sqrt{p\rho} + \frac{A_b(t)r_b}{V(t)}\left(\rho_p - \rho\right)p^n + \frac{\tilde{d}_1}{V(t)}\xi_1(t), \tag{8.27}$$

$$\frac{dR}{dt} = r_b p^n + \tilde{d}_3\xi_3(t), \quad \dot{V} = A_b(t)r_b p^n.$$

Here the burning area $A_b(t) = f(R(t))$ is a given (by the design of the propellant grading) function of R. The dynamics of the SRM operation is now characterized by the parameters averaged over the length of the combustion camera. For example, the cracking and bore chocking faults will affect $A_b(t)$ and $V(t)$, while the nozzle failure or the case breach faults may affect $A_t(t)$. Note that the growing intensity of the noise can also be an important indication of the fault. The dynamics of the fault in these settings is characterized by an unknown time-dependent fault function $f_{fault}(t)$ that in many cases of interest can be assumed to be a polynomial function of time with unknown coefficients. We now demonstrate that abrupt changes of the model parameters can be accommodated within introduced Bayesian inferential framework of on-board FD&P system for SRMs.

8.5.2 Estimation of the Parameters of Nozzle Blocking

In our first example, we consider inference of the nozzle blocking fault modeled as a stepwise change in the nozzle throat area A_t. An additional rationale behind this example is that the significant stepwise change in the parameters of nonlinear models is one of the plausible reasons for the failure of alternative Kalman filter–based algorithms. The nozzle blocking is modeled as a sudden change of the nozzle area to the effective value $A_{et} = A_t + \Delta A_t$ (cf. [33]). The corresponding geometry change of the nozzle is shown in Figure 8.5a. To estimate the change of the nozzle area we notice that it is the coefficient $a = \frac{c_0\gamma\Gamma A_t(t)}{V(t)}$, which is proportional to A_t. We also notice that time series $R(t)$, $V(t)$, and $A_b(t)$ are completely determined by

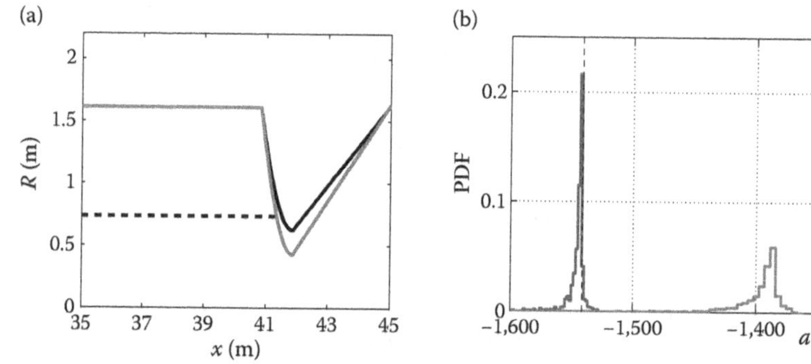

FIGURE 8.5 (a) Geometry of the nozzle blocking model. The contour of the combustion chamber before the fault is shown by the top solid line and after the fault by the bottom line. The grain location is shown by the dashed line. (b) Estimation of the value of the parameter a before (left curve) and after (right curve) the fault with $t = 0.5$ sec, $h = 0.001$ sec, $N = 500$. The dashed line shows the actual value of the parameter.

the time-trace of pressure $p(t)$ and can be excluded from the calculations. The results of dynamical inference obtained following this approach are shown in Figure 8.5b and summarized in Table 8.4.

It can be seen that the dynamical inference algorithm provides an accurate estimation of abruptly changing SRM parameters and that parameter's PDFs become sharply peaked about the true value.

8.5.3 Predicting "Misses" in the Fault Detection

To illustrate prediction of "misses" in on-board FD&P systems let us consider a situation when small pressure-induced deviation from the nominal value persists for a few second prior to the rapid crossing the "alarm" level. Such fault dynamics is typical for combustion instabilities in solid propellant motors [34] and represents an archetypal example of "misses."

TABLE 8.4 Results of the Parameter Estimation of Model 8.25

Parameter	Value	Estimate	Relative Error
$2\gamma\rho_p$	2575.34	2572	1.1%
$-(c_0\gamma\Gamma A_t)/V$	-1101.75	-1103	1.1%
d_1^2	0.0902	0.0906	0.4%
d_2^2	0.0902	0.0906	0.4%

Note: The total time of the measurements in this test was $t = 0.5$ sec, the sampling time was $h = 0.001$ sec, and the number of measured points was $N = 500$. The propellant grain has cylindrical geometry.

This situation is illustrated in Figure 8.6, where measured pressure signal (black solid line) crosses the alarm level (dashed line) and initiates the alarm at approximately t_A = 15 sec. The overpressure fault occurs at t_F = 17 sec and the time window between the alarm and a "catastrophic" event becomes too short, which can be considered as a model of "miss" situation. We model the "misses" by introducing fault-induced changes to the nozzle throat area and assume that the time evolution of the nozzle fault is highly nonlinear and can be described by a polynomial function

$$A_{et} = A_{t0} - (\alpha\tau + \beta\tau^2 + \delta\tau^3). \tag{8.28}$$

This polynomial approximation of the fault dynamics is substituted into Equation 8.27. It is the presence of high powers of time that causes rapid deviation of pressure from the nominal value at t_F = 17 sec. The earlier detection of "misses" in this context is reduced to a two-step procedure: (i) detection of the fault initiation and (ii) earlier detection of high-power terms in the polynomial fit to the time evolution of the fault. Accordingly, the pressure time traces have to be followed continuously online and fast algorithm of dynamical inference has to be employed for on-board estimation of the fault parameters.

The burn distance, burning area, and the combustion chamber volume become known functions of the pressure time trace if one neglects random

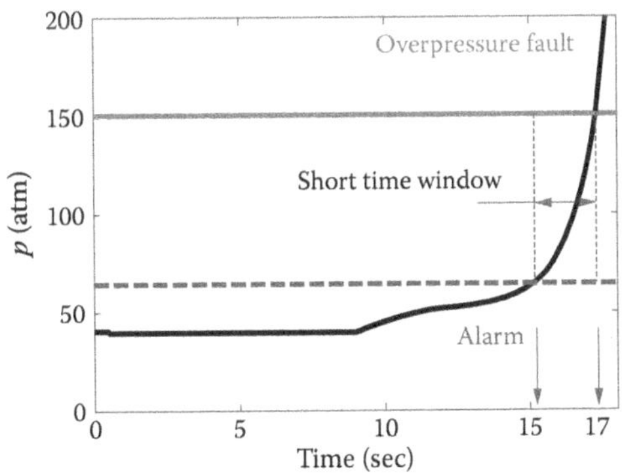

FIGURE 8.6 Sketch of the pressure fault (black line) representing a possible "miss" situation. The dashed and solid horizontal lines indicate the "alarm" and the "catastrophe" levels, respectively. The alarm onset and the overpressure events are indicated by arrows.

terms in the equation for the burn distance and assumes that the thrust curve is neutral

$$R(t)=r_b\int_0^t p^n(t')dt'; \quad A_b(t)=A_{b0}=const; \quad V(t)=V_0+A_{b0}R(t). \quad (8.29)$$

In this case the equations of the SRM internal ballistics take the form

$$\frac{dp}{dt}=-\frac{c_0\gamma\Gamma A_{et}(t)}{r_bV(t)}p\sqrt{p/\rho}+\frac{p^n}{R(t)}\left(\gamma\rho_p-p\right)+d_2'\xi_2(t), \quad (8.30)$$

$$\frac{d\rho}{dt}=-\frac{c_0\Gamma A_{et}}{r_bV(t)}\sqrt{p\rho}+\frac{p^n}{R(t)}\left(\rho_p-\rho\right)+d_1'\xi_1(t).$$

Here we have introduced dimensionless time $t\to r_bt$ and appropriately scaled constant unknown noise intensities d_1' and d_2'. Effective throat area $A_{et}(t)$ given by Equation 8.28, and $R'(t) = V_0/A_{b0} + R(t)$.

To apply an algorithm of dynamical inference we rewrite the factorized vector field in the form $f(x,t)=\hat{C}\phi$ with the set of the base function chosen in the form

$$\{\phi_i(t)\}=\left\{\frac{p^n}{R'},\frac{p^{n+1}}{R'},\frac{p^n\rho}{R'},\sqrt{\frac{p}{\rho}}\frac{p}{R'},\sqrt{\frac{p}{\rho}}\frac{p\tau}{R'},\ldots\right.$$

$$\left.\sqrt{\frac{p}{\rho}}\frac{p\tau^2}{R'},\sqrt{\frac{p}{\rho}}\frac{p\tau^3}{R'},\sqrt{\frac{p}{\rho}}\frac{\rho}{R'},\sqrt{p\rho}\frac{\tau}{R'},\sqrt{p\rho}\frac{\tau^2}{R'},\sqrt{p\rho}\frac{\tau^3}{R'}\right\},$$

and the matrix $\hat{C}$ written as follows

$$\hat{C}=\begin{bmatrix}\gamma\rho_p & -1 & 0 & -a\gamma & -a\gamma\alpha' & -a\gamma\beta' & -a\gamma\delta' & 0 & 0 & 0 & 0\\ \rho_p & 0 & -1 & 0 & 0 & 0 & 0 & -a & -a\alpha' & -a\beta' & -a\delta'\end{bmatrix}.$$

Here $a = (c_0\Gamma A_{t0})/(r_bV_0)$ and coefficients α', β', and δ' are equal to the coefficients α, β, and δ in Equation 8.28 scaled by A_{b0}/V_0.

The results of the detecting "misses" are shown in Figure 8.7. To detect fault initiation, parameters of the system are monitored in real time. Once fault is detected at time t_d the algorithm is continuously updating the inferred values of parameters estimated on increasing intervals Δt of time elapsed from t_d. The inferred values of parameters are used to generate a set of trajectories predicting pressure dynamics and to calculate pressure distributions ahead of time and to determine the most probable time of the case burst. Example of trajectories generated for two different Δt are shown in Figure 8.7a by jiggling lines. Examples of the PDFs of the time moment of the overpressure fault are shown in Figure 8.7b. It can be seen from the figure that predicted PDFs converges to the correct value 2.1 sec after the fault initiation at 9 sec, extending the time window between the "alarm" and the fault almost three fold (cf. with Figure 8.6) and reducing the probability of the "misses." Further detailed discussion and examples of application to an analysis of the results of the ground

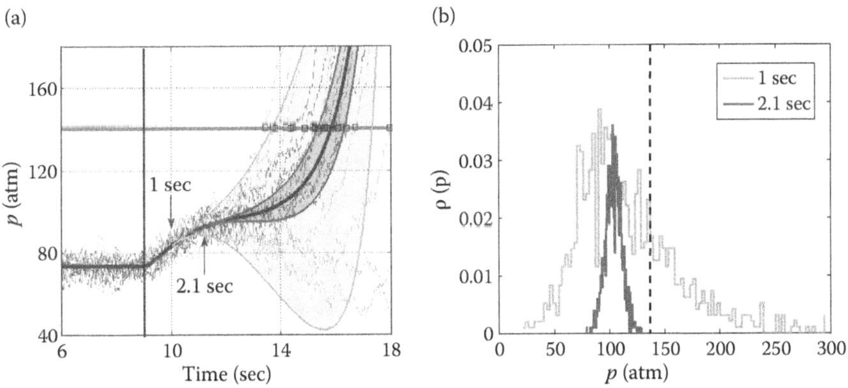

FIGURE 8.7 (a) Time evolution of the pressure build up after the nozzle blocking fault is shown by the solid line. The results of the predictions build 1 and 2.1 sec after the fault are shown by light-gray and dark-gray areas correspondingly. Examples of the pressure time traces predicted ahead of time using inferred values of the fault parameters are shown by the jiggling lines. The lines bounding shaded areas indicate standard deviations of the predictions. The time moments of the predicted overpressure faults used to build the PDF of the case burst times as shown by the circles on the horizontal margin line. Fault occurs at $t = 9$ sec. (b) The PDF of the predicted values of pressure at $t = 14$ sec build 1 sec (gray line) and 2.1 sec (black lines) after the fault. The dashed vertical line shows the dangerous level of the pressure.

firing test and to a case breach FD&P in a large segmented SRM are given in [35].

8.6 MODAL DYNAMICS BASED DAMAGE DETECTION

Metallic and composite plates are common members of airspace structures. As the size and complexity of the space hardware grows, the structural weight control becomes crucial. Weight requirement usually involves the use of lighter weight structures such as composites. All composite structures have one basic drawback in common: structural failures will occur internally, out of view of normal visual means of inspection. Internal debonds, delaminations, cracks, and/or buckles are typical. For heavily loaded composites, such as the heavy lift launcher will undoubtedly employ, this type of damage represents a major safety concern especially on man-rated vehicles. The large area composite structures like Payload Shroud, the Interstage, the Core Intertank, Storage Fuel Tank, Crew Composite Modulus, etc. are under broad investigation for usage in space vehicles. The peculiarities of some of these structures you can find in reports [36–38].

In this section, based on the classical Kirchhoff thin-plate vibration theory (see the seminal work of Lessia [39]), we discuss the damage detection method and show how these results can be used for the dynamical inference and development of structural health monitoring methods.

8.6.1 Mathematical Model of Pristine Plate

Assuming the midsurface of a plate as a reference, we shall use in-plane Cartesian coordinates. Let w be the deflection of the middle surface of the sandwich plate and then the governing differential equation of the isotropic thin plate can be found using the classical thin plate theory. For the transversely vibrating plates, we consider the following equation [39]

$$D\Delta^2 w + \rho\frac{\partial^2 w}{\partial t^2} = f(t,x,y), \tag{8.31}$$

where D is the flexural rigidity (bending stiffness), and ρ is the mass density per unit area of the plate. Operator $\Delta^2 = \nabla^2\nabla^2$ is a biharmonic operator, $f(t,x,y)$ term is the lateral force per unit area responsible for external load or PZT excitation. In the case of sandwich panel, stiffness is [40]

$$D = E_f t_f^3/6 + E_c t_c^3/12 + E_f t_f (t_f + t_c)^2/2 = 2D_f + D_c + D_0. \tag{8.32}$$

Here Young modulus for facesheets is E_f and modulus of the core E_c, t_f, and t_c are the thicknesses of the core and facesheets, correspondingly. The sandwich plate has a thickness $h = 2t_f + t_c$. Each term on the right hand side is denoted by its own capital letter D, $\rho = (2\rho_f t_f + \rho_c t_c)/h$, ρ_f, ρ_c are the densities of the facesheets and the core. In the Expression 8.32 we have $D_f \ll D_0$, $D_c \ll D_0$.

The displacement field $w(x,y,t) = w_0 \exp(i(\omega t - kx))$ describes transverse plane wave traveling in the x direction. As a result, by solving 8.31 for the wavenumber we get two real roots, which describe two waves propagating in opposite directions:

$$k = \pm\sqrt[4]{\omega^2 \rho / D}, \tag{8.33}$$

while the two imaginary roots correspond to evanescent fields. The phase speed v_p of the flexural wave with frequency ω in this case is a function of the plate stiffness and the mass per unit area $v_p = \omega/k = \pm\sqrt[4]{\omega^2 D/\rho}$.

Let us consider the generation and propagation of the flexural waves based on the results of the eigenfunctions and eigenfrequencies obtained for a rectangular plate. The collection of solutions for rectangular plates can be represented by deflection functions $w(x,y)$ as the product of beam functions

$$w(x,y) = \sum_m \sum_n c_{mn} X_m(x) Y_n(y), \tag{8.34}$$

where $X_m(x)$ and $Y_n(y)$ are chosen as the fundamental modes satisfying corresponding boundary conditions along the edges $x,y = 0$ and $x = L_x$, $y = L_y$. In this case the eigenfrequencies ω are given by the formula [18]

$$\omega^2 = \frac{\pi^4 D}{L_x^4 \rho}\left[G_x^4 + G_y^4 \frac{L_x^4}{L_y^4} + 2\frac{L_x^2}{L_y^2}[\nu H_x H_y + (1-\nu) J_y J_y] \right], \tag{8.35}$$

where G, H, and J are the given functions of natural numbers determined by the specific boundary conditions and L_x, L_y are the rectangular plate dimensions (Table 4.1 [39]). As it was noted by Leissa, there are 21 combinations of simple boundary conditions (i.e., clamped (C), simply supported (SS), or free (F)) for rectangular plates. Frequency parameters are expressed in terms of $\omega L_x^2\sqrt{\rho D}$, where L_x is the length dimension. For example, for SS boundary conditions of a rectangular plate we have

$$w(x,y)=\sum_{m,n} c_{mn}\sin(m\pi x/L_x)\sin(n\pi y/L_y), \tag{8.36}$$

where m,n are neutral numbers. As a result, from 8.35 eigenfrequencies are

$$\omega_{mn}^2=(\pi^4 D/\rho)\left[(m/L_x)^2+(n/L_y)^2\right]^2. \tag{8.37}$$

The evolution of the plate under external load can be found by considering amplitudes c_{mn} in the Equation 8.34 as functions of t. Substituting 8.34 into 8.31, multiplying both sides by $X_{m'}(x)Y_{n'}(y)$, integrating over the area of the plate and using the orthogonality condition, we obtain

$$\ddot{c}_{mn}(t)+\omega_{mn}^2 c_{mn}=f_{mn}, \tag{8.38}$$

where $f_{mn}(t)=\dfrac{\langle f(t,x,y)X_m(x)Y_n(y)\rangle}{\|X_m(x)Y_n(y)\|^2}, \langle\cdot\rangle=\dfrac{4}{L_xL_y}\displaystyle\int_0^{L_x}\int_0^{L_y}(\cdot)dx\,dy.$

For example, consider the decaying oscillating pulse is applied to the point (x_s,y_s) of the plate [41]

$$f_{mn}(t)=f_{mn}^0 e^{-t/\tau}\cos\omega t, \tag{8.39}$$

where τ is the time of the decay and ω is the excitation frequency, term $f_{mn}^0=\dfrac{4}{L_xL_y}X_m(x_s)Y_n(y_s)$ determines localization of the source characterized by spatial delta function. The excitation signal 8.39 makes it possible to present the solution of the system as [41]

$$\begin{aligned} c_{mn} &= f_{mn}^0\frac{\varpi_{mn}^2\cos\omega_{mn}t+(\tau\omega_{mn})^{-1}(\varpi_{mn}^2+2\omega^2)\sin\omega_{mn}t}{\varpi_{mn}^4+4\omega^2/\tau^2} \\ &+ f_{mn}^0 e^{-t/\tau}\frac{\varpi_{mn}^2\cos\omega t-2(\omega/\tau)\sin\omega t}{\varpi_{mn}^4+4\omega^2/\tau^2}, \end{aligned} \tag{8.40}$$

where $\varpi_{mn}^2 = \omega_{mn}^2 + 1/\tau^2 - \omega^2$. In the next subsection, we consider how a wave propagating from a point source can interact with the local damage.

8.6.2 Mathematical Model of Damaged Plate

Local stiffness reduction can be caused by different faults. The loss of bending stiffness can be caused by a debond, delamination, honeycomb crash or the reduction of the modulus of facesheet due to an impact. Any of these structural failures will lead to the change of the stiffness and mass parameters. As a result, in framework of the plate theory, we can generalize damage as a domain with different structural properties than the rest of the plate.

Let us consider the damage model as a decrease in structural local stiffness by a function $g(x - x_d, y - y_d)$, which reflects changes in the local structural material elastic coefficients. The term on the left-hand side of Equation 8.31 is modified

$$Dg(x - x_d, y - y_d)\Delta^2 w + \rho \frac{\partial^2 w}{\partial t^2} = f(t,x,y), \tag{8.41}$$

where the function describing damage is centered $g(x - x_d, y - y_d)$ at a certain point (x_d, y_d) characterizing local change in stiffness. We transform the Equation 8.41 to the form

$$D\Delta^2 w + \rho \frac{\partial^2 w}{\partial t^2} = \varepsilon f(t,x,y) + D\varepsilon(x,y)\Delta^2 w, \tag{8.42}$$

where the right hand side value $\varepsilon(x,y) = 1 - g(x - x_d, y - y_d)$ is not equal to zero in a small domain of the damage.

Considering that the value $\varepsilon(x,y)$ in 8.42 is sufficiently small we can apply the perturbation technique: we find the solution for $\varepsilon(x,y) = 0$ and then substitute the displacement into the right hand side of the expansion system through the normal modes form 8.34. In this case, the damage term works like a source of secondary waves generated in the structure.

Applying the same procedure of the expansion solutions system of equations for mode amplitudes is presented as

$$\ddot{c}_{mn}(t) + c_{mn}(t)\omega_{mn}^2 = f_{mn}(t) + \varepsilon\Lambda_{mn}(x_d, y_d, t), \tag{8.43}$$

where for calculation of the term $\varepsilon\Lambda_{mn}$

$$\varepsilon\Lambda_{mn} = \frac{4}{L_x L_y}\sum_{k,l} < \varepsilon(x,y)\omega_{kl}^2 c_{kl} X_m(x) Y_n(y) X_{m'}(x) Y_{n'}(y) > \Theta(t > r/v_p), \tag{8.44}$$

where $\Theta(t > r/v_p)$ is a Heaviside step function, we have to know the exact distribution of the value $\varepsilon(x,y)$.

To calculate 8.44, we consider $c_{kl}(t)$ at the damaged region to be given by the solution of Equation 8.43 with $\varepsilon = 0$. Let us consider the simplest model of damage when in a local square we have a reduced stiffness by value $\varepsilon = const$

$$\varepsilon(x,y) = \begin{cases} \varepsilon & \text{for } |x - x_d| \le d, |y - y_d| \le d, \\ 0 & \text{otherwise.} \end{cases} \tag{8.45}$$

In the case of simply supported boundary conditions we have sine eigenfunctions 8.36 and the damaged term in 8.43 can be expressed as

$$\varepsilon\Lambda_{mn} = \frac{\Theta(t > r/v_p)}{\pi^2 L_x L_y \omega_{mn}^2}\sum_{k,l}\omega_{kl}^2 c_{kl}[a_{km}(x_d) - a_{km}(y_d)][a_{ln}(x_d) - a_{ln}(y_d)], \tag{8.46}$$

where $a_{km}(x_d) = \frac{1}{k-m}\sin(k-m)\frac{\pi x}{L_x}\Big|_{x_d-d}^{x_d+d}$, $a_{km}(y_d) = \frac{1}{k+m}\sin(k+m)\frac{\pi y}{L_y}\Big|_{y_d-d}^{y_d+d}$,

$a_{ln}(x_d) = \frac{1}{(l-n)}\sin(l-n)\frac{\pi x}{L_x}\Big|_{x_d-d}^{x_d+d}$, $a_{ln}(y_d) = \frac{1}{(l+n)}\sin(l+n)\frac{\pi y}{L_y}\Big|_{y_d-d}^{y_d+d}$. The solution is plotted on Figure 8.8a and 8.8b where the first graph represents the total solution and the plot (b) reflects the difference in displacements for the healthy structure and the damaged one.

For computer simulation, the force excitation point S is kept fixed at the center of the plate ($L_x/2, L_y/2$) and the damage is located at a certain distance (point x_d, y_d) from the corner (0,0). Some of the results for the transverse displacement are plotted in Figure 8.9 as a function of (x,y) for a different value of t and the simply supported boundary conditions along

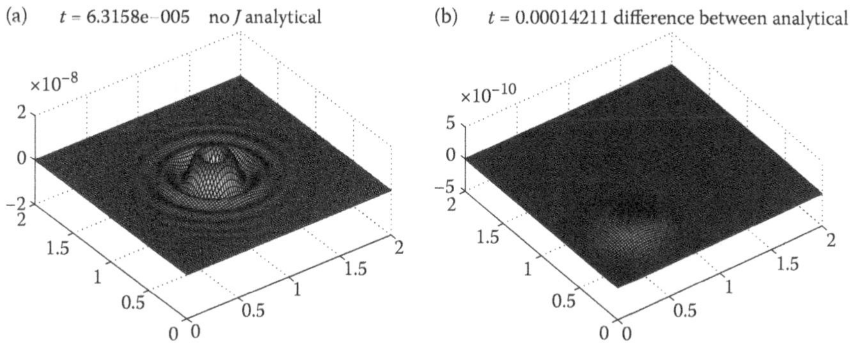

FIGURE 8.8 Analytical solution for w out of plane displacement for (a) $t = 0.0002$ sec and (b) difference between the damaged and pristine signal ($x_d = y_d = 0.5$ m, $d = 0.025$ m, $x_s = y_s = 1$ m, $L_x = L_y = 2$ m, $N_x = N_y = 20$). Input signal is determined by formula 8.40, $\tau = 0.00001$ sec, $f = 10$ kHz, $\varepsilon = 0.1$, $\omega = 2\pi f$.

the edges. In the simulation, we used Hanning windowed signal with a number of pulses $N = 3.5$.

The variation of the scattered signal as a function of the wavelength of the excitation signal and the size of the damage were analyzed in numerical simulations and found to be in a good agreement with the theoretical predictions. The simulation was made for a different number of the modes N_x,N_y. The mode number affects the accuracy of calculation. By increasing the time of the simulation, the incident pulse interferes with the reflected one and the patterns of wave distribution are much more complicated. The loss of the local stiffness changes the amplitude of the scattered signal but does not change

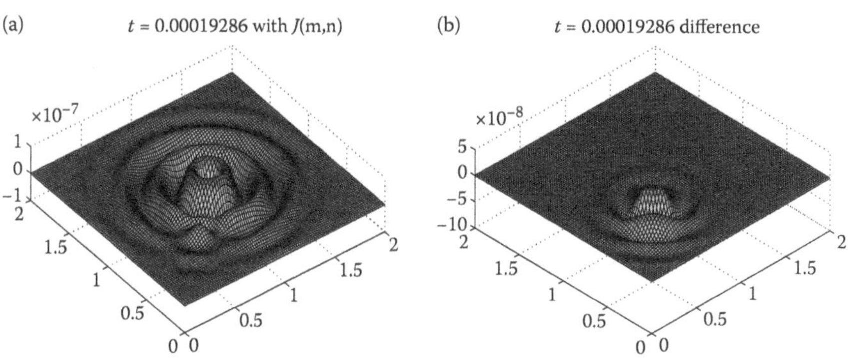

FIGURE 8.9 (a) Out-of-plane displacement $w(x,y)$ at $t = 2 \cdot 10^{-4}$ sec and (b) difference between the damaged and pristine value of $w(x,y)$ at $t = 2 \cdot 10^{-4}$ sec. Input signal is Hanning windowed signal with $N = 3.5$, $f = 10$ kHz.

the general picture. We can observe that the damaged region plays the role of the scatterer generating scattering waves. Two distributions are plotted just to show the evolution of the field $w(x,y)$. A clear indication of the damage was shown for the time when the wave reaches the damaged region (Figure 8.9a and 8.9b). The difference in the $w(x,y)$ between the damaged and pristine signal is depicted in Figure 8.9b.

This study clearly shows that the normal mode expansion method is capable of capturing the physics of wave propagation in plate structures. We see a scattering pattern that in general closely resembles the interaction of a wave with the damaged region. Using the results obtained earlier, it is analytically shown and verified by the computer simulation that the normal function expansion method describes the damage state of the plate by changing the dynamics of several modes. It is shown that at least ten terms in the expansion should be used for the DI method to grasp the main characteristic features of the damage presented in the plate.

8.7 DYNAMICAL INFERENCE OF A SET OF COUPLED OSCILLATORS

It was shown that several important problems of the IHM in aerospace applications render themselves into a set of coupled stochastic differential equations ideally suited for analysis in the framework of dynamical inference. In particular, an open problem of the on-board SHM of damage in composite materials can be reduced to the DI of a set of oscillators representing the dynamics of the eigenmodes. Furthermore fast online inference of a set of oscillators with time varying parameters is of general importance across a wide range of interdisciplinary applications including, e.g., neurophysiology [24] and active control of combustion instability in liquid motors [42]. In this section, we illustrate how such inference can be performed using proposed DI algorithm.

8.7.1 General Inferential Framework for a Set of Coupled Oscillators

Consider dynamics of L nonlinear oscillators with coordinates q_j and velocities v_j driven by random force with components ξ_j, which are mixed by a diffusion matrix with elements D_{ij}. A set of equations governing this dynamics can be written in the form

$$\begin{aligned} \dot{v}_j &= \alpha_j v_j + b_j v_j^2 + c_j v_j^3 - q_j + \eta_j + \sqrt{D_{ij}}\,\xi_j, \\ \dot{q}_j &= -\beta q_j + \gamma_j\, v_j, \quad \langle \xi_j(t)\xi_i(t')\rangle = \delta_{ij}\,\delta(t-t'), \quad j=1:L. \end{aligned} \tag{8.47}$$

This form of equation models a wide range of nonlinear phenomena (see, e.g., [43]) and can be reduced, in particular, to a set of linear oscillators obtained in previous sections in the context of online SHM problems.

To model measurements of the dynamical variables $v_j(t)$ we use a measurement matrix X, which is frequently found in theory of dynamical inference [19]

$$y_i = X_{ij} v_j. \tag{8.48}$$

Here y_i are measured variables, which are related to v_j by linear transformation with an unknown matrix X. All the accessible information is contained in y_i. The problem is, therefore, to learn model parameters $\mathcal{M} = \{\eta_i, \alpha_i, q_i(0), \gamma_i, D_{ij}, X_{ij}\}$ from time series data $\{y_i\}$.

To develop a general inferential framework for a solution of this problem we notice that the second linear ordinary differential equation in Equation 8.47 connecting coordinates and velocities can be integrated explicitly (see [24,44]) to obtain

$$q_j(t) = \gamma \int_0^t d\tau e^{-\beta(t-\tau)} v_j(\tau) + e^{-\beta t} q_j(0). \tag{8.49}$$

On substituting 8.49 into the first equation in 8.47, we have

$$\begin{aligned} \dot{v}_j &= \alpha_j v_j + b_j v_j^2 + c_j v_j^3 + \eta_j \\ &\quad - \gamma_j \int_0^t d\tau e^{-\beta(t-\tau)} v_j(\tau) - e^{-\beta t} q_j(0) + \sqrt{D_{ij}} \xi_j. \end{aligned} \tag{8.50}$$

Here $j = 1,\ldots,L$ and $q_j(0)$ is a set of initial coordinates for unobservable variable $q_j(t)$. Thus the reconstruction of unobservable variables $q_j(t)$ is reduced to inference of L initial conditions $q_j(0)$.

Variables $v_j(t)$ can also be excluded from further consideration by using Equation 8.48. Indeed, on substituting $v = X^{-1}y$ into 8.50 we obtain in vector notations:

$$\begin{aligned} \dot{y} &= X\alpha\left(X^{-1}y\right) + Xb\left(X^{-1}y\right)^2 + Xc\left(X^{-1}y\right)^3 e^{-\beta t} X q_0 \\ &\quad - \int_0^t e^{\beta(t-\tau)} X\gamma\left(X^{-1}y\right) d\tau + X\eta + X\sqrt{D}\xi(t), \end{aligned} \tag{8.51}$$

where α, *b*, and *c* are vectors with components $\{\alpha_j\}$, $\{b_j\}$, and $\{c_j\}$ respectively, $q_0 = q(t = 0)$ and

$$(X^{-1}y)^n = \begin{pmatrix} \left(\sum_{i=1}^{L} \tilde{x}_{1i} y_i\right)^n & \cdots & 0 \\ \vdots & \ddots & \vdots \\ 0 & \cdots & \left(\sum_{i=1}^{L} \tilde{x}_{Li} y_i\right)^n \end{pmatrix}.$$

Equation 8.51 can be rewritten in a more general form

$$\begin{aligned} \dot{y}_i &= \tilde{\alpha}_{ij} y_j + \tilde{b}_{ijk} y_j y_k + \tilde{c}_{ijkl} y_j y_k y_l + \tilde{\eta}_i + \\ &- \Phi_i \tilde{\gamma}_{il} y_l - e^{-\beta t} \tilde{z}_i + \sqrt{\tilde{D}_{ij}}\, \xi_j(t), \end{aligned} \tag{8.52}$$

where the meaning of the parameters with ~ is clear from the comparison of 8.51 and 8.52; see also [24,44].

Note that presentation 8.51 covers the whole model space of a set of coupled nonlinear oscillators with polynomial base functions of the power 3. This presentation can also be used to infer unknown parameters of a system of coupled nonlinear oscillators in the case when oscillator velocities y_i are measured directly.

To infer the set of unknown parameters

$$\tilde{\mathcal{M}} = \{\tilde{\eta}_i, \tilde{\alpha}_{ij}, \tilde{b}_{ijk}, \tilde{c}_{ijkl}, \tilde{\gamma}_{ij}, \tilde{z}_i, \tilde{D}_{ij}\}$$

of the system 8.52 of *L* coupled oscillators 8.51 within our inferential framework one can introduce the following base functions

$$\begin{aligned} \phi(x) = \{&1, y_1, \ldots, y_L, y_1^2, y_1 y_2, \ldots, y_1 y_L, y_2^2, y_2 y_3, \ldots, y_2 y_L, \ldots, y_1^3, \\ &y_1^2 y_2, \ldots, y_1^2 y_L, y_2^3, y_2^2 y_1, \ldots, y_2^2 y_L, \ldots, y_L^2 y_{L-1}, y_L^3, \Phi_1, \ldots, \Phi_L, e^{-\beta t}\} \end{aligned} \tag{8.53}$$

where $\Phi_i \equiv \int_0^t y_i(\tau) e^{\beta(\tau - t)}$.

The total number of unknown parameters is $N_{tot} = (2 + 2L + L^2) \times L + L(L + 1)^2/2$ and is increasing as L^3 with the dimension of the system. These parameters can be inferred directly from 8.52 using Equations 8.8 and 8.9 if the measurement matrix is known. In the case of unknown measurement matrix one has to infer additional L^2 coefficient of unknown matrix X_{ij}, which is not symmetric in general. To solve this problem, we notice that in practice the number of the coefficients of the original system is always significantly smaller than the full set of equations N_{tot}, because of the symmetry, which is always present in real systems. For example, for the sets of linear oscillators that appear in eigenmode expansion of the SHM problem in the on-board analysis of damage of the composite materials the second equation in 8.47 has the form $\dot{q}_j = v_j$, i.e., $\beta = 0$, $\gamma_j \equiv 1$, and all $\tilde{b}_{ijk}$ and $\tilde{c}_{ijkl}$ are also zero. Furthermore, the eigenfrequencies of expansion in eigenmodes of every specific structural component that requires on-board SHM must be learned beforehand from extensive preliminary ground and in-flight tests, which significantly simplifies the problem. With this in mind let us consider a numerical example with a set of two oscillators 8.50 where all the coefficients of 8.50 but β are assumed unknown together with unknown coefficients of the measurement matrix $\{x_{ij}\}$.

8.7.2 Numerical Example

In the case of two oscillators, the set $\tilde{\mathcal{M}}$ of variables of the transformed dynamics 8.51 corresponds to the following set of the base functions

$$\phi(x) = \{1, y_1, y_2, y_1^2, y_2^2, y_1 y_2, y_1^3, y_1^2 y_2, y_1 y_2^2, y_2^3, \Phi_1, \Phi_2, e^{-\beta t}\}.$$

In general it should be possible to reconstruct all unknown coefficients of the original system for any number of FHN oscillators as long as we can establish the connection between the set $\tilde{\mathcal{M}}$ and the set of original unknown variables of Equation 8.47

$$\mathcal{M} = \{\eta_i, \alpha_i, b_i, c_i, \gamma_i, q_i(0), D_{ij}, X_{ij}\}.$$

This connection for the case of two oscillators ($L = 2$) can be written explicitly as follows

$$X^{-1}\begin{bmatrix} \eta_1 \\ \eta_2 \end{bmatrix} = \begin{bmatrix} \tilde{\eta}_1 \\ \tilde{\eta}_2 \end{bmatrix}, \quad \begin{bmatrix} q_{0,1} \\ q_{0,2} \end{bmatrix} = X^{-1}\begin{bmatrix} \tilde{q}_1 \\ \tilde{q}_2 \end{bmatrix}, \quad \tilde{D}X^{-1} = X^{-1}D. \quad (8.54)$$

$$\begin{bmatrix} \gamma_1 & 0 \\ 0 & \gamma_2 \end{bmatrix} X^{-1} = X^{-1} \begin{bmatrix} \tilde{\gamma}_{11} & \tilde{\gamma}_{12} \\ \tilde{\gamma}_{21} & \tilde{\gamma}_{22} \end{bmatrix}, \tag{8.55}$$

$$\begin{bmatrix} \alpha_1 & 0 \\ 0 & \alpha_2 \end{bmatrix} X^{-1} = X^{-1} \begin{bmatrix} \tilde{\alpha}_{11} & \tilde{\alpha}_{12} \\ \tilde{\alpha}_{21} & \tilde{\alpha}_{22} \end{bmatrix}. \tag{8.56}$$

The unknown elements x_{ij} of the inverse measurement matrix X^{-1} and parameters with tilde are the model parameters of the transformed system 8.51 that can be inferred directly using time series data $\{y_i\}$. Relations 8.54 through 8.58 allow one to reconstruct 15 unknown parameters of the original system, including elements of the noise and measurement matrixes. Note, however, that coefficients $(1 + \alpha_i)$ can also be assumed unknown in general and the following relations can be used to reconstruct them

$$\begin{bmatrix} 1+\alpha_1 & 0 \\ 0 & 1+\alpha_2 \end{bmatrix} \begin{bmatrix} x_{11}^2 & 2x_{11}x_{12} & x_{12}^2 \\ x_{21}^2 & 2x_{21}x_{22} & x_{22}^2 \end{bmatrix} = X^{-1} \begin{bmatrix} \tilde{b}_{111} & \tilde{b}_{112} & \tilde{b}_{122} \\ \tilde{b}_{211} & \tilde{b}_{212} & \tilde{b}_{222} \end{bmatrix}. \tag{8.57}$$

Similarly, the relation between the coefficients for polynomials of power 3 are given by

$$\begin{bmatrix} -1 & 0 \\ 0 & -1 \end{bmatrix} \begin{bmatrix} x_{11}^3 & 2x_{11}^2x_{12} & 2x_{11}x_{12}^2 & x_{12}^3 \\ x_{21}^3 & 2x_{21}^2x_{22} & 2x_{21}x_{22}^2 & x_{22}^3 \end{bmatrix} = X^{-1} \begin{bmatrix} \tilde{c}_{111} & \tilde{c}_{112} & \tilde{c}_{121} & \tilde{c}_{122} \\ \tilde{c}_{211} & \tilde{c}_{212} & \tilde{c}_{221} & \tilde{c}_{222} \end{bmatrix}. \tag{8.58}$$

We now analyze the convergence of the method in the case when all the parameters of the reduced model 8.51, including elements of the measurement matrix, are unknown. The sampling rate was 35 kHz, we used 9 blocks of data with 5000 points in each block, and these blocks of data were generated at random 1000 times to analyze the statistics of the convergence.

To reconstruct both the mixing matrix X and the parameters of the original system $\mathcal{M}$ from the inferred parameters $\tilde{\mathcal{M}}$ of the transformed system 8.51, we have to solve equations (8.54–8.58) with respect to elements

TABLE 8.5 Values of Some of the Transformed and Original Coefficients Inferred Using 30,000 Points Obtained from Measurement Matrix and Real Parameters Reconstruction

Parameter	Actual Value	Estimate	Error
$\tilde{\eta}_1$	0.9200	0.924384	0.022624
$\tilde{\eta}_2$	0.3500	0.351001	0.009063
$\tilde{b}_{222}$	1.7550	1.758011	0.037047
$\tilde{b}_{112}$	-2.1086	-2.114731	0.068268
X_{11}	1.7	1.686459	0.796526
X_{12}	0.8	0.794263	0.717092
X_{21}	0.2	0.196746	1.626811
X_{22}	0.9	0.898222	0.197610
η_1	0.4	0.406227	1.556788
η_2	0.3	0.302462	0.820660
α_1	−0.35	−0.351992	0.569082
α_2	−0.2	−0.200376	0.188228
b_1	1.35	1.357427	0.550145
b_2	1.2	1.203863	0.321885
c_1	−1.0	0.999520	0.047957
c_2	−1.0	−0.999114	0.088582

Note: The actual values (second column) are compared with the inferred values (third column); relative errors are given in the last column in %.

of $\tilde{\mathcal{M}}$. We stress that the present technique is not restricted to the $L = 2$ case and can be employed in the general case of L oscillators; however, an additional optimization over the set of initial values may be required.

The results of the inference of the transformed and original parameters of a set of oscillators are summarized in Table 8.5. It can be seen from the table that the DI method allows us to reconstruct both the elements of the measurement matrix and the parameters of the transformed 8.52 and original 8.50 systems. Our analysis shows that the relative error of inference better than 2% is achieved in less than 1 sec of measurements.

We therefore conclude that the general inferential framework for a set of coupled oscillators provided by Equations 8.8 and 8.9, Equations 8.52 and 8.53 is one of the primary candidates for on-board SHM in aerospace applications.

8.8 CONCLUSION

It was shown that several important applications of IHM in aerospace render themselves into a problem of online inference of a set of coupled stochastic differential equations. A convenient approach to a solution of this problem was introduced within the framework of dynamical inference. It was shown that in a wide range of practical cases this problem can be solved analytically using a path-integral approach to the maximum likelihood estimation. The robustness of the solution in the presence of strong dynamical noise was demonstrated in application to the inference of the archetypal nonlinear stochastic Lorenz system. The ability of the method to infer simultaneously discrete and continuous parameters was further illustrated by the application of the DI to the fast online estimation of parameters and detection of leaks in stochastic hybrid three tank system, which is often considered as a standard engineering control benchmark.

In the remaining part of the chapter a progress in development of some selected FD&P systems for in-flight structural health monitoring was discussed. In the first example a dynamical model of the SRM ballistics in nominal and off-nominal regimes was presented. It was shown that the DI approach can be used to detect abrupt fault-induced changes in the parameters of the SRM. It was shown further that the DI method can be used to reduce the probability of "misses" in detection of the SRM faults on-board by inferring coefficients of the diverging terms in polynomial fit to small deviations of pressure from the nominal regime and generating an earlier warning when the inference results begin to converge.

In the next example a problem of in-flight structural health monitoring of a composite sandwich plate was analyzed. It was shown that a normal mode expansion method is capable of capturing the physics of wave propagation in plate structures and may describe damage state of the plate. As a result, the dynamic inference method is very promising for SHM of the plates. It was demonstrated in this context that it is sufficient to keep in the expansion up to ten terms to grasp the main characteristic dynamical features corresponding to the damage present in the plate.

Finally, the general inferential framework for fast online reconstruction of the parameters of a set of coupled oscillators was presented in the context of the SHM problem. It was shown that the method allows for an accurate reconstruction of the system parameters even when the velocities of the oscillators are mixed by the linear measurement matrix with unknown elements and coordinates of the oscillators are not accessible for measurements.

The ability of the proposed technique to provide nonbiased analytical estimation of discrete and continuous model parameters of stochastic nonlinear hybrid dynamical systems makes the DI method one of the primary candidates for on-board SHM in aerospace applications. It was shown recently [45,46] (see also Section 8.7) that the method can be extended to infer hidden (unobservable) dynamical variables and parameters of the measurements model. Note that the case of unknown non-white noise represent a specific example of hidden dynamical variables when additional variable $\eta(t)$ appears in Equation 8.1, whose unobservable dynamics is governed by stochastic differential equation (e.g., $\dot{\eta} = a\eta + \xi(t)$ for exponentially correlated noise) with unknown parameters. Therefore a trivial extension of the proposed technique to the case of colored noise is possible.

With these extensions the method can be useful for a wide range of problems outside the aerospace domain including, e.g., applications to diagnostic and prognostic of cardiovascular systems [13,23] and ecological [45,46] and physiological [24,44] systems. Because of the fast convergence and accuracy the method is especially attractive for real time applications to streaming data; see a detailed discussion and examples of this application in [24,44].

Note, however, the tradeoffs of the method related to the fact that the method requires dense sampling of the measured time-series data and calculations of second derivatives of the base functions with respect to all dynamical variables in case of hidden variables. The number of the base functions is usually a fast growing function of the size of the system, e.g., as a number of terms in multinomial series for polynomial base functions. Therefore, for effective real-time onboard applications one should try to limit the size of the performance models. For offline applications with sparse time-series data a traditional particle filter and/or MCMC approaches should be considered as a simple and efficient alternative. For complex optimization problems with hidden dynamical variables a combination of the proposed technique and MCMC method can be used to improve convergence in the space of unknown dynamical trajectories. Detailed comparisons of the method with alternative DI and MCMC algorithms are given, e.g., in [13,45].

The proposed method can be extended and improved in a number of ways to encompass in particular the case of multiplicative noise and a general case of colored noise. The corresponding work is currently in progress. We are also planning to provide free online access to the key algorithms of the method and to a large number of examples.

Ultimately, IHM research will yield integrated, multidisciplinary analysis and optimization capabilities that enable system-level designs providing graceful recovery from in-flight failures and computationally efficient tools for in-flight prognosis of aircraft health.

REFERENCES

[1] A. Patterson-Hine, G. Aaseng, G. Biswas, S. Narasimhan, K. Pattipati. A review of diagnostic techniques for ishm applications, 2008. http://citeseerx.ist.psu.edu/viewdoc/summary?doi=10.1.1.118.953.

[2] K. Visscher, M.J. Schnitzer, S.M. Block. Single kinezin molecules studied with a molecular force clamp. *Nature*, 400:184–189, 1999.

[3] D.J.D. Earn, S.A. Levin, P. Rohani. Coherence and conservation. *Science*, 290(5495):1360–1364, 2000.

[4] M. Salita. Modern srm ignition transient modeling. I—Introduction and physical models. In *27th Joint Propulsion Conference and Exhibit, 2001 AIAA*, 2001.

[5] J. Christensen-Dalsgaard. Helioseismology. *Review of Modern Physics*, 74 (4):1073–1129, 2002.

[6] P.E. McSharry, L.A. Smith. Better nonlinear models from noisy data: attractors with maximum likelihood. *Physical Review Letters*, 83:4285–4288, 1999.

[7] J. Gradisek, S. Siegert, R. Friedrich, I. Grabec. Analysis of time series from stochastic processes. *Physical Review E*, 62(3):3146–3155, 2000.

[8] J.P.M. Heald, J. Stark. Estimation of noise levels for models of chaotic dynamical systems. *Physical Review Letters*, 84(11):2366–2369, 2000.

[9] R. Meyer, N. Christensen. Fast bayesian reconstruction of chaotic dynamical systems via extended kalman filtering. *Physical Review E*, 65:016206, 2001.

[10] J.-M. Fullana, M. Rossi. Identification methods for nonlinear stochastic systems. *Physical Review E*, 65:031107, 2002.

[11] M. Siefert, A. Kittel, R. Friedrich, J. Peinke. On a quantitative method to analyze dynamical and measurement noise. *Europhysics Letters*, 61(4):466–472, 2003.

[12] F. Watanabe, H. Konno, S. Kanemoto. Identification method of stochastic nonlinear dynamics using dynamical phase analysis—application to for SMARK data. *Annals of Nuclear Energy*, 31(4):375–397, 2004.

[13] V.N. Smelyanskiy, D.G. Luchinsky, A. Stefanovska, P.V.E. McClintock. Inference of a nonlinear stochastic model of the cardiorespiratory interaction. *Physical Review Letters*, 94(9), 2005.

[14] S. Siegert, R. Friedrich, J. Peinke. Analysis of data sets of stochastic systems. *Physical Letters A*, 253:275–280, 1998.

[15] E.L. Ionides, C. Breto, A.A. King. Inference for nonlinear dynamical systems. *Proceedings of the National Academy of Sciences*, 103(49):18438–18443, 2006.

[16] S. Cauchemez, N.M. Ferguson. Likelihood-based estimation of continuous-time epidemic models from time-series data: application to measles transmission in London. *Journal of the Royal Society Interface*, 5(25):885–897, 2008.

[17] C.L. Bremer, D.T. Kaplan. Markov chain Monte Carlo estimation of nonlinear dynamics from time series. *Physica D*, 160:116–126, 2001.

[18] R. Meyer, N. Christensen. Bayesian reconstruction of chaotic dynamical systems. *Physical Review E*, 62:3535–3542, 2000.

[19] P. Congdon. *Bayesian statistical modelling*. Wiley series in probability and statistics. Wiley, Chichester, 2001.
[20] V.N. Smelyanskiy, D.G. Luchinsky, D.A. Timucin, A. Bandrivskyy. Reconstruction of stochastic nonlinear dynamical models from trajectory measurements. *Physical Review E*, 72(2), 2005.
[21] R. Graham. Path integral formulation of general diffusion processes. *Zeitschrift für Physik B*, 26:281–290, 1977.
[22] M.I. Dykman. Large fluctuations and fluctuational transitions in systems driven by colored Gaussian noise—a high frequency noise. *Physical Review A*, 42:2020–2029, 1990.
[23] D.G. Luchinsky, M.M. Millonas, V.N. Smelyanskiy, A. Pershakova, A. Stefanovska, P.V.E. McClintock. Nonlinear statistical modeling and model discovery for cardiorespiratory data. *Physical Review E*, 72(2):021905, 2005. Part 1 021905.
[24] D.G. Luchinsky, V.N. Smelyanskiy, A. Duggento, P.V.E. McClintock. Inferential framework for nonstationary dynamics part I: Theory. *Physical Review E*, 77:061105, 2008.
[25] V.V. Osipov, D.G. Luchinsky, V.N. Smelyanskiy, D.A. Timucin. Case breach fault model and diagnostic for the solid rocket. AIAA 2007-5823. 2007. 43rd Joint Propulsion Conference & Exhibit, 8–11 July 2007, Cincinnati, OH.
[26] V.V. Osipov, D.G. Luchinsky, V.N. Smelyanskiy, S.-H. Lee, C. Kiris, D.A. Timucin. Bayesian framework for in-flight SRM data management and decision support. In D.G. Luchinsky, editor, *Aerospace Conference, 2007 IEEE*, pp. 1–16, 2007.
[27] J. Wu, G. Biswas, S. Abdelwahed, E. Manders. A hybrid control system design and implementation for a three tank testbed. In *Control Applications. 2005 IEEE*, pp. 645–650, 2005.
[28] A.S. Lincon, D. Sivakumar, J. Prakash. State and fault parameter estimation applied to three-tank bench mark relying on augmented state Kalman filter. *ICGST-ACSE Journal*, 7:33–41, 2007.
[29] A. Zolghadri, D. Henry, M. Monsion. Design of nonlinear observers for fault diagnosis: a case study. *Control Engineering Practice*, 4:1535–1544, 1996.
[30] D.G. Luchinsky, V.V. Osipov, V.N. Smelyanskiy, A. Patterson-Hine, B. Hayashida, M. Watson, J. McMillin, D. Shook, M. Johnson, S. Hyde. Model-based diagnostics and prognostics for solid rocket motors. In *Annual Conference of the Prognostics and Health Management Society, 2009 PHM*, 2009.
[31] E. Sorkin. *Dynamics and Thermodynamics of Solid-Propellant Rockets*. Wiener Bindery, Jerusalem, 1967.
[32] M. Sailta. Verification of spatial and temporal pressure distribution in segmented solid rocket motors. In *27th Aerospace Science Meeting 1989 AIAA*.
[33] J.E. McMillin. Scaling equations for ballistic modeling of solid rocket motor case breach. In *2nd Joint Propulsion Conference and Exhibit*, 2006.
[34] W.G. Brownlee. *An Experimental Investigation of Unstable Combustion in Solid Propellant Rocket Motors*. PhD thesis, Dissertation, California Institute of Technology, 1959.

[35] D.G. Luchinsky, V.V. Osipov, V.N. Smelyanskiy, I. Kulikov, A. Patterson-Hein, B. Hayashida, M. Watson, D. Shook, M. Johnson, S. Hyde, J. Shipley. *Integrated Vehicle Health Management for Solid Rocket Motors*, INTECH, Croatia, 2010.

[36] B.A. Bednarcyk, S.M. Arnold, C.S. Collier, P.W. Yarrington. Preliminary structural sizing and alternative material trade study for CEV crew module. In *48th Structures, Structural Dynamics, and Materials Conference*, Honolulu, HI, April 2007.

[37] B.A. Bednarcyk, S.M. Arnold, D.A. Hopkins. Design of fiber reinforced foam sandwich panels for large ARES V structural applications. In *51st Structures, Structural Dynamics, and Materials Conference*, pp. 1–19, Orlando, FL, April 2010.

[38] D.W. Sleight, D. Paddock, J. Jeans, J. Hudeck. Structural design and analysis of the upper pressure shell section of a composite crew module. In *11th ASCE Aerospace Division International Conference (Earth and Space 2008)*, Long Beach, CA, 2008.

[39] A.W. Leissa. *Vibration of Plates*. Number NASA SP-160. U.S. Government Printing Office, Washington, U.S. government printing office edition, 1969.

[40] D. Zenkert, editor. *The Handbook of Sandwich Construction*. EMAS, Warley, 1997.

[41] L.D. Landau, E.M. Lifshitz. *Mechanics*. Pergamon, London, 1976.

[42] A. Annaswamy, A. Ghoniem. Active control of combustion instability: Theory and practice. *IEEE Control Systems Magazine*, 22(6):37–54, 2002.

[43] B. Linder, L. Schimansky-Geier. Analytical approach to the stochastic Fitzhugh-Nagumo system and coherence resonance. *Physical Review E*, 60(6):7270–7276, 1999.

[44] A. Duggento, D.G. Luchinsky, V.N. Smelyanskiy, I.A. Khovanov, P.V.E. McClintock. Inferential framework for nonstationary dynamics. ii. application to a model of physiological signalling. *Physical Review E*, 77:061106, 2008.

[45] D. Luchinsky, V. Smelyanskiy, M. Millonas, P. McClintock. Dynamical inference of hidden biological populations. *European Physical Journal B Condensed Matter and Complex Systems*, 65:369–377, 2008.

[46] V.N. Smelyanskiy, D.G. Luchinsky, M.M. Millonas, P.V.E. McClintock. Recovering 'lost' information in the presence of noise: application to rodent-predator dynamics. *New Journal of Physics*, 11(5):053012, 2009.

CHAPTER 9

Model-Based Tools and Techniques for Real-Time System and Software Health Management

Sherif Abdelwahed

Abhishek Dubey

Gabor Karsai

Nagabhushan Mahadevan

CONTENTS

9.1 INTRODUCTION

The ultimate challenge in system health management is the theory for and application of the technology to *systems*, for instance to an entire vehicle. The main problem the designer faces is complexity; simply the sheer size of the system, the number of data points, anomalies, and failure modes can be overwhelming. Furthermore, systems are heterogeneous and one has to have a systems engineer's view to understand interactions among systems. Yet, system-level health management is crucial, as faults increasingly arise from system-level effects and interactions. While individual subsystems tend to have built-in redundancy or local anomaly detection, fault management, and prognostics features, the system integrators are required to provide the same capabilities for the entire vehicle, across different engineering subsystems and areas.

The heterogeneity of subsystems necessitates the use of techniques that are domain-independent and applicable in mechanical and hydraulic systems, as well as to electronics, and even in software. The most common, high-level concept one can find across these domains is the concepts of (1) observable degradations, anomalies, discrepancies caused by failure modes, (2) their propagation, and (3) their temporal evolution toward system-level faults. Such propagation and evolution effects are clearly cross-domain and cross-subsystem, and if they are understood by the designer, they can be used in system-level diagnostics and prognostics.

In this chapter we present a model-based technology for system-level health management based on the concept of timed fault propagation graphs (TFPG, for short). We present the modeling language, fault diagnostics, and prognostics algorithms that are applicable to complex systems, and how these techniques can be used for the health management of software systems.

We argue that the reasoning approach described here can be called as "real time" in two ways. First, it can operate in real time, online, and is capable of producing the new fault diagnostics results as soon as new evidence becomes available (or as time elapses). This will be elaborated in the section on the reasoner. Second, the operation of the reasoner is bounded in space (memory footprint) and time (response time). The bounds depend on the complexity and structure of the particular model used. Note that we use a model-based reasoner; hence, its operation and behavior is ultimately determined by the model used. While we cannot at this time analytically estimate the worst-case response time of the reasoner simply from the model, we can predict it using the specific model in the reasoner and applying it to a "worst-case" input sequence of alarms.

The model-based approach described in this chapter could be the next logical step after applying machine learning and/or data mining algorithms to data collected during operations. What the data mining techniques discover can be used in the following ways. One, the data mining can provide assistance to the modeler who constructs the TFPG models: correlations discovered in the data may serve as clues to building the causal models. Second, the data mining results can be used to build the discrete monitors (fault detectors) needed by the TFPG reasoner in operation. These monitors detect the anomalies in the system and will trigger the higher-level TFPG reasoner. Third, the data mining can play an important role in refining and improving the models. We will discuss this last point in a separate section.

The chapter is organized as follows. First we review system-level fault diagnostics techniques. Next, we describe the TFPG model, the reasoning algorithms used, and present examples. This is followed by a section about the application of the TFPG in software health management. Then, we describe the application of TFPG in system prognosis. Finally, we conclude with a discussion on how data mining techniques can be used to improve a TFPG-based system and some observations for future research. We also included a number of examples, with specific test results. The tests were conducted on an Intel Xeon (W5320) 2.67 GHz Processor with 4 GB of RAM.

9.2 RELATED WORK

9.2.1 Failure Propagation Models

Diagnostic reasoning techniques share a common process in which the system is continuously monitored, and the observed behavior is compared with the expected one to detect abnormal conditions. In many industrial systems, diagnosis is limited to signal monitoring and fault identification via threshold logic, e.g., detecting if a sensor reading deviates from its nominal value. Failure propagation is modeled by capturing the qualitative association between sensor signals in the system for a number of different fault scenarios. Typically, such associations correspond to relations used by human experts in detecting and isolating faults. This approach has been effectively used for many complex engineering systems. Common industrial diagnosis methods include fault trees [1–4], cause-consequence diagrams [5,6], diagnosis dictionaries [7], and expert systems [8,9].

Model-based diagnosis (see [10–12] and the references therein), on the other hand, compares observations from the real system with the predictions from a model. Analytical models, such as state equations [13], finite state machines [14], hidden Markov models [15], and predicate/temporal logic [16] are used to describe the nominal system behavior. In the case of a fault, discrepancies between the observed behavior and the predicted normal behavior occur. These discrepancies can then be used to detect, isolate, and identify the fault depending on the type of model and methods used. In consistency-based diagnosis, the behavior of the system is predicted using a nominal system model and then compared with observations of the actual behavior of the system to obtain the minimal set of faulty component that is consistent with the observations and the nominal model. Consistency-based diagnosis was introduced in a logical framework in [16] and was later extended in [17]. The approach has been applied to develop diagnosis algorithms for causal systems [18,19] and temporal causal systems [20,21].

The diagnosis approach presented in this chapter is conceptually related to the temporal causal network approach presented in [21]. However, we focus on incremental reasoning and diagnosis robustness with respect to sensor failures. The causal model presented in this chapter is based on the timed failure propagation graph (TFPG) introduced in [22,23]. The TFPG model is closely related to fault models presented in [24–26] and used for an integrated fault diagnosis and process control system [27]. The TFPG model was extended in [28] to include mode dependency constraints on the propagation links, which can then be used to handle failure scenarios in hybrid and switching systems. TFPG modeling and reasoning tool has been developed and used successfully in an integrated fault diagnoses and process control system [27].

The temporal aspects of the TFPG model are closely related to the domain theoretic notion of temporal dependency proposed in [29]. However, there are several major differences between the two approaches. In particular, TFPG-based diagnosis implements a real-time incremental reasoning approach that can handle multiple failures including sensor/alarm faults. In addition, the underlying TFPG model can represent a general form of temporal and logical dependency that directly incorporates the dynamics of multimode systems.

9.2.2 Fault Recovery

The fault recovery aspect of this work has been inspired by our previous work on reflex and healing architecture. Reflex and healing (RH) [30,31] is a biologically inspired two-stage mechanism for recovering from faults in large distributed real-time systems. The first stage is composed of primary building blocks of fault management components called reflex engines (also referred to as managers) that are arranged in a hierarchical management structure. A reflex engine integrates several fault management devices expressed as timed state machine models. While some of these state machines are used as observers that generate "fault events" (discussed in next section), others are used as mitigators to perform time-bounded reactive actions upon occurrence of certain fault events.

At the system-level, fault recovery is done using a planned reconfiguration of the system. This involves a planning step that searches over a set of candidate models obtained using a set of prespecified goals and a list of constraints [32]. This step is typically multiobjective in nature and is dependent on several factors such as system goals, resilience to future faults, and performance. This architecture has been successfully demonstrated for a large-scale software infrastructure for a particle accelerator [33] and computing cluster infrastructure used for scientific workflows [34].

9.2.3 Fault Detection and Health Management of Software

Conmy et al. presented a framework for certifying Integrated Modular Avionics software applications build on ARINC-653 platforms in [35]. Their main approach was the use of "safety contracts" to validate the system at design time. They defined the relationship between two or more components within a safety critical system. However, they did not present any details on the nature of these contracts and how they can be specified. We believe that a similar approach can be taken to formulate acceptance criteria, in terms of "correct" value domain and temporal domain properties that will let us detect any deviation in a software component's behavior.

Nicholson presented the concept of reconfiguration in integrated modular systems running on operating systems that provide robust spatial and temporal partitioning in [36]. He identified that health monitoring is critical for a safety-critical software system and that in the future it will be necessary to trade off redundancy-based fault tolerance for the ability of "reconfiguration on failure" while still operational. He described that a possibility for achieving this goal is to use a set of lookup tables, similar to the health monitoring tables used in ARINC-653 system specification, that maps trigger event to a set of system blueprints providing the mapping functions. Furthermore, he identified that this kind of reconfiguration is more amenable to failures that happen gradually, indicated by parameter deviations. Finally, he identified the challenge of validating systems that can reconfigure at run time.

Goldberg and Horvath have discussed discrepancy monitoring in the context of ARINC-653 health management architecture in [37]. They describe extensions to the application executive component, software instrumentation, and a temporal logic run-time framework. Their method primarily depends on modeling the expected timed behavior of a process, a partition, or a core module—the different levels of fault protection layers.

All behavior models contain "faulty states," which represent the violation of an expected property. They associate mitigation functions using callbacks with each fault.

Sammapun et al. describe a run-time verification approach for properties written in a timed variant of LTL called MEDL in [38]. They described an architecture called RT-MaC for checking the properties of a target program during run-time. All properties are evaluated based on a sequence of observations made on a "target program." To make these observations, all target programs are modified to include a "filter" that generates the interesting event and reports values to the event recognizer. The event recognizer is a module that forwards the events to a checker that can check the property. Timing properties are checked using watchdog timers on the machines executing the target program. The main difference in this approach and the approach of Goldberg and Horvath outlined in previous paragraph is that RT-MaC supports an "until" operator that allows specification of a time bound where a given property must hold. Both of these efforts provided valuable input to our design of run-time component-level health management.

9.3 FAULT DIAGNOSTICS USING TIMED FAILURE PROPAGATION GRAPHS

9.3.1 The Timed Failure Propagation Graph Model

A TFPG is a labeled directed graph where nodes represent either failure modes, which are fault causes, or discrepancies, which are off-nominal conditions that are the effects of failure modes. Edges between nodes in the graph capture the effect of failure propagation over time in the underlying dynamic system. To represent failure propagation in multimode (switching) systems, edges in the graph model can be activated or deactivated depending on a set of possible operation modes of the system. Formally, a TFPG is represented as a tuple $(F,D,E,M,\mathrm{ET},\mathrm{EM},\mathrm{DC})$, where

- F is a nonempty set of failure nodes.
- D is a nonempty set of discrepancy nodes.
- $E \subseteq V \times V$ is a set of edges connecting the set of all nodes $V = F \cup D$.
- M is a nonempty set of system modes. At each time instance t the system can be in only one mode.

- ET : $E \rightarrow I$ is a map that associates every edge in E with a time interval $[t_1,t_2] \in I$.
- EM : $E \rightarrow \mathcal{P}(M)$ is a map that associates every edge in E with a set of modes in M. We assume that EM(e) $\neq \emptyset$ for any edge $e \in E$.
- DC : $D \rightarrow$ {AND,OR} is a map defining the class of each discrepancy as either AND or an OR node.
- DS : $D \rightarrow$ {A,I} is a map defining the monitoring status of the discrepancy as either A for the case when the discrepancy is active (monitored by an online alarm) or I for the case when the discrepancy is inactive (not monitored).*

In the above model, the map ET associates each edge $e \in E$ with the minimum and maximum time for the failure to propagate along the edge. For an edge $e \in E$, we will use the notation e.min and e.tmax to indicate the corresponding minimum and maximum time for failure propagation along e, respectively. That is, given that a propagation edge is enabled (active), it will take at least (most) tmin (tmax) time for the fault to propagate from the source node to the destination node. The map EM associates each edge $e \in E$ with a subset of the system modes at which the failure can propagate along the edge. Consequently, the propagation link e is enabled (active) in a mode $m \in M$ if and only if $m \in$ EM(e). The map DC defines the type of a given discrepancy as either AND or OR. An OR type discrepancy node will be activated when the failure propagates to the node from any of its parents. On the other hand, an AND discrepancy node can only be activated if the failure propagates to the node from all its parents. We assume that TFPG models do not contain self-loops and that failure modes are always root nodes, i.e., they cannot be a destination of an edge. Also, a discrepancy cannot be a root node, that is, every discrepancy must be a successor of another discrepancy or a failure mode.

Figure 9.1 shows a graphical depiction of a failure propagation graph model. Rectangles in the graph model represent the failure modes, while circles and squares represent OR and AND type discrepancies, respectively. The arrows between the nodes represent failure propagation. Propagation edges are parameterized with the corresponding interval, [e.tmin,e.tmax], and the set of modes at which the edge is active. Figure 9.1 also shows a

* In this chapter, we will use the terms *alarms* and *monitored discrepancies* interchangeably, as they mean the same thing.

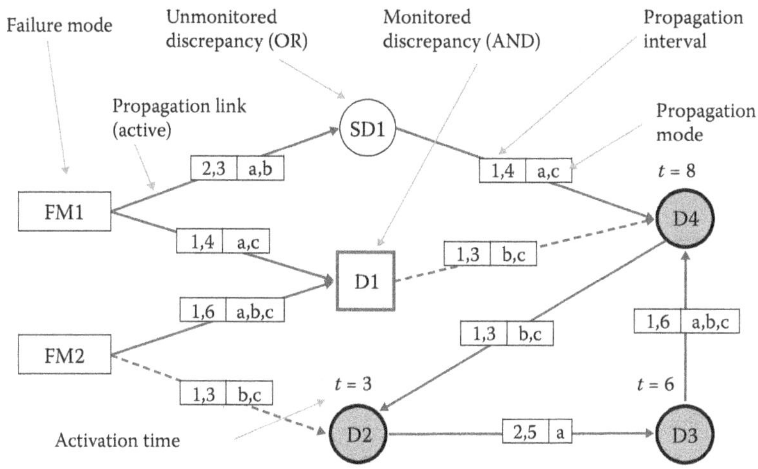

FIGURE 9.1 TFPG model ($t = 10$, Mode = A $\forall t \in [0,10]$).

sequence of active discrepancies (alarm signals) identified by shaded discrepancies. The time at which the alarm is observed is shown above the corresponding discrepancy. Dashed lines are used to distinguish inactive propagation links.

The TFPG model captures observable failure propagations between discrepancies in dynamic systems. In this model, alarms capture state deviations from nominal values. The set of all observed deviations corresponds to the monitored discrepancy set in the TFPG model. Propagation edges, on the other hand, correspond to causality (for example, as defined by energy flow) in the system dynamics. Due to the dynamic nature of the system, failure effects take time to propagate between the system components. Such time in general depends on the system's time constants as well as the size and timing of underlying failure. Propagation delay intervals can be computed analytically or through simulation of an accurate physical model.

Failure propagation in a TFPG system has the following semantics. The state of a node indicates whether the failure effects reached this node. For an OR type node v' and an edge $e = (v,v') \in E$, once a failure effect reaches v at time t it will reach v' at a time t' where $e.\text{tmin} \leq t' - t \leq e.\text{tmax}$. On the other hand, the activation period of an AND alarm v' is the composition of the activation periods for each link $(v,v') \in E$. For a failure to propagate through an edge $e = (v,v')$, the edge should be active throughout the propagation, that is, from the time the failure reaches v to the time it reaches v'.

An edge e is active if and only if the current operation mode of the system, m_c is in the set of activation modes of the edge, that is, $m_c \in \mathtt{EM}(e)$. When a failure propagates to a monitored node v' ($\mathtt{DS}(v') = \mathtt{A}$) its physical state is considered to be ON; otherwise, it is considered to be OFF. If the link is deactivated any time during the propagation (because of mode switching), the propagation stops. Links are assumed to be memory less with respect to failure propagation so that current failure propagation is independent of any (incomplete) previous propagation. Also, once a failure effect reaches a node, its state will change permanently and will not be affected by any future failure propagation.

9.3.2 Reasoning Algorithm

The reasoning algorithm for TFPG model diagnosis is based on a consistency relationship defined using three state mappings for the graph nodes of the TFPG model: physical, observed, and hypothetical.

A *physical system* state corresponds to the current state of all nodes in the TFPG model. At any time t the physical state is given by a map $AS_t : V \rightarrow \{\mathtt{ON},\mathtt{OFF}\} \times \mathbb{R}$, where V is the set of nodes in the TFPG model. An ON state for a node indicates that the failure (effect) reached this node, otherwise it is set to OFF. The physical state at time t is denoted $AS_t(v)$.state, while $AS_t(v)$.time denote the last time at which the state of v is changed. Failure effects are assumed permanent, therefore, the state of a node once changed will remain constant after that. A similar map is used to define the state of edges based on the current mode of the system.

The *observed state* at time t is defined as a map $S_t : D \rightarrow \{\mathtt{ON},\mathtt{OFF}\} \times \mathbb{R}$. Clearly, observed states are only defined for discrepancies. The observed state of the system may not be consistent with the failure propagation graph model temporal constraints, due to potential alarm failures. However, we assume that monitored discrepancy signals are permanent so that once the observed state of a discrepancy has changed, it will remain constant after that.

The aim of the diagnosis reasoning process is to find a consistent and plausible explanation of the current system state based on the observed state. Such explanation is given in the form of a valid hypothetical state. A *hypothetical state* is a map that defines node states and the interval at which each node changes its state. Formally, a hypothetical state at time t is a map $H_t^{V'} : V' \rightarrow \{\mathtt{ON},\mathtt{OFF}\} \times \mathbb{R} \times \mathbb{R}$ where $V' \subseteq V$. Similar to actual states, hypothetical states are defined for both discrepancies and failure modes. The estimated earliest (latest) time of the state change is denoted $H(v)$.terl ($H(v)$.tlat).

A hypothetical state is an estimation of the current state of all nodes in the system and the time period at which each node changed its states. Estimation of the current state is valid only if it is consistent with the TFPG model. State consistency in TFPG models is a node-parents relationship that can be extended pairwise to arbitrary subsets of nodes. Formally, let $\Pr(v)$ denote the set of parents of v in a TFPG model G. We can define *observable consistency* at time t as a relation $\text{OCons}_t \subset \mathcal{P}(V) \times V$ such that $(V',v) \in \text{OCons}$ if and only if $V' = \Pr(v)$ and the observable state of v is consistent with that of all its parents V' based on the map S_t and the failure propagation semantics. The observable state consistency relationship can be directly extended to any set of nodes representing a subgraph of G. In this case, we overload the relationship OCons so that $\text{OCons}_t \subseteq \mathcal{P}(V)$, where for each $V' \subseteq V$:

$$V' \in \text{OCons}_t \quad \Leftrightarrow \quad \forall v \in V' \ (\Pr_{V'}(v),v) \in \text{OCons},$$

where $\Pr_{V'}(v)$ is the parents of v restricted to V'. The set of maximally consistent set of nodes is denoted as Φ_t where $V' \in \Phi_t$ if and only if

$$V' \in \text{OCons}_t \text{ and } \forall V'' \subseteq V \ V' \subset V'' \Rightarrow V'' \notin \text{OCons}_t.$$

The set Φ_t can be efficiently computed incrementally based on Φ_{t-1} based on a new event e_t. Rvent e_t corresponds to either a new triggered monitored discrepancy or a time-out event generated when a sensor alarm is not observed at state ON while it is supposed to be based on its current hypothetical state. The underlying procedure will be denoted $\texttt{UpdateMCO}(\Phi_{t-1},e_t)$. Note that initially $\Phi_0 = \{V\}$.

Based on the semantics of failure propagation, it is possible to define a constructive notion of *hypothetical consistency* such that given a consistent hypothetical state $H_t^{V'}$ it is possible to extend this map forward (procedure $\texttt{BProp}(H_t^{V'},v)$) by assigning the maximal hypothetical state of the node v based on the hypothetical state of its parents in V', or backward (operation $\texttt{FProp}(H_t^{V'},v)$) by assigning the maximal hypothetical state for v' based on the state of its children in V'. The following algorithm outlines the incremental reasoning procedure.

The above diagnosis algorithm returns a set of new hypotheses that can consistently explain the current observed state of the TFPG system. A failure report is then generated from the computed set of hypotheses HS_t. The failure report enlists the set of all consistent state assignments that maximally matches the current set of observations. Any observed state that does not match the current hypothesis is considered faulty. A detailed description and analysis of the diagnosis algorithm can be found in [39].

9.3.2.1 Hypothesis Ranking

The quality of the generated hypotheses is measured based on three independent factors:

- *Plausibility* is a measure of the degree to which a given hypothesis group explains the current fault signature. Plausibility is typically used as the first metric for sorting the hypotheses, focusing the search on the failure modes that explain the data that is currently being observed.
- *Robustness* is a measure of the degree to which a given hypothesis is expected to remain constant. Robustness is typically used as the second metric for sorting the hypotheses, helping to determine when to take action to repair the system.
- *Failure rate* is a measure of how often a particular failure mode will occur.

Algorithm 1. The Diagnosis Procedure `Diag`(Φ_{t-1}, e_t)

$\Phi_t \leftarrow$ `UpdateMCO`(Φ_{t-1}, e_t)
$HS_t \leftarrow \varnothing$
define
`In`$(X) := \{v \in X | (\forall v' \in X)\ (v,v') \notin E\}$
`PSet`$(X) := \{v \in V - X | (\exists v' \in \text{In}(X))\ (v,v') \in E\}$
`ODC`$(X) := \cup_{v \in X}$`Reach`$(v) - X$
`TSet`$(X) := \{v \in V - X | \text{ODC}(X) \times v \cap E = \varnothing\}$
`CSet`$(X) := \{v \in \text{TSet}(X) | (\exists v' \in X)\ (v,v') \in E\}$
end define
for all $V' \in \Phi_t$ **do**
  $H \leftarrow S_t|_{V'}$
  while `PSet`$(V') \neq \varnothing$ **do**
    select v from `PSet`(V')
    $H \leftarrow$ `BProp`(H,v)
    $V' \leftarrow V' \cup \{v\}$
  end while
  while `CSet`$(V') \neq \varnothing$ **do**
    select v from `CSet`(V')

```
    H ← FProp(H,v)
    V′ ← V′ ∪ {v}
  end while
  for all v ∈ V − V′ do
    H(v).state ← OFF
    H(v).terl, H(v).terl ← 0
  end for
  HS_t ← HS_t ∪ {H}
end for
return Φ_t,HS_t
```

The plausibility metric considers two independent factors, namely, alarm consistency and failure mode parsimony. The alarm consistency factor is defined as the ratio of the active consistent alarms to that of all (currently) identified alarms. The failure mode factor is defined as the ratio of identified failure modes (according to the underlying hypothesis) to the total number of failure modes in the system. This factor is a direct representation of the parsimony principle (hypothesis with fewer failures is more plausible). Hypotheses plausibility metrics are ordered lexicographically (alarm factor is more dominant).

The diagnoser selects the current set of hypothesis incrementally in an attempt to improve the current plausibility measure. In other words, the diagnoser will update a given hypothetical state map only if such an update can increase the plausibility of the underlying hypothesis. In addition, changes are restricted so that the updated hypothesis remains valid.

9.3.2.2 Reasoner Performance

The reasoning algorithm described above is an online algorithm as it continually updates the active hypothesis set as new evidence (alarms and mode changes) becomes available. Furthermore, it can be triggered based on the elapse of time (with no changes in the alarms) and it will revise the hypotheses as needed. This latter capability is made possible by the propagation time interval on the edges: if propagation does not happen, it may facilitate the preference of one failure mode over another in the hypothesis. From a strictly pragmatic point of view, the two most important metrics of reasoner performance are the worst-case response time, and the dynamic memory footprint used. We argue that for a particular model (with a given topology, failure modes, alarms, etc.) these two metrics are always bounded.

The dynamic memory footprint is based on the number and the size of the hypotheses the reasoner has to manage during an update. This worst case can be induced by an alarm input sequence that activates alarms in the reverse, that is, alarms farthest from the failure modes are activated first, then the alarms before them, etc. As the number of alarms is bounded, this process results in a bounded use of dynamic memory. In fact, we have created a tool that predicts the dynamic memory footprint of the reasoner based on this technique and simulating its memory usage patterns.

The worst-case response time of the reasoner depends on the size and complexity of the graph and the size of the active hypothesis set. As both of them are bounded, the worst-case response time is bounded as well. Similarly to the previous case, one can predict the worst-case response time based on worst-case scenarios, on the actual reasoner implementation.

Note that the worst-case response time of the reasoner might be much worse than the average or typical response time. It is an open research question how to analytically estimate the average-case response time. In our experience, presented below, we found that for realistic models and realistic scenarios, the reasoner performs well.

9.3.3 TFPG Examples

This section describes example of TFPG models and the results of the experiment performed using these models. Each test case involved loading the appropriate TFPG model into the centralized TFPG reasoner, feeding the reasoner with a timed sequence of events. Each test case typically contained at least 10 events or more—depending on the nature of the test case. These test cases included cases with alarms firing in the correct sequence as per the graph (for single and multiple simultaneous failures), false alarms, intermittent alarms, mode changes. The tests were conducted on an Intel Xeon (W5320) 2.67-GHz processor with 4-MB RAM.

Figure 9.2 shows a trivial nonhierarchical TFPG model. It shows the root causes of the failure (failure modes FM1 and FM2) and the anomalies (discrepancies RD1, D1, SD12, D12, RD2, and D2) that would be triggered when one or more of these failures were to occur. All observable discrepancies have an alarm associated with them (e.g., alarm MRD1 for RD1, M1 for D1). The links capture the failure propagation starting from the failure modes to discrepancies and to subsequent discrepancies downstream. Some of these links depict additional constraints related to activation and timing for failure propagation. The activation condition (a Boolean expression over the modes) captures when failure can propagate over a

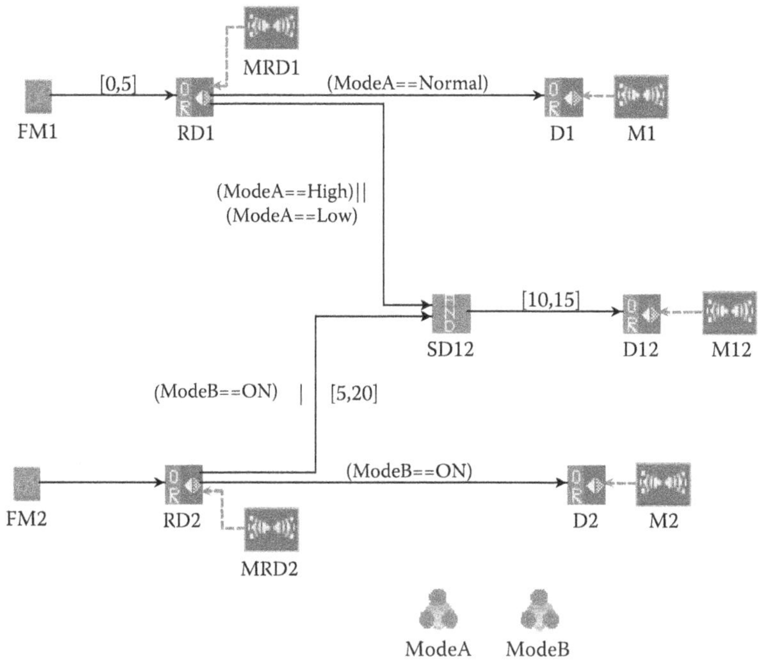

FIGURE 9.2 An illustrative TFPG example.

link. The timing constraint expresses the time bounds within which the failure effect is expected to propagate over that link. These constraints are not specified when the failure can propagate over the link at any time or in any mode. While this model is flat, it can be made hierarchical by composing a TFPG model (at any level) from a TFPG model of components. Overall, this model has 2 failure modes, 6 discrepancies (5 observable), 7 failure propagation links, and 2 modes (with 2 states each).

The next two examples describe typical TFPG models of subsystems in real applications. However, visual rendering of these models is not possible because of their complexity and the space limitations.

Example 1: Pump and Valve

This example refers to a pump and valve (PV) subsystem. Table 9.1 presents a summary. It captures a hierarchical TFPG model of the faults, anomalies, and failure effect propagation in a simple example consisting of a pump, a valve, two sensors, and the associated interface hardware. The operation (and hence the failure propagation) of

TABLE 9.1 TFPG Diagnosis Experiments

			Model Parameters											
		Example	**#C**	**#FM**	**#D**	**#FP**	**#A**	**#M**	**#IP**	**#OP**	**#AG**	**Mem**	**Avg. Update Time (seconds)**	**#TC**
Centralized	1	Pump and Valve	11	36	120	174	27	3	NA	NA	29	18.5	0.0128	87
	2	Generic Fuel Transfer System	153	481	1973	3409	270	9	NA	NA	244	275	0.21	550
Distributed	3	Global	0	0	2				1	9	9	30	0.11 (0.06)	110
		AREA1	17	34	11	43	11	0	0	9	11	7.5	0.012	
		AREA2	50	153	62	590	62	0	1	0	87	25.4	0.08	
	4	Global	21	52	92	176	18	3	28	18	38	74.5	0.3 (0.18)	223
		AREA1/AREA2	39	131	268	539	33	3	3	2	54	47.5	0.14	
		AREA3/AREA4	34	87	119	266	13	0	11	7	24	24.5	0.096	
Software diagnosis (centralized)	5	GPS System	1	19	78	172	9	22			12	12	0.063	10

Note: #C, number of components; #FM, number of failure modes; #D, number of discrepancies; #FP, number of failure propagations; #A, number of alarms; #M, number of modes; #IP, number of input ports; #OP, number of output ports; #AG, number of ambiguity groups; #TC, number of test cases; Mem, worst-case memory used (MB).

For the global reasoners, the time noted in parentheses is the actual time spent in reasoning for creating and updating global hypotheses. Additional time is spent in the global reasoner for assimilating the information from local and global hypotheses to create the set of systemwide consistent failure hypotheses (see Section 9.3.5.1 for further details).

the PV subsystem is governed by 3 modes-associated with pump (on/off), valve(open/close), and the subsystem (active/standby/shutdown). Apart from capturing the failure modes and their propagation in the pump, valve, and their sensors, it also captures the failure effects originating from the power distribution and remote interface units associated with these devices. The failure propagation in the system depends on the operating of the system and the control actions (modes) issued to the devices (pump, valve). Further offline analysis of the model revealed that given the causal linkage expressed between the failures and the observed discrepancies in the TFPG model, the number of ambiguity groups in terms of failure modes was 29. The average processing time to update the result was 0.0128 second. The estimate of worst case memory usage is around 18.5 MB.

Example 2: Generic Fuel Transfer System

Another example, comparable to a real-life application, is a hierarchical TFPG model for an aircraft fuel transfer subsystem. The subsystem is symmetrically divided into left and right parts with supply tanks (left and right fuselage tanks, left and right wing tanks) feeding to a central transfer manifold, which then feeds to engine feed tanks (left and right feed tanks). The supply tanks and the feed tanks are full initially. The controller tries to maintain a constant supply to the engines while maintaining the center of gravity of the aircraft. The hierarchical TFPG model of this subsystem includes failure propagation across the main subsystem elements (the tanks and the manifold) as well as the power and control elements. The properties of this TFPG model are captured in Table 9.1. Offline analysis revealed that there were a total of 244 failure ambiguity groups. The average processing time for an update was 0.22 second. The estimate of worst-case memory usage is around 275 MB.

9.3.4 Distributed Reasoning

A centralized reasoning approach (TFPG-based or otherwise) is not well-suited for the online diagnosis of very large systems that are made up of many subsystems with limited failure interactions across the subsystem boundary. In such cases, a single centralized diagnosis engine might not scale to provide the desired response time. In a centralized system, all alarms and mode changes have to be routed to the central reasoner that would have to operate large models. It might be more pragmatic to use multiple reasoners to perform diagnosis of different areas/subsystems. These area reasoners can be hosted on the same or different processing nodes and can respond faster than a single systemwide reasoner as their search spaces are much smaller and they can operate in parallel. However, this approach of splitting the task among multiple independent area reasoners does not address cascading fault effects from one subsystem to another. Additionally, it cannot provide a coherent systemwide knowledge of the fault status of the entire system. This section discusses a TFPG-based distributed reasoning approach that handles these problems.

9.3.4.1 Overview

The TFPG-based distributed reasoning approach employs a global TFPG reasoner and multiple local TFPG reasoners. The TFPG model of the global reasoner captures only the failure interactions between subsystems monitored by the local reasoner(s). The TFPG model of the local reasoner contains a detailed TFPG model of the subsystem monitored by the local reasoner(s). The local reasoners operate in parallel and autonomously, each of them reasoning over the events in its specific subsystem. The local reasoner(s) communicate updates of any potential fault cascade to the global reasoner. The global reasoner, on its part, ensures that this information is transferred only to relevant local reasoners whose subsystem could be potentially affected by the fault cascade. Using the knowledge obtained from the local reasoner of the potential fault cascades and their individual diagnosis results, the global reasoner builds a globally consistent systemwide diagnosis report. This approach ensures that there is a relatively quick diagnosis result from the local reasoners that, if required, could be refined or updated by the global reasoner that yields systemwide consistent result.

Figure 9.3 captures a model of the local reasoners and the global reasoner. Note that the global reasoner does not have a detailed view of the subsystems. That is the purview of the local reasoners. The following subsections list extensions to the original TFPG model and algorithms. Readers are referred to [40] for further details.

9.3.4.2 Extensions to the TFPG Model

The original TFPG model—a tuple of (F, D, E, M, ET, EM, DC)—needs to be extended to be able to express the global and subsystem TFPG models.

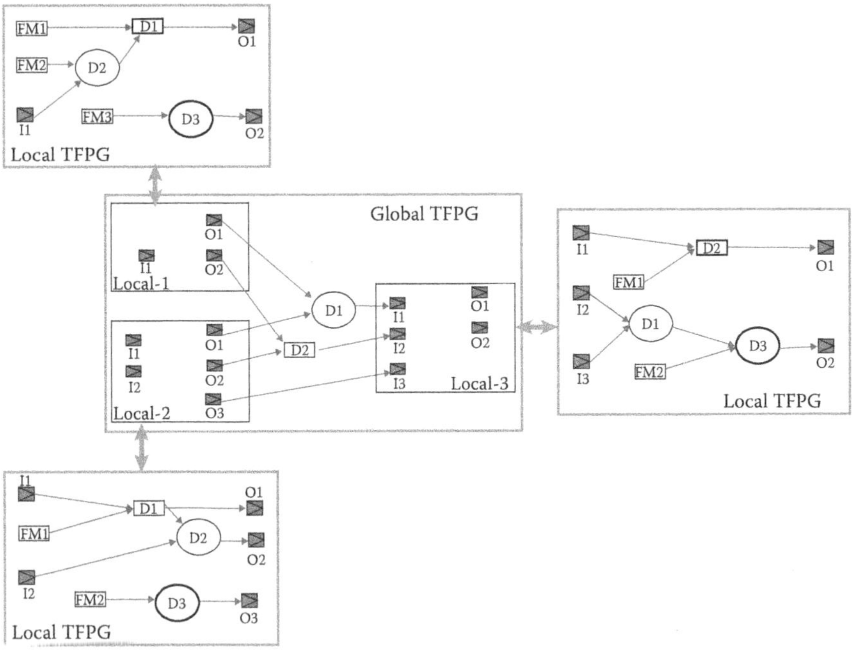

FIGURE 9.3 Distributed TFPG model with three local TFPG models and one global TFPG model.

The global and the subsystem model include two additional sets of elements

- (*IP*): a set of input ports
- (*OP*): a set of output ports

The input and output ports are the interface points for connecting a subsystem model (monitored by a local reasoner) with the rest of the model. Any failure propagation into the subsystem model has to go through an input port. Likewise, any failure propagation out of the subsystem model has to go through an output port. In the TFPG failure propagation semantics, these input and output port extension elements are similar to unmonitored (silent) OR discrepancies. While both the subsystem models and the global model are aware of the input and output ports associated with each subsystem model, there is a difference in their knowledge of the failure propagation interactions associated with these ports. While each subsystem model is aware of only the failure propagation interactions within the subsystem, the global model is aware of only the failure propagation interaction outside of the subsystem models.

The global model is a TFPG model that is composed of multiple subsystem models and the failure propagation interaction between these subsystem models. As is evident from the model depicted in Figure 9.3, the global model is not aware of the detailed TFPG model inside a subsystem. It is aware of just the input and output ports of a subsystem model.

9.3.4.3 Extensions to the Reasoner

The TFPG Algorithm 1 described earlier is central to the reasoning in the global and local reasoners as well. This algorithm is applied to reason upon any external physical event update—changes in the state of the alarm or modes. Additionally, in a local reasoner, if the update resulted in a change to the hypothetical state of any of the interface nodes (input and output ports), then the update needs to be communicated to the global, then to the local reasoner, which shares the interface element. In the reasoner receiving the update information, the update to the hypothetical state of the input/output port is treated like an external physical event update and the reasoning Algorithm 1 is applied to reason about it. This process of applying the reasoning algorithm and communicating updates to the input/output port from the local (global) to the global (local) reasoner continues, until the propagation stops. At this point, the metrics associated with the hypothesis in the local reasoner are recomputed and transmitted to the global reasoner, which then recomputes the metrics of the hypothesis in the global reasoner.

All through the update propagation, the local and global reasoners are responsible for maintaining the hypothetical states of the nodes (including input/output ports) in their respective TFPG models. In addition, the hypothesis in the local reasoner have a reference to a parent hypothesis in the global reasoner. Likewise, the hypothesis in the global reasoner has references to a hypothesis from each of the local reasoners. This reference tracking and maintaining of consistent hypothetical states of the shared interface elements (input/output ports) ensures that the hypothesis in the local and global reasoners are consistent with one another. This enables the hypothesis in the global reasoner to provide a systemwide consistent diagnosis result. This knowledge of a systemwide consistent result in the global allows the global reasoner to find the best explanation, i.e., the fault source, for the sequence of alarms observed in the local reasoner(s). This information can be fed back to the local reasoner to help in prioritizing its hypotheses.

An essential aspect of the distributed diagnosis is its reliance on the communication between the reasoners. It should be noted that this algorithm permits a local reasoner to communicate only with the global reasoner. The communication messages transferred between global and local reasoner(s) include

- (*Command*): Commands issued by the global reasoner to the local reasoner. This could include commands to initiate a diagnosis update, command to report diagnosis result or event update, etc.
- (*Response*): Message sent the local reasoner in response to a command from the global reasoner.
- (*Update*): Updates to the hypothetical state of an input/output port sent in either direction between the global and its associated local reasoners.

The previous paragraphs provided an outline of the steps associated with processing event updates in distributed reasoning approach. The following paragraphs describe two specific strategies that were applied to handle reasoning about new events. Throughout the course of this discussion, it is assumed that the reasoner(s) and the event generator(s)/event reporter(s) share the same clock.

9.3.4.4 Synchronous Event Processing

In this strategy, when an external physical event (alarm/mode updates) is received, the reasoner immediately reports it to the global reasoner without updating the local state. The global reasoner stores all such external events (in any reasoner) in a queue, which is sorted by the event occurrence (or detection) time. This strategy assumes that the external physical events are reported to the global reasoner as soon as they are reported to the local reasoner. This ensures that the events are processed in the correct order. The global reasoner takes up the first event in the queue and commands the appropriate reasoner (where the event occurred) to process it. The reasoner applies the TFPG algorithm 1 to process the event. In case any updates on the hypothetical states of the interface elements are detected, the reasoner performs the updates and communicates them as described in the Section 9.3. Once this step is finished, the global reasoner directs each of the local reasoners to report their local hypothetical results. It fuses these results to get a global consistent result and commands the local reasoners to update/synchronize their hypothesis with the global. This could involve elimination of unnecessary local hypotheses. The global reasoner takes up the next event in the queue and directs the appropriate reasoner to process it. If the queue is empty, the global reasoner waits for a new physical event to be reported (either triggered directly in the global reasoner or reported from the local reasoner). Thus, in this strategy, the global reasoner directs the execution of events across the areas.

9.3.4.5 Asynchronous Event Processing

In the asynchronous strategy, the local reasoner(s) does not wait for the global reasoner's command to process/reasoner about an external physical event. As soon as an external physical event is reported to a local reasoner(s), it informs the global reasoner about this event. It also stores the event in a local sorted event queue. If the local reasoner is not processing any command from the global reasoner and its local event queue is not empty, it starts processing these events from the queue. It stores the hypothesis updates based on these events in a separate list (the `transient-hypothesis-list`). Each local reasoner continues to maintain a primary hypothesis list that is synchronized with the global reasoner (the `stable-hypothesis-list`).

The global reasoner also maintains a sorted event queue. It processes the events in the queue in the similar manner as synchronous event processing. When it is time to process the next event in the queue, it directs the appropriate reasoner to start processing it. Since the local reasoner could have already processed this event, it checks its transient hypothesis list for updates relating to the event. If available, it does not process the event but updates the stable-hypothesis-list based on the results in the transient-hypothesis-list. If not available, it processes the event and updates the stable-hypothesis-list. It then communicates the updates to the global reasoner. Next, this leads to propagation of updates across the system, which continues until the states stabilize. The global reasoner collects the updated hypothesis results from the local reasoners and creates an updated globally consistent result that is used to prune the stable-hypothesis-list in local reasoners.

Whenever the stable-hypothesis-list is updated in the local reasoner, hypotheses in its transient-hypothesis-list need to be checked if they need to be reevaluated. It is possible that none of them need to be updated or some or all of them need to be updated because of lack of consistency with the global stable state.

The two strategies described above have their own advantages and disadvantages. While with asynchronous updates, one can receive a quick update on a local event from a local reasoner. This is not possible in the synchronous mode of processing where the event is processed only when all events (across all reasoners) that occurred before it are processed. This is particularly inefficient if the local event was purely the result of a local fault or a cascading fault whose effect was already stored in a stable manner in the local reasoner. On the other hand, the asynchronous processing algorithm is much more complicated and involves a lot more bookkeeping in the local reasoner. It could suffer from repeated recomputation if the stable-hypothesis-list involving the events are repeatedly updated by events currently being processed by the global reasoner.

The distributed reasoning approach described above has been implemented using a distributed software platform, and evaluated on numerous examples, but further refinements are subject of active research.

9.3.5 Distributed TFPG Examples

This section describes examples of distributed TFPG models and the results of the experiment performed using these models. Each reasoner (global and subsystem) was hosted on a separate process and the communication between the processes was realized through a CORBA based middleware. Each test case involved starting the associated reasoner processes (global and subsystems) with the appropriate model and feeding the timed sequence of events to the appropriate reasoner (based on which subsystem the event triggered in). The total number of events triggered (across all subsystems and global) in each test case depended on the nature of the test case. Some test cases triggered the alarms associated with one or more failure modes in the correct sequence as per the graph. Others altered these sequences by introducing false alarms, intermittent alarms, mode changes, etc.

Figure 9.4 shows a trivial distributed TFPG model. It shows a distributed TFPG model with a global model and three subsystem (area) models. The global model is aware of the presence of subsystem models and the interface element (input or output port) in each of the subsystem models. The global model is not aware of the internal detailed TFPG model in each subsystem. This model illustrates how failure effects from AREA1 (failure mode–FM11) and AREA3 (failure mode–FM31) propagate out (through output ports OP11 and OP31, respectively), enter the failure propagation graph in the global model, and finally reach into AREA2 (through input

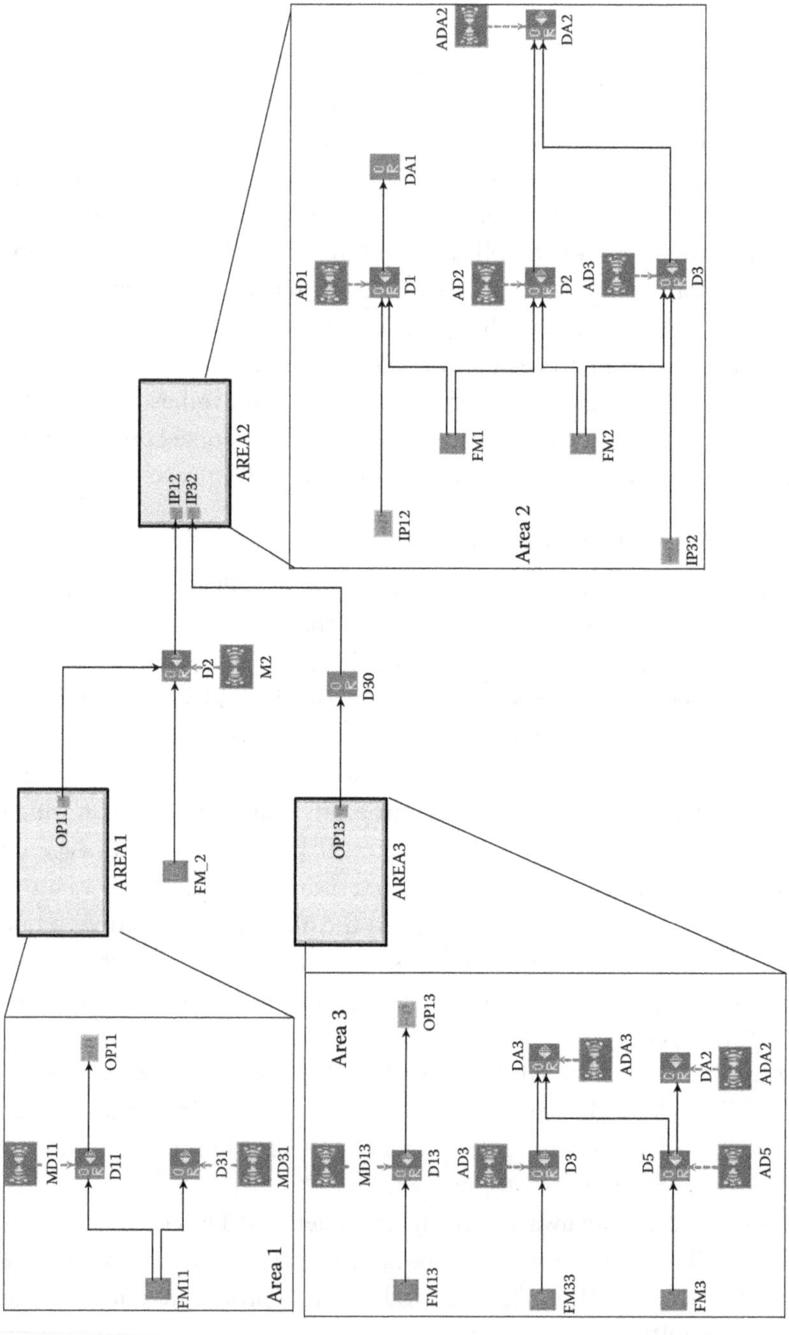

FIGURE 9.4 **An illustrative distributed TFPG example.**

ports IP21 and IP22). It shows that there are portions of the TFPG model in some subsystems that do not participate in the cascading failure effects across subsystem boundaries.

The above example illustrates a simple distributed TFPG model with 1 global model, 3 subsystem (area) models, with 1 or 2 input and output ports in each of the areas, 1–3 failure modes per subsystem, 2–5 discrepancies, and 3–10 failure propagation links. The next examples are distributed TFPG models developed for a small portion of real-life systems. Due to proprietary restrictions, these models can be presented only in an abstract format.

Example 3: Distributed System 1

The first example is a distributed TFPG model that includes a global (Global model) with two subsystems (AREA1 and AREA2). Here, the failure propagation across subsystem boundaries is strictly from AREA2 to AREA1. Table 9.1 describes the properties of the three TFPG models (Global, AREA1, AREA2). Offline analysis revealed that there were a total of 9 failure ambiguity groups in the global model, 11 failure ambiguity groups in AREA1, and 87 failure ambiguity groups in AREA2. This includes the ambiguities introduced by the secondary failure modes (the ports that carry in failure effects from other subsystems). The estimated worst case memory usage for global was around 30 MB, AREA1 is 7.5 MB, and AREA2 is 25.4 MB.

The tests with this example were run with the 3 reasoners (2 subsystem reasoners and 1 global reasoner) hosted on separate processes. The average time to update the hypothesis in the subsystem reasoners was 0.01 second in AREA1, 0.08 second in AREA2, while in the global reasoner it took about 0.11 second. This time includes the reasoning time as well as time to compute systemwide results assimilated from the set of local and global hypotheses. It was found that the latter took a significant share of the total update time. The average reasoning time was around 0.065 second.

Example 4: Distributed System 2

The next example of a distributed TFPG model (again of a portion of a realistic system) is much richer in the failure propagation cascades across the subsystem model boundaries. Again, since the names or the exact nature of the system cannot be revealed these subsystems would be referred to as AREA1, AREA2, AREA3, and AREA4. The symmetric nature of the system introduces two sets of subsystem models that are similar to one another. The global model has 4 subsystem models. The TFPG models of subsystem AREA1 resemble that of subsystem AREA2; AREA3 model is similar to AREA4.

Table 9.1 describes the properties of the five TFPG models (Global, AREA1/AREA2, AREA3/AREA4). Offline analysis revealed that there were a total of 38 failure ambiguity groups in the global model, 54 failure ambiguity groups in AREA1 (and AREA2), and 24 failure ambiguity groups in AREA3 (and AREA4), in each case including the ambiguities introduced by the secondary failure modes (the ports that carry in failure effects from other subsystems). The estimated worst case memory usage for global is around 74.5 MB, AREA1/AREA2 is 47.5 MB, and AREA3/AREA4 is 24.5 MB. The tests were conducted in the same fashion as the previous example on the distributed reasoner. The difference is that in this example there were 2 more processes to account for the two additional subsystem reasoners. The time to update the hypothesis in the subsystem reasoners was 0.096–0.14 second, while in the global reasoner it took about 0.3 second. As stated in the previous example, this time includes the reasoning time, 0.18 second, as well as time taken to compute final systemwide results.

9.3.5.1 Discussion

The experimental results for the centralized and distributed reasoners were presented. While it was found that in general the reasoner performed quite well, it could certainly be improved. As it can be expected, the update time was more when the event could not be explained consistently with existing hypotheses and new explanations (multiple failures) had to be found that were consistent with the current event as well as the past events.

In both distributed examples, the tests revealed that the update time was almost the same in both synchronous and asynchronous modes of processing. The difference was that in the asynchronous mode of operation, the transient result was available with the same rate as the average update time for a subsystem reasoner. Some other interesting observations were made regarding the update times in the global reasoner. The global reasoner (in the current setup) performs two tasks. The first task deals with reasoning on events and updating global hypotheses to keep a consistent state across all subsystem reasoners. The second task involves computing the detailed results from all the local reasoners for reporting purposes, which involves the assimilation of global hypotheses and all local hypotheses. From our results, we infer that significant time is spent on the second task during each update. It is our opinion that this task is more relevant to report generation than reasoning. Hence, Table 9.1 shows two times for the global reasoner update, one that includes the total time of the two tasks and the other in parentheses that includes just the reasoning time (first task).

Another aspect that should be mentioned is that the total time spent in reasoning (in a global or local model) was influenced by local events as well as the hypothetical updates received from the input and output ports. It was observed that the distributed reasoners (global and subsystem) had an average update time that was longer than a centralized reasoner operating on a model of similar size. This can be attributed to the issue that the global and subsystem reasoners have an extra load associated with the fault cascades coming into the subsystem (from other subsystems). Whenever there is significant traffic of fault cascades across the system boundaries, more time is spent in updating the local hypothesis relative to the boundary events. It should be noted that enough care and offline analysis should be performed while designing the boundaries of these subsystems such that they have low coupling across the subsystem boundary and heavy cohesion within the subsystem.

9.4 APPLICATION OF TFPG FOR DIAGNOSING SOFTWARE FAILURES

Modern cyber-physical systems, such as aircraft, are increasingly becoming reliant on software for core functions and system integration [41,42]. Increase in the scope of functions covered by software tends to increase the complexity of software, which often tends to increase the potential of latent software defects. These latent defects can escape the existing rigorous testing and verification techniques but manifest under exceptional circumstances. These circumstances may include faults in the hardware system, including both the computing and noncomputing hardware. Often, systems are not prepared for such faults. Such problems have led to a number of failure incidents in the past, including but not limited to those referred to in these reports: [43–47].

One way to counter these challenges is to systematically extend classical software fault tolerance techniques and apply the system health management for complex engineering systems to software systems. System health management typically includes anomaly detection, fault source identification (diagnosis), fault effect mitigation (in operation), maintenance (offline), and fault prognostics (online or offline) [48,49].

Our research has focused on extending our diagnosis techniques described earlier in this chapter and apply it to software systems. It is essential to identify the boundaries of fault propagation in software systems to apply systematic diagnostic techniques such as TFPG. One approach is to ensure that software systems are built from *software components*, where each component is a reusable unit. Components (see Figure 9.5) encapsulate (and generalize) objects that provide functionality, and have well-defined ports for communication. We expect that these components are well-specified, independently developed, verified, and tested. Furthermore, all communication and synchronization among components is facilitated by a component framework that provides services for all component interactions with well-defined semantics, and no component interactions happen through "out-of-band" channels. This component framework acts as a middleware, provides composition services, and facilitates all messaging and synchronization among components, and is used to support fault management.

Having well-specified interactions enable deduction of behavioral dependencies and failure propagation across a component assembly. Similar approaches exist in [50,51]. The key differences between those and this work are that we apply an online diagnosis engine. The next section briefly describes an implementation of such a component framework and uses it to describe the application of online diagnosis to software systems.

9.4.1 ARINC Component Framework

The ARINC Component Framework (ACF) is a runtime software layer that implements the ARINC-653 component model (ACM) [52]. ACM

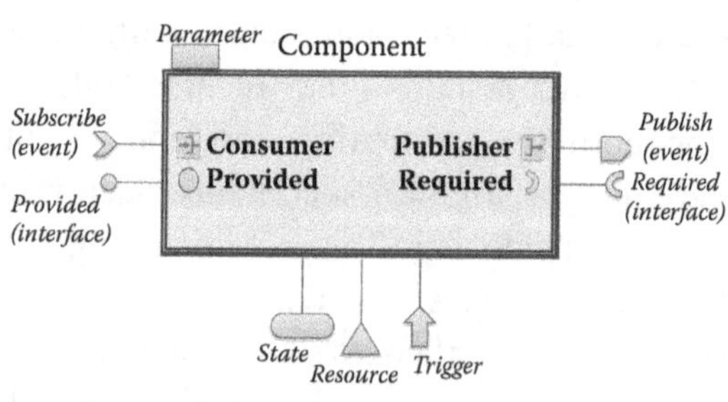

FIGURE 9.5 The ARINC component model.

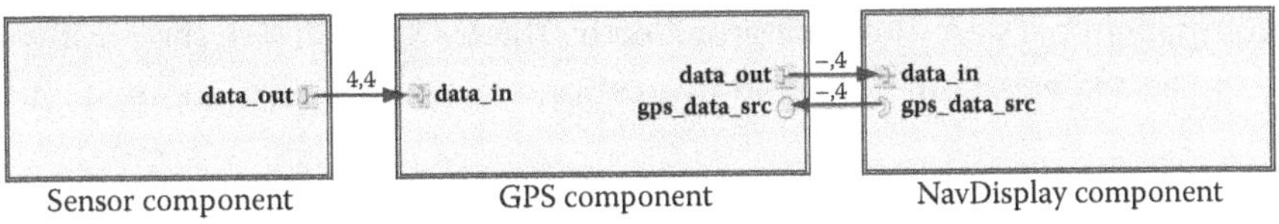

FIGURE 9.6 Example for component interactions. Each interface is annotated with its (periodicity, deadline) in seconds.

borrows concepts from other software component frameworks, notably from the CORBA Component Model (CCM) [53], and is built upon the capabilities of ARINC-653 [54], the state-of-the-art operating system standard used in integrated modular avionics. The key to ARINC-653 is the principle of spatial and temporal isolation among partitions. Discussion of abilities of ARINC-653 is out of the scope of this chapter. However, interested readers are suggested to refer to [54].

In ACM, a component can have four kinds of external ports for interactions: publishers, consumers, facets (provided interfaces*), and receptacles (required interfaces) (see Figure 9.5). Each port has an interface type (a named collection of methods) or an event type (a structure). The component can interact with other components through synchronous interfaces (assigned to provided or required ports) and/or asynchronous events (assigned to event publisher or consumer ports). Additionally, a component can host internal methods that are periodically triggered. Each port can be periodic (i.e., time triggered) or aperiodic (i.e., event triggered). It is statically bound to a unique ARINC-653 process. This binding is defined and configured during initialization. Given that a provided interface can have more than one method, every method is allocated to a separate process.

Figure 9.6 shows a simple example assembly of components. The sensor component contains an asynchronous publisher interface (source port) that is triggered periodically (every 4 seconds). The event published by this interface is consumed by a periodically triggered asynchronous consumer/event sink port on the GPS component (every 4 seconds). Note that the event sink process is periodically released, and each such invocation reads the last event published by the Sensor. The consumer process in the GPS, in turn, produces an event that is published through the GPS's event publisher port. This event triggers the aperiodic consumer/event sink port on the Navigation Display component. Upon activation, the display

* An interface is a collection of related methods.

component uses an interface provided by the GPS to retrieve the position data via a synchronous method invocation call into the GPS component.

9.4.2 Health Management in ACM

With this framework, there are various levels at which health management techniques can be applied, ranging from the level of individual components to the whole system.

9.4.2.1 Component-Level Health Management

Component-level health management (CLHM) provides localized and limited functionality for managing the health of one component by detecting anomalies, mitigating its effects using a reactive timed state machine, on the level of individual components. It also reports to higher-level health manager(s): the system health manager.

9.4.2.2 System-Level Health Manager

System-level health manager (SLHM) manages the overall health of the system, i.e., assembly of all components. The CLHM processes hosted inside each of the components report their input (monitored alarms) and output (mitigation actions) to the SLHM. It is important to know the local mitigation action because it could affect how the faults cascade through the system. Thereafter, the SLHM is responsible for the identification of root failure source(s)—multiple failure mode hypotheses are allowed. Once the fault source is identified, appropriate mitigation strategy is employed.

9.4.2.3 Component-Level Detection

The ACM framework allows the system designer to deploy monitors that can be configured to detect deviations from expected behavior, violations in properties, constraints, and contracts of an interaction port or component. Figure 9.7 summarizes various places where a component's behavior can be monitored. These monitors check properties associated with the resource utilization (locks, deadline violation), data (validity, precondition, and postcondition violation) and user code (exception or user-code violation). The placement of a monitor is dependent on the property violation that it monitors. The monitors associated with resource usage, data validity, and deadline violation are implicitly handled and triggered by the framework. All other monitors have to be configured and enabled at design time. For all ports, exceptions in the user provided functional code are abstracted by the framework as an error triggered by a monitor

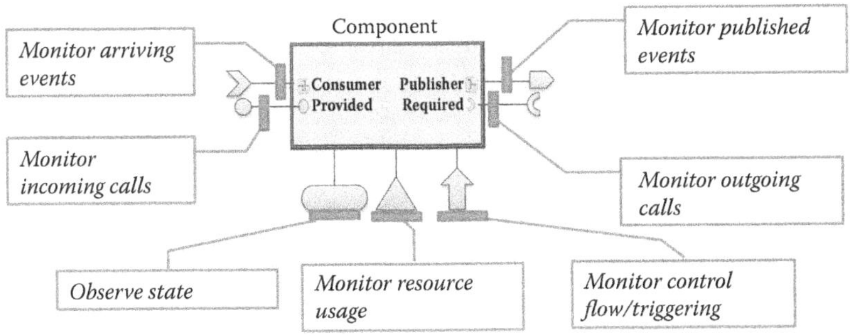

FIGURE 9.7 Approach to component monitoring.

associated with the user code. The monitors checking the precondition and postcondition violation on method calls are evaluated before and after the user provided functional code. These conditions are expressed over the current value or the historical change in the value, or rate of change of value of variables such as (a) the event data of asynchronous calls, (b) function parameters of synchronous calls, and (c) state variables of the component. Table 9.2 summarizes these monitors. While the monitors associated with resource usage are run in parallel by framework, other monitors are evaluated in the same thread executing the component port. When any monitor reports a violation, the status is communicated to its component-level health manager (CLHM) and then possibly to the system-level health manager (SLHM).

TABLE 9.2 Monitoring Specification

<**PreCondition**>::=<Condition>

<**PostCondition**>::=<Condition>

<**Deadline**>::=<double value>/* *from the start of the process associated with the port to the end of that method* */

<**Data_Validity**>::=<double value>/* *Max age from time of publication of data to the time when data is consumed*/

<**Lock Time Out**>::=<double value>/* *from start of obtaining lock*/

<Condition>::=<Primitive Clause><op><Primitive Clause>|<Condition><logical op><Condition>| !<Condition> | True| False

<Primitive Clause>::=<double value>| Delta(Var)| Rate(Var)|Var /* *A Var can be either the component State Variable, or the data received by the publisher, or the argument of the method defined in the facet or the receptacle*/

<op>::= < | > | <= | >= | == | !=

<logical op>::=&& | ||

Note: Comments are in italics.

TABLE 9.3 CLHM Mitigation Actions

CLHM Action	Semantics
IGNORE	Continue as if nothing has happened
ABORT	Discontinue current operation, but operation can run again
USE_PAST _DATA	Use most recent data (only for operations that expect fresh data)
STOP	Discontinue current operation
START	Re-enable a STOP-ped periodic operation
RESTART	STOP followed by a START for the current operation

9.4.2.4 Component-Level Mitigation

Once a local discrepancy is detected, the framework reports the problem to the CLHM. CLHM for each component is deployed on a high-priority ARINC process that is triggered when any of the associated ARINC processes (hosting the Component ports) or the underlying ACM framework report any violations detected by monitors. In a blocking call, the reporting process waits for a response/mitigation action from the CLHM. It is important to note that a local reactive action changes the state of the system and is therefore important information that is required to ascertain the diagnosis. Table 9.3 summarizes all possible local mitigation actions.

Example 5: CLHM Example

In this scenario, we inject a fault in the functional code written by us for the GPS example (see Figure 9.6) such that between 10 and 20 seconds since its first launch, the GPS get data process sends out bad data when queried by the navigation display. The bad data is defined as the rate of change of GPS data being greater than a threshold. This fault simulates an error in the filtering algorithm in the GPS such that it loses track of the actual position.

Figure 9.8 shows a snapshot of the sequence of events in the experiment. The fault is injected approximately 18 seconds after the start of the experiment. The navigation display component retrieves the current GPS data using the remote procedure call. Then, the post condition check of the remote procedure call is violated. This violation is defined by a threshold on the delta change of current GPS data compared to past data (last sample). The navigation display component raises an error,

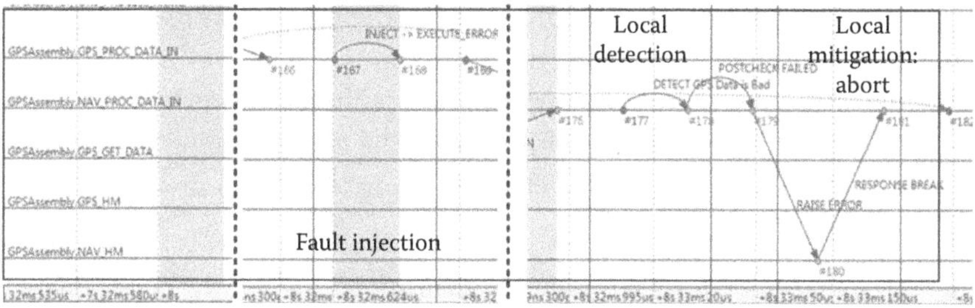

FIGURE 9.8 Example of a component level health management action. The fault was injected in the GPS component to publish bad data. This was detected via the post condition of the navigation component's required method. The specified reactive action is to abort the call. Note the time scale is nonlinear.

which is received by the local health manager. Consequently, it receives an ABORT response from the health manager. Notice that the execution of navigation display process is preempted till it receives a response from the health manager. The ABORT response means that the process that detected the fault should immediately stop processing and return. The effect of this action is that the navigation's GPS coordinates are not updated as the remote procedure call finished with an error.

Having described the local monitors and component level reactive action, we can discuss the TFPG model for a given software assembly. This model is used by the SLHM to diagnose failures. To facilitate this diagnosis, all component level managers are required to report their input (alarm/monitor) and output (response/mitigation action) to the SLHM. As more information (alarms or alarms plus responses) becomes available, the SLHM (using the diagnosis engine) improves its fault hypothesis, which can then be potentially used to drive the mitigation strategy at the system level.

9.4.3 Software Fault Propagation Model

As described in previous sections, in this framework the software assemblies are composed of components, which are in turn composed of specific kinds of ports—publisher, consumer, provider interface, and requires interface. While these interaction ports can be customized by the event-data-types published/consumed, interfaces/methods exposed, periodicity, deadline, etc., their fundamental behaviors and interaction patterns are well defined. This implies that it should be possible to identify specific faults and fault propagation pattern that are common to each kind of interaction pattern, which could result in a generic TFPG model for each interaction pattern (connection between two ports of different components). Thus, the failure propagation across the component boundaries can be captured from the assembly model. The generic TFPG model for a specific interaction port captures the following information:

1. Health monitor alarms and the component health manager's response to these alarms are captured as observable discrepancies
2. Failures originating from within the interaction port and the effect of their propagation as failure modes
3. Effect of failures originating from other entities as discrepancies
4. Cascading effects of failures within the interaction port as discrepancies
5. Effect of failures propagating to other entities as cascades

Additionally, a data and control flow model about the component internals (between the component processes) assists in capturing the failure propagation within the component. In principle, this approach is similar to the failure propagation and transformation calculus described by Wallace [51]. That paper showed how architectural wiring of components and failure behavior of individual components can be used to compute failure properties for the entire system.

Example 6: Generic TFPG for a Periodic Consumer

Figure 9.9 shows a generic TFPG model for a periodic consumer port. The list below explains the failure effects that are cap tured in a generic TFPG model through the example of the consumer TFPG model. The consumer's TFPG model 9 is presented in terms of the failure propagations captured in the context of the observed alarms—LOCK_TIMEOUT, VALIDITY_FAILURE, PRE-CONDITION Violation, USER-CODE Failure, Deadline Violation. Each of these subgraphs covers most of the points described above.

FIGURE 9.9 TFPG template showing five (out of six possible) fault propagation patterns for a periodic consumer.

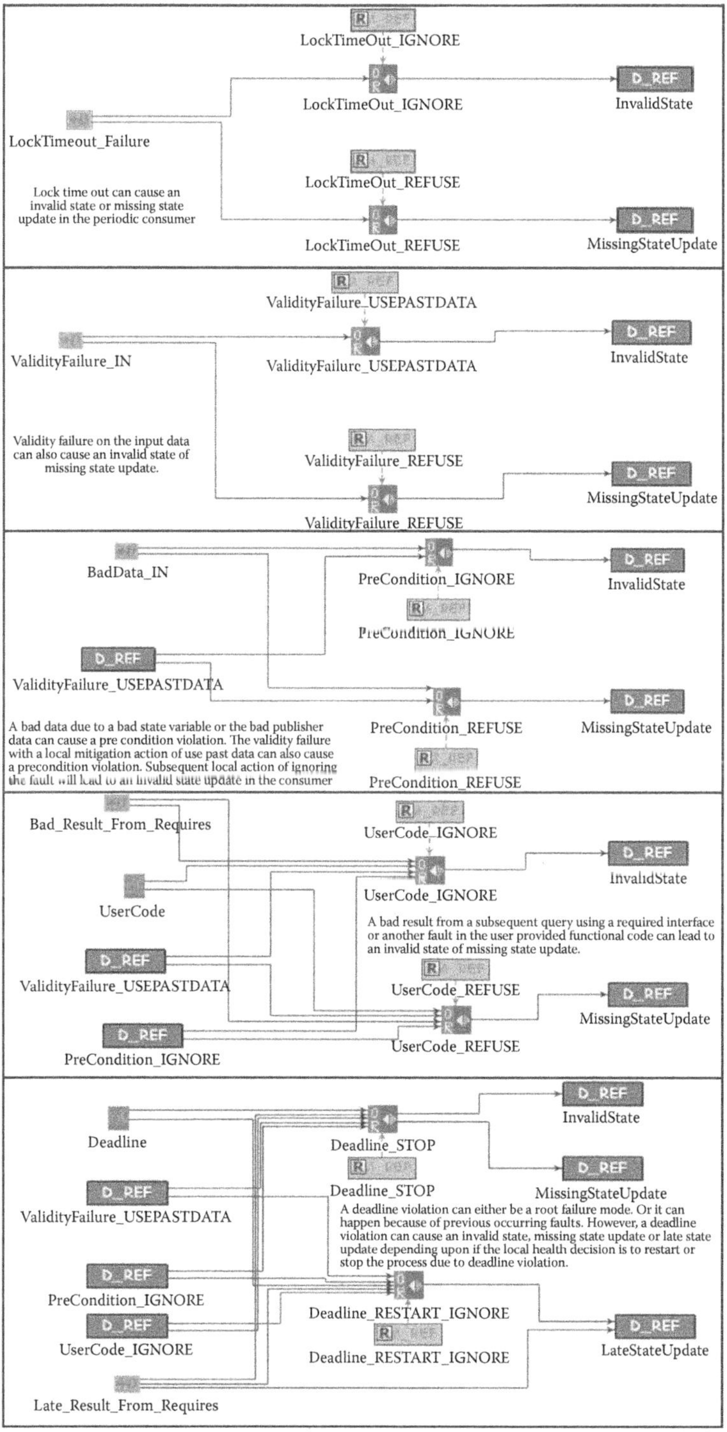
LockTimeOut_IGNORE
LockTimeOut_IGNORE
InvalidState
LockTimeout_Failure
LockTimeOut_REFUSE
Lock time out can cause an invalid state or missing state update in the periodic consumer
LockTimeOut_REFUSE
MissingStateUpdate
ValidityFailure_USEPASTDATA
ValidityFailure_IN
ValidityFailure_USEPASTDATA
InvalidState
Validity failure on the input data can also cause an invalid state of missing state update.
ValidityFailure_REFUSE
MissingStateUpdate
ValidityFailure_REFUSE
BadData_IN
PreCondition_IGNORE
InvalidState
PreCondition_IGNORE
ValidityFailure_USEPASTDATA
A bad data due to a bad state variable or the bad publisher data can cause a pre condition violation. The validity failure with a local mitigation action of use past data can also cause a precondition violation. Subsequent local action of ignoring the fault will lead to an invalid state update in the consumer
PreCondition_REFUSE
MissingStateUpdate
PreCondition_REFUSE
Bad_Result_From_Requires
UserCode_IGNORE
UserCode
UserCode_IGNORE
InvalidState
A bad result from a subsequent query using a required interface or another fault in the user provided functional code can lead to an invalid state of missing state update.
ValidityFailure_USEPASTDATA
UserCode_REFUSE
MissingStateUpdate
PreCondition_IGNORE
UserCode_REFUSE
Deadline
InvalidState
Deadline_STOP
Deadline_STOP
MissingStateUpdate
ValidityFailure_USEPASTDATA
A deadline violation can either be a root failure mode. Or it can happen because of previous occurring faults. However, a deadline violation can cause an invalid state, missing state update or late state update depending upon if the local health decision is to restart or stop the process due to deadline violation.
PreCondition_IGNORE
Deadline_RESTART_IGNORE
UserCode_IGNORE
Deadline_RESTART_IGNORE
LateStateUpdate
Late_Result_From_Requires

1. LOCK_TIMEOUT—This is caused by problems in obtaining the component lock. Being a real-time system, any attempt to obtain a lock is bounded by a maximum deadline. In case of timeout, the fault is observed as a discrepancy with an anomaly of LOCK_TIMEOUT, resulting in a CLHM response of either IGNORE or REFUSE/ABORT. Based on the response, its failure effect can lead to an invalid or a missing state update and affect entities downstream.
2. VALIDITY_FAILURE—This is caused when the "age" of the data fed to the consumer is not valid. It is observed as a discrepancy with an anomaly of VALIDITY_FAILURE, resulting in a CLHM response of either USE_PAST_DATA or REFUSE/ABORT. Based on the response, its failure effect can lead to an invalid or a missing state update and affect entities downstream. The failure effect could also cascade into one or more of the anomalies described below.
3. PRE-CONDITION_FAILURE—This is caused when the precondition to the consumer process is not satisfied. This anomaly could also be observed as a result of a VALIDITY_FAILURE followed by a response to USE_PAST_DATA. It is observed as a discrepancy with an anomaly of PRE-CONDITION_FAILURE, resulting in a CLHM response of either IGNORE or REFUSE/ABORT. Based on the response, its failure effect can lead to an invalid or a missing state update and affect entities downstream. The failure effect could also cascade into user-code and/or deadline violation.
4. USER-CODE_FAILURE—This is caused when there is a failure in the user-code (e.g., exception). This anomaly could be observed as a result of a USER-CODE failure (captured as a failure mode) or a cascading failure effect propagation from VALIDITY_FAILURE and subsequent health management response of USE_PAST_DATA or PRE-CONDITION violation followed by an IGNORE response. Once observed, depending on the local CLHM response its failure effect can lead to an invalid or a missing state update and affect entities downstream. The failure effect could also cascade into a deadline violation.
5. DEADLINE_FAILURE—This is caused when the process deadline is violated. This anomaly could be observed as a result

of a deadline failure (captured as a failure mode) or a cascading failure effect propagation from VALIDITY_FAILURE followed with a local health management response of USE_PAST_DATA or PRE-CONDITION violation with a local CLHM response of IGNORE or USER_CODE. It is observed as a discrepancy with an anomaly of DEADLINE_FAILURE, resulting in a CLHM response of either STOP (stopping the process in case of hard-deadline violation) or IGNORE (in case of soft-deadline violation) or RESTART. Based on the response, its failure effect can lead to an invalid/missing/late state update and affect entities downstream.

It should be noted that sometimes it might not be possible to monitor some of the failures/alarms mentioned above. In such cases, these observed discrepancies would turn into unobserved discrepancies and the fault effect would propagate through the discrepancy without raising any observation (alarm).

Example 7: TFPG for the GPS Assembly

Figure 9.10 shows the failure propagation link created for interaction between a publisher and a periodic consumer. The publisher and consumer boxes encapsulate the detailed TFPG model of the publisher and the consumer entities. The failure propagation in-to or out-of the ports captures the failure effect propagation across the entity boundary. Any failure in the Publisher entity that could lead to a discrepancy of NoDataPublished/LateDataPublished could possibly cascade to the consumer entity through a VALIDITY_FAILURE in its input data. Likewise, a failure in the publisher leading to

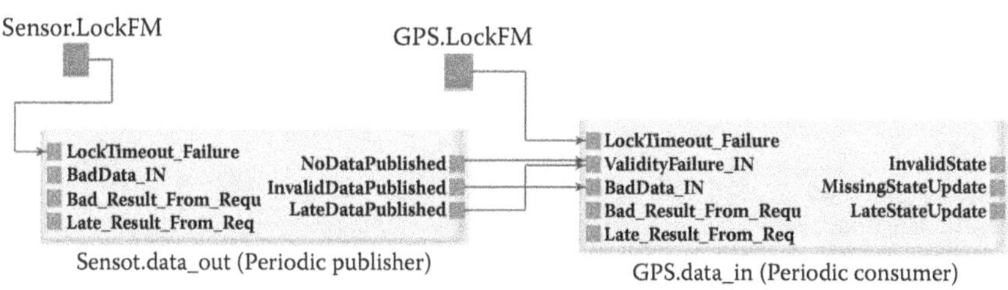

FIGURE 9.10 TFPG model for publisher–consumer interaction.

InvalidDataPublished discrepancy could produce anomalies in the consumer through the triggering of Bad-Data-Input discrepancy. The model also captures the possibility that a Failure-Mode of a problem in the Component lock could lead to discrepancies associated with Lock_Timeout in the Component's processes/interaction ports.

Once we can create the models to capture the failure propagation across all interactions, we can essentially create TFPG model for the full assembly. In practice, the assembly level TFPG model is generated by instantiating the appropriate TFPG model for the Component interaction ports and connecting the failure propagation links between the discrepancy ports. The data and control flow within and across the component can be used to generate the failure propagation links across the instantiated TFPG models. This information can either be obtained by static analysis of the user-level code for the component or relying on the designer to provide this information. Currently, we take the latter approach.

9.4.3.1 Complexity of the Generated Model

Table 9.4 describes the formulae for calculating the metrics for the size and complexity of failure propagation graphs generated from a given software assembly. These metrics are based on the templates of generic fault propagation across all possible intercomponent and intracomponent interactions in the ACM framework.

Figure 9.11 shows the high-level TFPG model for the system/assembly described in Figure 9.6. In this assembly, there were 3 components, 6 methods, 2 consumer ports, 1 requires port, and 2 post conditions. Hence, the number of failure modes totaled 20 and the number of alarms also totaled 20. The detailed TFPG model specific to each interaction pattern is contained inside the respective TFPG component model (brown box). Figure 9.11 shows failure propagation between the sensor publisher* and GPS consumer,† the GPS publisher‡ and NavDisplay consumer,§ the "requires" method in NavDisplay¶ and the "provides" method in GPS,** the effect of the bad updates on state variables and the entities updating or reading the

* (Sensor_data_out).
† (GPS_data_in).
‡ (GPS_data_out).
§ (NavDisplay_data_in).
¶ (NavDisplay_gps_data_src_getGPSData).
** (GPS_gps_data_src_getGPSData).

TABLE 9.4 Metrics for Automatically Generated TFPG from ACM Component Assembly

$\#FM = \#Comp + \#Interactions * 3 - \#ReqPorts$	(1.1)
$\#Alarm = \#Interactions * 2 + \#Cons * 2 + \#Post * 2 + \#Pre * 2$	(1.2)

Note: FM, failure modes; Comp, components; Interactions, all intracomponent component and intercomponent interactions; ReqPorts, required interface ports; Alarms, all observable TFPG discrepancies; Cons, consumer ports (validity monitors); Pre, preconditions; Post, postconditions.

state variables. Table 9.1 summarizes the worst case memory consumed by the reasoner and average time taken to update the hypothesis.

It should be noted that while the interaction pattern between a publisher port and a consumer port produces a fault propagation in the direction of data and event flow, i.e., from the publisher to the consumer, the interaction pattern between a "requires" interface and its corresponding "provides" interface involves fault propagation in both directions. The fault propagation within a component is captured through the propagations across the bad updates on the state variables within the component. Currently, in the framework, there is no instrumentation to deploy a monitor to specifically observe violations in state variable updates. These could be captured indirectly as precondition or postcondition monitors on the interfaces/interactions ports that update or read from these state variables.

9.4.4 The Diagnosis Process

The TFPG diagnosis engine is hosted inside a separate component. An alarm aggregator component is responsible for aggregating all the alarms and passing it to the diagnosis engine. These two components are hosted in a separated ARINC-653 partition. When the diagnosis engine receives the first alarm from a fault scenario, it generates all hypotheses that could have possibly triggered the alarm. Each hypothesis lists its possible failure modes and their possible timing interval, the triggered alarms that are supportive of the hypothesis, the triggered alarms that are inconsistent with the hypothesis, the missing alarms that should have triggered, and the alarms that are expected to trigger in future. Additionally, the reasoner computes hypothesis metrics such as plausibility and robustness, which provide a means of comparison. The higher the metrics, the more reasonable it is to expect the hypotheses to be the real cause of the problem. As more alarms are produced, the hypotheses are further refined. If the new alarms are supportive of existing sets of hypotheses, they are updated to reflect the refinement in their metrics and alarm list. If the

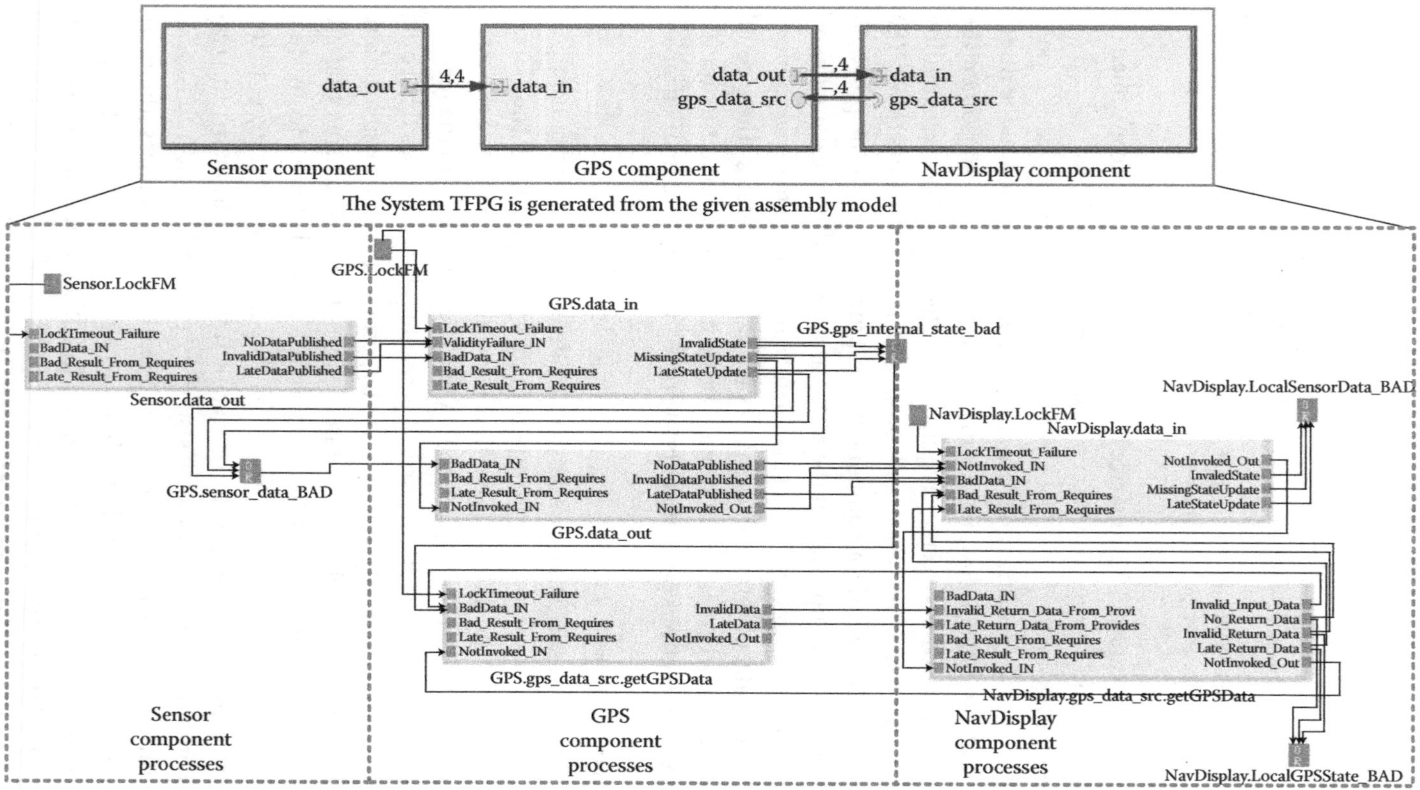

FIGURE 9.11 TFPG model for the component assembly.

new alarms are not supportive of any of the existing hypotheses with the highest plausibility, then the reasoner refines these hypotheses such that hypotheses can explain these alarms.

Figure 9.12 shows the diagnosis results for a specific scenario wherein the Sensor (Figure 9.6) stops publishing data. This results in a failure effect that cascades across component boundaries. The initial alarm is generated because of data validity violations in the consumer of the GPS component. When this alarm was reported to the local component health manager, it issued a response directing the GPS component to use past data (USE_PAST_DATA). While the issue was resolved local to the GPS component, the combined effect of the failure and mitigation action propagated to the navigation display component. In the navigation display component, a monitor observing the postcondition violation on a required interface was triggered because the GPS data validated its constraints. These two alarms were sent to the system-level health manager and processed by the TFPG diagnoser.

As can be seen from the results, the system correctly generated two hypotheses. The first hypothesis blamed the sensor component lock to be the root problem. The second hypothesis blamed the user level code in the sensor publisher process to be the root failure mode. In this situation the second hypothesis was the true cause. However, because we currently treat the lock time out monitors as unmonitored discrepancies the diagnoser was not able to reasonably disambiguate between the two possibilities.

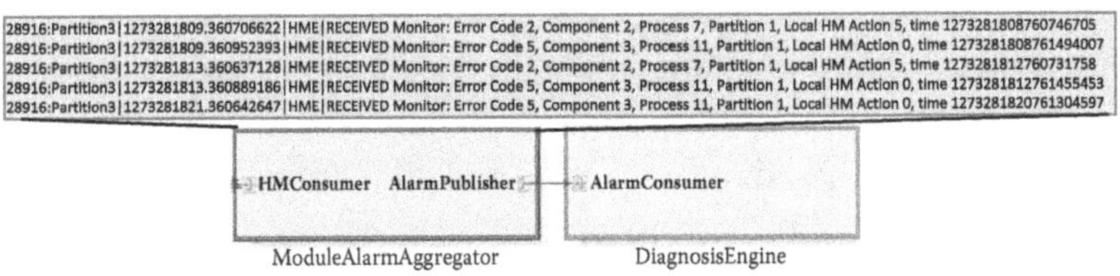

====[Hypothesis Group 1]=====
Fault: FM_Sensor_data_out_USER_CODE time: [0.000000,24.341104]
Supporting Alarms : VALIDITY_FAILURE (GPS Dataln),
POSTCONDITION_FAILURE (NavDisplay GetGPSData)
Plausibility: 100.000000 **Robustness**: 100.000000 **FRMetric**: 0

====[Hypothesis Group 2]=======
Fault: Sensor_LOCK_PROBLEM time: [0.000000, 24.341104]
Supporting Alarms : VALIDITY_FAILURE (GPS Dataln),
POSTCONDITION_FAILURE (NavDisplay GetGPSData)
Plausibility: 100.000000 **Robustness**: 100.000000 **FRMetric**: 0

FIGURE 9.12 Diagnosis results from the TFPG reasoner.

9.5 APPLICATION OF TFPG FOR PROGNOSTICS OF IMPENDING FAULTS

In general, the aim of failure prognosis is to estimate the system reliability, given a set of conditions and observations, by assessing how close the system is to a critical manifestation of current failures. The reliability estimation can then be used to reconfigure the system, change the operating settings, or schedule specific maintenance procedures targeting the faulty components. In the TFPG modeling and reasoning settings, the prognosis problem and the associated reliability measure can be defined based on three main factors: failure criticality levels, diagnosis or hypothesis plausibility, and the time distance from the current state to the critical failure.

The first factor addresses the common situation in which different sections of the system may correspond to different levels of criticality with respect to system functionality. These sections can be identified using a measure of criticality assigned to all discrepancies in the systems. The second factor addresses the current estimated (diagnosed) condition of the system and the plausibility of the corresponding hypothesis. The third factor addresses the timing proximity of the current estimated state relative to a given critical region of the system. All these factors directly contribute to the reliability of the system at a given time. We will discuss them in detail in the rest of this section.

9.5.1 Failure Criticality

In typical practical situations, failure progresses starting from the initial failure modes into several stages with increasing level of criticality. To capture this situation, we define the map $\mathrm{CL} : D \rightarrow \mathbb{N}$ that assign to each discrepancy, $d \in D$, a criticality level $\mathrm{CL}(d)$. The lowest criticality level, 0, is reserved for noncritical discrepancies and all failure modes. To capture the increasing criticality with respect to propagation depth, we assume that

$$(\forall d',d \in D) \quad (d',d) \in E \rightarrow \mathrm{CL}(d') \leq \mathrm{CL}(d)$$

The above condition states that if d' is a parent of d in a TFPG model, then the criticality level of d' should be less than or equal to that of d. As a consequence, the criticality levels along any given path in a TFPG model form a monotonically increasing sequence. Note that we only assign a criticality level to all monitored and nonmonitored discrepancies D. Failure modes are assigned a criticality level of 0 by default.

Based on the definition of criticality levels, we can define *criticality fronts* associated with a given TFPG model by the map, $\mathtt{CF} : \mathbb{N} \rightarrow \mathcal{P}(D)$, where,

$$(\forall d \in D) \quad d \in \mathtt{CF}(n) \leftrightarrow \mathtt{CL}(d) \geq n \text{ and } (\forall d',d) \in E)\ \mathtt{CL}(d') \leq n.$$

The set of criticality fronts are essentially the codomain of the above map, and the set of criticality front levels CFL are the set $\{n \in \mathbb{N}\ |\mathtt{CF}(n) \neq \emptyset\}$. It can be shown that CFL corresponds bijectively to the codomain of CL. Based on the above definitions, a criticality front level, $n \in \mathbb{N}$, corresponds to a graph cut of the TFPG model in which the nodes on one side of the cut have criticality levels less than n and the remaining nodes have criticality levels greater than or equal to n. Figure 9.13 shows an example TFPG model with assigned criticality levels and the corresponding criticality fronts.

Criticality levels are typically assessed based on the requirements for system operation and functionality. In particular, the criticality value for a given discrepancy depends on the operation cost associated with the fault reaching and progressing from this discrepancy. This will include the cost of maintenance, reconfiguration, and recovery when applicable. However, in some situations, it is not possible to have a precise value for the criticality of a sensor. In such situations, an enumeration of criticality levels (e.g., low, medium, high) can be used to distinguish between sensors with

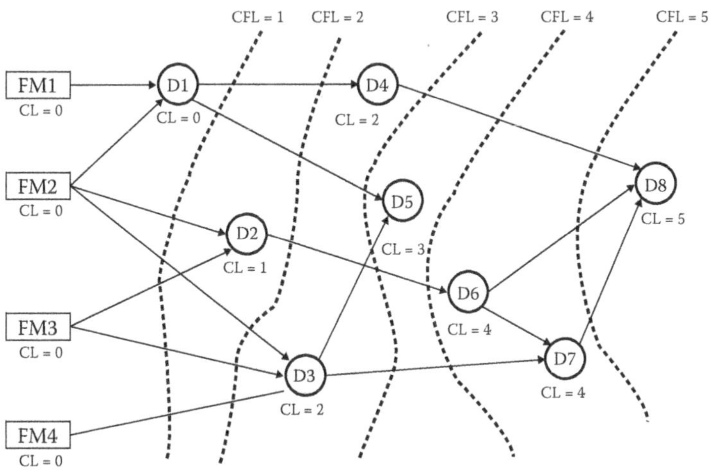

FIGURE 9.13 TFPG model with assigned criticality levels and the corresponding criticality fronts.

respect to fault severity. Such enumeration can be assigned an approximate integer value, that reflects its relative importance, which can be used later in this section to compute a reliability measure for the system, in terms of the remaining useful life (RUL) or the time to criticality.

9.5.2 State Estimation Plausibility

As discussed in the previous section, the TFPG reasoning algorithms relies on sensor signals (alarms) and the TFPG model structure to identify the most plausible estimates of the current system condition as a set of state hypotheses. The plausibility of each hypothesis is defined based on the number of supporting sensor signals (alarms) versus the inconsistent and missing ones. We will write $A(H)$ for the set of discrepancies that are presumed active (ON) according to H and $I(H)$ for the set of discrepancies that are presumed inactive (OFF) according to H. That is,

$$A(H) = \{d \in D | H(d).\text{state} = \texttt{OFF}$$

The state front of a hypothesis H is denoted $\texttt{SF}_H$ and is defined as a set of discrepancies $D' \subseteq D$ such that $(\forall d \in \texttt{SF}_H)$

$$d \in A(H) \text{ and } (\exists(d,d') \in E) d' \in I(H).$$

Accordingly, the state front $\texttt{SF}_H$ is the set of discrepancies that are currently active as estimated by H, but some of their children discrepancies are not active according to H. Given that any discrepancy in D can either be in $A(H)$ or $I(A)$ but not both. The set $\texttt{SF}_H$ is well-defined and the boundary line between D' and $D - D'$ forms a graph cut for the underlying TFPG model.

The intuitive meaning of the state front for a hypothesis is that all the discrepancies on this front have the same likelihood of being active at the current time and they are forming the front of fault propagation in the sense that they are the discrepancies that could become active based on the next set of alarms as the fault propagation continues to progress.

The plausibility of a state front is equal to the plausibility of the underlying hypothesis. It is possible that several hypotheses can have the same plausibility level and therefore several state fronts may have the same plausibility level.

9.5.3 Time Proximity

The time proximity factor measures how close the current state of the system is to a future failure. As discussed earlier, future failures are identified by their criticality level front as defined by the map CF. Each front is defined as a set of discrepancies at the boundary of a graph cut for the TFPG model. Similarly, a state estimation is defined by a hypothesis (with its plausibility level) and is identified by the discrepancies at the boundaries of the cut formed by the underlying hypothesis level state front. Accordingly, the time proximity factor is a measure for the temporal distance between two fronts (graph cuts) each corresponding to a set of discrepancies in the TFPG model.

To define such distance, consider two sets of discrepancies D_1, $D_2 \subseteq D$ such that all discrepancies in D_1 are either ancestors of some discrepancies in D_2 or not connected to any discrepancy in D_2. In this case, we write $D_1 \prec D_2$. We define the propagation time between D_1 and D_2 with respect to a hypothesis H, denoted $t_H(D_1,D_2)$, as the minimum time to trigger discrepancy in D_2 as a result of failure propagation from discrepancies in D_1.

To compute $\hat{t}_H(D_1,D_2)$, we first consider the set of all discrepancies that are children of D_1. We then compute the earlier propagation time to these discrepancies based on the activation times of their parent nodes according to H. The computation of the earliest propagation time for all subsequent nodes is done based on the earliest propagation times available for their parents. This incremental computation continues until the earliest propagation time is computed for all the nodes in D_2. The minimum time is selected as the output. Algorithm 2 outlines the computation procedure.

Algorithm 2. The Propagation Time Procedure $\hat{t}_H(D_1,D_2)$

assumption: $D_1 \prec D_2$
if $D_1 \cap D_2 \neq \emptyset$ **then**
  return 0
end if
define `RSet`$(X) := \{d \in D - X | \forall d' \in D)\ (d',d) \in E \rightarrow d' \in X\}$
`TNodes` $= \{(d,H(d).\text{terl}) | d \in A(H)\}$
$t_{\min} = \infty$
while $D_2 \not\subset$ `TNodes`.nodes **do**
  Select d from `RSet`(D_1)
  compute terl(`TNodes`,d)
  if $d \in D_2$ **then**
    $t_{\min} = \min(t_{\min}, \text{terl}(d))$
  end if
end while
return $t_{\min}$

In Algorithm 2, for a given subset of nodes X in the TFPG model, RSet(X) is the set of discrepancies outside of X where all their parents belong to X. The function (TNodes,d) computes the earliest time for d to become activated based on the earliest time the parents of d are activated. This function can be directly computed based on the semantics of failure propagation in TFPG models.

9.5.4 Time to Criticality

Given a set of criticality levels, the associated criticality fronts are computed directly from the earlier definition. The state front for a given hypothesis can be directly computed based on the given definition. Let Y be the set of hypothesis with the highest plausibility value at a time t. We define the time to criticality level n at a given time t, denoted TTC(Y,n), as follows

$$\mathtt{TTC}(Y,n) = \min\{\hat{t}_H(\mathtt{SF}_H, \mathtt{CF}(n)) \mid H \in Y\},$$

that is, the time to criticality level n is the minimum of all (earliest) propagation times for all hypotheses with the maximum plausibility. In practice, there are typically few enumerated criticality levels. The time to criticality, therefore, follows the increasing order of the criticality. That is, the time to reach a high criticality level is usually longer than the time to reach a lower criticality level, as expected. However, this is not always the case as shown in Figure 9.14. In this example, there are three different paths from the state (estimation) front $\mathtt{SF}_H$ to the criticality front level 1

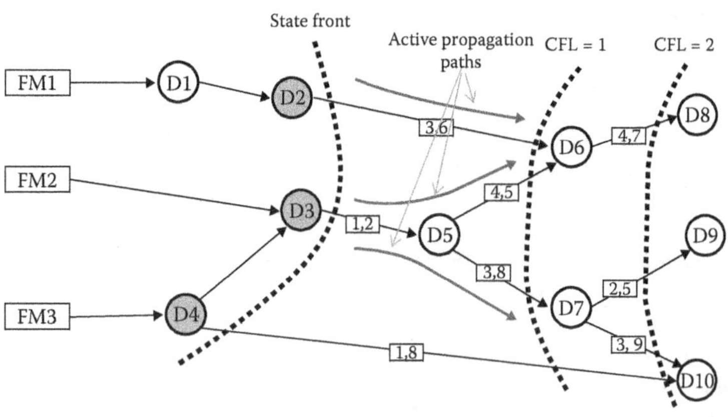

FIGURE 9.14 Example for time to criticality.

(CFL = 1), where H is the most plausible hypothesis in which D2, D3, D4 are assumed active and D1 is assumed faulty. In this example, the time to criticality to the first level is 3 (time units) while the time to criticality for the next higher level is 1.

9.6 RELATION TO MACHINE LEARNING AND DATA MINING

The TFPG model for a specific physical system, as described above, is the result of an engineering process that is based on the engineer's knowledge of the system, first principles, and system connectivity. For large systems this model construction can be quite complex and will rarely be perfect on the first try. Hence, a model maturation process is needed where the TFPG model is refined over time, based on operational experience. Arguably, machine learning and/or data mining techniques can be used to assist in this maturation, as discussed below.

Imperfections in TFPG models come in two main forms: (1) a failure propagation edge is missing and (2) a failure propagation edge is present where it should not exist. Assume that we operate the reasoner using such a degraded model. If the input and the output of the reasoner are logged, one can apply data mining techniques to the input, with the goal of discovering the weak points in the model. For validation, we will also need the result of the maintenance activity that repaired the system, i.e., the knowledge of the real, physical failure mode.

If a failure propagation edge is missing from the model, then the reasoner will likely produce degraded results and mark the downstream alarms as false alarms, indicating that they are not on a valid propagation path. However, if the data mining algorithm tells that there is a strong correlation between the two alarms and such correlation can be supported because of a physical connection in the system, then it is likely that an edge is missing and adding it would improve the performance of the reasoner.

If a failure propagation edge is present but it should not be, then the reasoner will again produce degraded results because it will misidentify failure modes as fault sources. Again, if the data mining indicates that there is very little correlation between the two alarms, and this can be corroborated by the lack (or the weakness) of physical connections between the components, then it is likely that the edge is superfluous and its removal would improve the diagnostic reasoning.

In either case, the change on the model shall be validated with the real physical failure mode. This can be done by supplying the same input sequence to the reasoner but now with the changed model, and observing if it is able to correctly produce the real failure mode. Once this validation is performed, the model can be fielded and used.

9.7 SUMMARY

We described a discrete event-based, graph-oriented approach to fault source isolation that is applicable in the context of system-level health management. We presented the modeling approach, the algorithms for the centralized reasoner, the algorithms for a distributed, master-slaves reasoner architecture. We have also illustrated how the same approach can be used for software health management and potentially integrating reasoners about the physical and the software components of a system. The modeling approach and the reasoner have been applied in various example systems and were found to provide satisfactory performance. The software health management application has been recently developed and tested only on small examples. The full, system-level integration and application of the reasoner approach are the subject of active research and development.

ACKNOWLEDGMENTS

This chapter is based upon work supported in part by NASA, the Boeing Company, and the DARPA Software Enabled Control program. Any opinions, findings, and conclusions or recommendations expressed in this material are those of the author(s) and do not necessarily reflect the views of the National Aeronautics and Space Administration, the Boeing Company, or DARPA. The authors would like to thank Paul Miner, Eric Cooper, and Suzette Person of NASA Langley Research Center and Stan Ofsthun, Tim Wilmering, Erik Fries, and John Wilson of the Boeing Company for their help and guidance.

REFERENCES

[1] D.M. Himmelblau. Fault detection and diagnosis in chemical and petrochemical processes. *Chemical Eng. Mon.*, 8, 1978.

[2] N. Viswanadham, T.L. Johnson. Fault detection and diagnosis of automated manufacturing systems. In *27th IEEE Conference on Decision and Control*, 1988.

[3] R. Hessian, B. Salter, E. Goodwin. Fault-tree analysis for system design, development, modification, and verification. *IEEE Transactions on Reliability*, 39(1):87–91, 1990.

[4] Y. Ishida, N. Adachi, H. Tokumaru. Topological approach to failure diagnosis of large-scale systems. *IEEE Trans. Syst., Man and Cybernetics*, 15(3):327–333, 1985.

[5] S.V. Nageswara Rao, N. Viswanadham. Fault diagnosis in dynamical systems: A graph theoretic approach. *Int. J. Systems Sci.*, 18(4):687–695, 1987.

[6] S.V. Nageswara Rao, N. Viswanadham. A methodology for knowledge acquisition and reasoning in failure analysis of systems. *IEEE Trans. Syst., Man and Cybernetics*, 17(2):274–288, 1987.

[7] J. Richman, K.R. Bowden. The modern fault dictionary. In *International Test Conference*, pp. 696–702, 1985.

[8] W.T. Scherer, C.C. White. A survey of expert systems for equipment maintenance and diagnostics. In S.G. Tzafestas, editor, *Knowledge-based system diagnosis, supervision and control*, pp. 285–300. Plenum, New York, 1989.

[9] S. Tzafestas, K. Watanabe. Modern approaches to system/sensor fault detection and diagnosis. *J. A. IRCU Lab.*, 31(4):42–57, 1990.

[10] P. Frank. Fault diagnosis in dynamic systems using analytical and knowledge-based redundancy—a survey and some new results. *Automatica*, 26(3):459–474, 1990.

[11] W. Hamscher, L. Console, J. de Kleer. *Readings in Model-Based Diagnosis*. Morgan Kaufmann Publishers Inc., San Francisco, CA, 1992.

[12] R. Patton. Robust model-based fault diagnosis: the state of the art. In *IFAC Fault Detection, Supervision and Safety for Technical Processes*, pp. 1–24, Espoo, Finland, 1994.

[13] R. Patton, P. Frank, R. Clark. *Fault Diagnosis in Dynamic Systems: Theory and Application*. Prentice Hall International, UK, 1989.

[14] M. Sampath, R. Sengupta, S. Lafortune, K. Sinnamohideen, D. Teneketzis. Failure diagnosis using discrete event models. *IEEE Trans. Contr. Syst. Technol.*, 4:105–124, 1996.

[15] A.N. Srivastava. Discovering system health anomalies using data mining techniques. In *Proceedings of the Joint Army Navy NASA Air Force Conference on Propulsion*, 2005.

[16] R. Reiter. A theory of diagnosis from first principles. *Artificial Intelligence*, 32(1):57–95, 1987.

[17] J. de Kleer, A. Mackworth, R. Reiter. Characterizing diagnoses and systems. *Artificial Intelligence*, 56, 1992.

[18] A. Darwiche. Model-based diagnosis using structured system descriptions. *Journal of Artificial Intelligence Research*, 8:165–222, 1998.

[19] A. Darwiche, G. Provan. Exploiting system structure in model-based diagnosis of discrete-event systems. In *Proc. of the Seventh International Workshop on Principles of Diagnosis*, pp. 95–105, 1996.

[20] J. Gamper. *A Temporal Reasoning and Abstraction Framework for Model-Based Diagnosis Systems*. PhD thesis, RWTH, Aachen, Germany, 1996.

[21] L. Console, P. Torasso. On the co-operation between abductive and temporal reasoning in medical diagnosis. *Artificial Intelligence in Medicine*, 3(6):291–311, 1991.

[22] A. Misra. *Sensor-Based Diagnosis of Dynamical Systems*. PhD thesis, Vanderbilt University, 1994.

[23] A. Misra, J. Sztipanovits, J. Carnes. Robust diagnostics: Structural redundancy approach. In *SPIE's Symposium on Intelligent Systems*, 1994.

[24] S. Padalkar, J. Sztipanovits, G. Karsai, N. Miyasaka, K.C. Okuda. Real-time fault diagnostics. *IEEE Expert*, 6(3):75–85, 1991.

[25] G. Karsai, J. Sztipanovits, S. Padalkar, C. Biegl. Model based intelligent process control for cogenerator plants. *Journal of Parallel and Distributed Systems*, 15:90–103, 1992.

[26] P.J. Mosterman, G. Biswas. Diagnosis of continuous valued systems in transient operating regions. *IEEE Trans. on Systems, Man and Cybernetics*, 29(6):554–565, 1999.

[27] G. Karsai, G. Biswas, S. Abdelwahed. Towards fault-adaptive control of complex dynamic systems. In T. Samad, G. Balas, editors, *Software-Enabled Control: Information Technology for Dynamical Systems*, chapter 17. IEEE publication, 2003.

[28] S. Abdelwahed, G. Karsai, G. Biswas. System diagnosis using hybrid failure propagation graphs. In *The 15th International Workshop on Principles of Diagnosis*, Carcassonne, France, 2004.

[29] V. Brusoni, L. Console, P. Terenziani, D.T. Dupre. A spectrum of definitions for temporal model-based diagnosis. *Artificial Intelligence*, 102(1):39–79, 1998.

[30] S. Neema, T. Bapty, S. Shetty, S. Nordstrom. Developing autonomic fault mitigation systems. *Journal of Engineering Applications of Artificial Intelligence Special Issue on Autonomic Computing and Grids*, 2004.

[31] A. Dubey, S. Nordstrom, T. Keskinpala, S. Neema, T. Bapty, G. Karsai. Towards a verifiable real-time, autonomic, fault mitigation framework for large scale real-time systems. *Innovations in Systems and Software Engineering*, 3:33–52, March 2007.

[32] T. Saxena, A. Dubey, D. Balasubramanian, and G. Karsai. Enabling self-management by using model-based design space exploration. *Engineering of Autonomic and Autonomous Systems, IEEE International Workshop on*, 0:137–144, 2010.

[33] S. Ahuja, et al. RTES demo system 2004. *SIGBED Rev.*, 2(3):1–6, 2005.

[34] L. Piccoli, A. Dubey, J.N. Simone, J.B. Kowalkowlski. LQCD workflow execution framework: Models, provenance and fault-tolerance. *Journal of Physics: Conference Series*, 219(7):072047, 2010.

[35] P. Conmy, J. McDermid, M. Nicholson. Safety analysis and certification of open distributed systems. In *International System Safety Conference*, Denver, 2002.

[36] Mark Nicholson. Health monitoring for reconfigurable integrated control systems. *Constituents of Modern System Safety Thinking. Proceedings of the Thirteenth Safety-Critical Systems Symposium*, 5:149–162, 2007.

[37] A. Goldberg, G. Horvath. Software fault protection with ARINC 653. In *Proc. IEEE Aerospace Conference*, pp. 1–11, March 2007.

[38] U. Sammapun, I. Lee, O. Sokolsky. Rt-MaC: runtime monitoring and checking of quantitative and probabilistic properties. In *Proc. 11th IEEE International Conference on Embedded and Real-Time Computing Systems and Applications*, pp. 147–153, 17–19, August 2005.

[39] S. Abdelwahed, G. Karsai, G. Biswas. A consistency-based robust diagnosis approach for temporal causal systems. In *The 16th International Workshop on Principles of Diagnosis*, Pacific Grove, CA, 2005.

[40] N. Mahadevan, S. Abdelwahed, A. Dubey, G. Karsai. Distributed diagnosis of complex causal systems using timed failure propagation graph models. In *IEEE Systems Readiness Technology Conference, AUTOTESTCON*, 2010.

[41] D.R. Morgan. Pave pace: system avionics for the 21st century. *Aerospace and Electronic Systems Magazine, IEEE*, 4(1):12–22, January 1989.

[42] Joint advanced strike technology program, avionics architecture definition appendix B. Technical report, August 1994.

[43] Ariane 5 inquiry board report. Technical report, Available at http://esamultimedia.esa.int/docs/esa-x-1819eng.pdf, June 1996.

[44] NASA. Report on the loss of the Mars Polar Lander and Deep Space 2 missions. Technical report, NASA, 2000.

[45] Australian Transport Safety Bureau. In-flight upset; 240km nw Perth, WA; Boeing co 777-200, 9m-mrg. Technical report, August 2005.

[46] Australian Transport Safety Bureau. Ao-2008-070: In-flight upset, 154 km west of Learmonth, WA, 7 october 2008, vh-qpa, airbus a330-303. Technical report, October 2008.

[47] W.S. Greenwell, J. Knight, J.C. Knight. What should aviation safety incidents teach us? In *SAFECOMP 2003, The 22nd International Conference on Computer Safety, Reliability and Security*, 2003.

[48] S. Ofsthun. Integrated vehicle health management for aerospace platforms. *Instrumentation Measurement Magazine, IEEE*, 5(3):21–24, September 2002.

[49] S. Johnson, editor. *System Health Management: With Aerospace Applications*. John Wiley & Sons, based on papers from First International Forum on Integrated System Health Engineering and Management in Aerospace, 2005. To Appear in 2011.

[50] Rogrio de Lemos. Analysing failure behaviours in component interaction. *Journal of Systems and Software*, 71(1–2):97–115, 2004.

[51] M. Wallace. Modular architectural representation and analysis of fault propagation and transformation. *Electron. Notes Theor. Comput. Sci.*, 141(3):53–71, 2005.

[52] A. Dubey, G. Karsai, R. Kereskenyi, N. Mahadevan. A real-time component framework: Experience with CCM and ARINC-653. *Object-Oriented Real-Time Distributed Computing, IEEE International Symposium on*, pp. 143–150, 2010.

[53] N. Wang, D.C. Schmidt, C. O'Ryan. Overview of the corba component model. *Component-based software engineering: putting the pieces together*, pp. 557–571, 2001.

[54] ARINC Specification 653-2: Avionics application software standard interface part 1—required services. Technical report.

III

Applications

CHAPTER 10

Real-Time Identification of Performance Problems in Large Distributed Systems

Moises Goldszmidt

Dawn Woodard

Peter Bodik

CONTENTS

10.1 INTRODUCTION

The large networked distributed systems that support Internet services, cloud computing services, etc. operate in datacenters hosting on the order of ten to hundreds of thousands of machines. These are the large-scale systems behind email, search, news, e-commerce, storage, computational services, and other services. These systems are subject to very stringent requirements of availability and responsiveness since competition is harsh and expectations on the side of the users are high. In these large systems, both availability and responsiveness goals are inevitably compromised by hardware and software problems [1]. During a performance *crisis*,* the operators' highest priority is to stabilize the system as soon as possible and avoid crisis escalation. They typically do this by inspecting collected system signals (telemetry), logs, and alarms and then matching what they see with previous cases to apply best practices and, in case of no matches, escalating the problem to a higher tier. The time it takes to match the crises, the accuracy of the match, and whether the issue can be solved by the operators or needs to be escalated have a direct and very significant economic impact on the cost of the datacenter operations.

We call the problem of matching a crisis to either a past instance of an already seen crisis or to a new one, the "crisis identification" problem. In this chapter, we describe an approach to automate this identification problem. The objective is to increase accuracy so that the proper mitigation procedures (i.e., repair scripts/actions) are applied, to decrease the unnecessary escalation of problems, and to decrease the time to identification from the start of the crisis.

The benefits of this technique are directly proportional to the frequency with which these crises recur. There are various reasons why these performance problems recur often enough in the context of very large distributed systems such as those operating in a datacenter or in cloud computing environments. We list a few: first, the bug fix for the underlying problem

* We will see in Section 10.2 that the operators have a formal and detectable definition of what constitutes a crisis, in terms of the percentage of machines that are not performing according to measurable performance objectives.

may have not yet been deployed (and may not be scheduled to be deployed soon enough) and also the fix could be based on a misunderstanding of the root cause [2,3]. Second, the problem may involve dependencies on other services (authentication, or data analysis) that are located in other datacenters and/or are part of a federated ecosystem, thus beyond the control of the datacenter in question. Third, these issues are amplified by emergent misbehaviors due to large scale and high utilization common in these environments.*

The approach we describe in this chapter can be summarized as follows: First, we automatically extract a *fingerprint* of each crisis. As will be shown in this chapter (Section 10.3), a fingerprint is an efficient and effective representation of a crisis that enables crisis matching. Second, we apply a clustering algorithm to the fingerprints that enables us to both group the crises and to consider in real-time the partial information from the occurring crisis to enable accurate identification. There is already a considerable literature on related work: As early as 2003, the authors of [4] proposed the use of compute-intensive modeling techniques to perform such automatic recognition. Since then, researchers have tried to identify operational problems by analyzing performance signals using machine learning [5–11], by identifying unusual or noteworthy sequences of events that might be indicators of unexpected behavior [12,13], and by manually instrumenting the system [14] and creating libraries of possible faults and their consequences [15]. Others have laid out general methodological challenges in using computers to diagnose computer problems [16,17]. The work closest in spirit to the one in this chapter is the "signatures" approach to identifying and retrieving the essential system state corresponding to previously seen crises [6]. Unfortunately, as set up in that paper, the "signatures" approach would be unlikely to scale effectively to a very large distributed system. Furthermore, as we showed empirically in [18], even if it is made to scale, its accuracy in terms of recognizing a crisis is significantly inferior to the approach we report here.

The three main contributions in this chapter are

- An efficient and effective representation of the state of a datacenter during a crisis in terms of the signals being monitored, which we call fingerprints. This representation in terms of a time series, scales to the requirements of large distributed systems.

* Jeff Dean, Google Fellow, keynote at LADIS 2009 workshop.

- A fully Bayesian estimation algorithm for clustering these time series and the related approximations to enable real time performance.
- The application and evaluation of the approach on real data from a small to mid-size commercial datacenter running a 24 × 7 enterprise-class application.

The rest of this chapter is organized as follows: Section 10.2 establishes some context and defines the problem. Section 10.3 describes the algorithms and procedures to automatically transform the raw signals collected from the machines in the datacenter to fingerprints. Section 10.4 is concerned with the algorithms and procedures necessary for the identification of a crisis. Section 10.5 reports on experiments and results on real data, and Section 10.6 provides some final remarks.

10.2 PROBLEM DEFINITION

A typical datacenter-scale user-facing application runs simultaneously on hundreds or thousands of machines. To detect performance problems and perform postmortem analysis after such problems, several *signals* are usually collected on each machine and logged to online storage. Since large collections of servers execute the same code, under normal load balancing conditions the values of these signals should come from the same distribution; as we will show, we use this intuition to capture the state of each signal and identify unusual behavior.

Each signal is usually measured once per aggregation *epoch*—typically a few minutes—and the measured values may represent a simple aggregate over the aggregation epoch, e.g., the mean. The signals correspond to hardware, OS, application, or runtime-level measurements, such as the size of the object heap or number of threads waiting in the run queue. Wide variation exists in what is collected and at what granularity with several commercial packages providing off-the-shelf starting points.

A small subset of the collected signals may be used as *key performance indicators* (KPIs) whose values form part of the definition of a contractual service-level objective (SLO) for the application. An SLO typically specifies a threshold value for each KPI and the minimum fraction of machines that have to satisfy the requirement over a particular time interval. For example, an SLO might require the end-to-end interactive response time to be below a certain threshold value for 99.9% of all requests in any 15-min interval.

A performance *crisis* is defined in our specific domain of application as a prolonged violation of one or more specified SLOs by a certain percentage of the machines in the datacenter.* Recovery from the crisis involves taking the necessary actions to return the system to an SLO-compliant state. If the operators can recognize that the crisis is of a previously seen type, a known remedy can be applied, reducing overall recovery time. Conversely, if the operators can quickly determine that the crisis does not correspond to any previously seen incident, they can immediately focus on diagnosis and resolution steps and record the result in case a similar crisis recurs.

Please note that the issue of *detecting* the beginning of a crisis is solved in this context and is not the concern of this chapter. Indeed, detection is performed by monitoring of violations of the SLO (through the KPIs) as described in the previous paragraph. Our goal is to automate the *crisis identification* process. To this end, we first automatically build a model of each crisis as a time series of a transformation of the signals and then propose an online clustering approach to identify when a crisis is similar to a known crisis to identify recurring problems.

10.3 FROM COLLECTED SIGNALS TO FINGERPRINTS

The goal of a fingerprint is to be a representation of a crisis that uniquely identifies that crisis. In this section, we describe how to automatically build a crisis fingerprint from the raw signals collected in the datacenter. In our experience, the number of signals is on the order of low to mid hundreds and they can represent the usual vital health information on the machines as the CPU, memory, disk, utilization, etc., as well as the sizes of different queues and even high-level alarms and events (usually designed by operators). The construction of a fingerprint involves three steps. In the first step, we transform the raw signals into a compact representation that (a) is easy to monitor by the operators, (b) summarizes the state of the data center for a particular time epoch, and (c) normalizes the different raw signals. In the second step, we include the dynamics of the monitored signals using a 1-Markov time series representation. The third step involves selecting the subset of the signals that are relevant for discriminating among different crises.

* Specific numbers are proprietary.

10.3.1 Summarizing the State of the Datacenter

In the first step, we compactly represent the values of each signal on all servers during a particular epoch. Because servers of the same application typically run the same code, we can view the measured values of the same signal as samples of a random variable whose distribution is unknown. We thus summarize the signal values across all servers using the median of the observed empirical cumulative distribution over an epoch. We use a quantile, such as the median, instead of other statistics, such as mean and variance, because quantiles are less susceptible to outliers. This summarization of the state of the signals does not grow as the number of machines increases, so the size of the fingerprint is only proportional to the number of signals being collected. In addition, there are well-known algorithms for estimating quantiles with bounded error using online sampling [19], which guarantee that the entries in the fingerprints can be computed efficiently. In our case study, which involved several hundred machines, we computed the values of the quantiles exactly.

In our initial exploratory data analysis phase, we observed that the main factor that differentiates between different types of crises are the different signal medians that take extreme values during each crisis [18,20]. In other words, compared with values of the (median of each) signal during normal periods with no crises, the values observed during a crisis are either too high or too low. Because the number of occurrences of each crisis type is relatively low, it is hard to build a robust model of the evolution of the medians during a crisis. Instead, we focused on capturing in the fingerprint the fact that the medians jump either to extremely high or low values. We achieve this by discretizing the value of each signal median to one of three states: extremely high (hot), extremely low (cold), or normal relative to its past values. For example, if the median of a signal is hot, the most recent value for that quantile is higher than normal. We remark that this representation is favored by the operators, as it is similar to standard monitoring techniques.

The computation of hot and cold thresholds is parameterized by p—which stands for the percentage of past values of a signal quantile that are considered extremely low or high during normal system operation. The cold threshold of a particular signal m (such as median of CPU utilization) is computed as the p/2th percentile of values of m in the past W days that exclude epochs with SLO violations. The hot threshold of m is computed as (100 – p/2)th percentile over the same period. For example, for $p = 4\%$

we use the 2nd and 98th percentiles. The particular choices of the values for the parameters and the sensitivity of the approach to this value are discussed and characterized in [18].

10.3.2 Crisis Modeling

As defined above, a crisis is represented by the evolution in time of the monitored median for each signal, where each monitored median has three states: 1 for *cold* (value of the median is too low), 2 for *normal*, and 3 for *hot* (value of the median is too high). It is natural then to use a time series model for the crisis evolution. Thus, denote the vector of signals for the ith crisis in the lth epoch after the start of the crisis by $Y_{il} = (Y_{il1}, \ldots, Y_{ilJ})$, where J indexes over the number of signals being monitored. We associate a *type* to each crisis and use k to denote this type. We assume that, given the crisis type, the initial state vector Y_{i1} is sampled from the appropriate discrete distribution, and that the state vector Y_{il} subsequently evolves according to a Markov chain.

Estimation of the full joint distribution of Y_{i1} and the full transition matrix is infeasible when the number of crises I is small and the number of signals J is moderate or large, as is typical in a reasonably well-behaved data center (in the data we examined, $I = 27$ and $J = 18$ after some feature selection, which we discuss in Section 10.3.3). For such small sample sizes, extremely parsimonious conditional independence structures have been found, both empirically and theoretically, to provide the best accuracy in estimation of a class variable [21–23]. In particular, naive Bayes models, which assume conditional independence of all attributes conditional on the class, and augmented naive Bayes models that assume only pairwise dependencies conditional on the class, have been found to have the best accuracy.

We therefore assume independence of all signals conditional on the crisis type (dependencies between pairs of signals can easily be accommodated by replacing the pair of signals, having three states each, with a single signal that has nine states). Conditional on k, Y_{i1j} then has a discrete distribution with probability vector $\gamma^{(jk)} = (\gamma_1^{(jk)}, \gamma_2^{(jk)}, \gamma_3^{(jk)})$, and Y_{ilj} evolves according to a Markov chain of order 1 over the three states. The elements of the row-stochastic Markov transition matrix are denoted by $T_{st}^{(jk)}$, where subscripts s, $t \in \{1,2,3\}$ indicate the states. The resulting complete-data likelihood function is as follows, where we condition on the unknown type indicators Z_i of each crisis $i = 1, \ldots I$ and where the values n_{ijst} are the number of transitions of the jth signal from state s to state t during crisis i:

$$\pi\left(\mathcal{D}|\{Z_i\}_{i=1}^{I},\{\gamma^{(jk)},T^{(jk)}\}_{j,k}\right)=\prod_{i,j,t}\left[\left(\gamma_t^{(jZ_i)}\right)^{1(Y_{i1j}=t)}\prod_{s}\left(T_{st}^{(jZ_i)}\right)^{n_{ijst}}\right]. \quad (10.1)$$

For simplicity, we will use π to indicate likelihood, prior, and posterior distributions, as distinguished by their arguments. As will be seen in Section 10.4, this parameterization and the natural (conjugate) choice of Dirichlet priors for the parameters will enable a closed form solution.

In our particular setting, we observed that to model a crisis we should use one or two epochs before the crisis actually starts and no more than four to six epochs into the crisis (when a lot of the signals tend to saturate).

10.3.3 Selecting the Relevant Signals

As discussed above, and as we corroborated experimentally in [18], using all the collected signals in the fingerprint decreases the accuracy of the identification. Ideally, to select the subset of relevant signals, we would like to propose and execute a set of algorithms that would find the signals that help the most in the identification process. An approach is to wrap a search around the identification process and use accuracy to select the signals that influence the process the most. Unfortunately, this is not only an expensive proposition (from the computational point of view) but it also requires that we have a continuously updated pool of labeled crises, which in turn implies that each crisis must be carefully diagnosed. Although this is the subject of future research, in this work, we propose a heuristic based on transforming the signal selection problem to that of classification with feature selection.

The setup is as follows. We consider each individual machine as providing a sample of the behavior of the signals. Recall that we can determine whether at a given epoch each machine is in crisis (class 1) or not (class 0) by simply looking at the state of the KPI (whether they violate or not the SLO).* As features, we used the raw signals from the machines. Notice that we are treating these as i.i.d. samples. Now, we have a regular classification

* We were careful to avoid those epochs that where at the apex of the crises—normally, in our domain, after an hour or so, since at that time all signals were high.

problem: given the signals as features, find a classifier that would predict whether a machine is in crisis.

In our work, we use logistic regression as a model for the classifier and L1 regularization for the feature selection. There is theoretical evidence and strong experimental evidence that this method works well even when the number of parameters is comparable to the number of samples [24]. It also typically produces sparse coefficient vectors in which many of the coefficients are 0 and can thus be used effectively for feature selection.

In our specific domain, to select the signals we will use in the fingerprints for the experiments, we build a classifier for each crisis over a period of 3 months prior to the data used for identification. We pick the S signals that were selected most frequently in each of the models. Note that once S is determined, the whole procedure of selecting the relevant signals is completely automated. In our experiments reported in this chapter, we set S = 18. Tradeoff on the specific number S used are shown experimentally in [18].

10.4 IDENTIFYING A CRISIS

In our proposed solution, we have an indicator label associated with each crisis. We use Z_i to denote the label associated with the crisis i. Thus, if $Z_i = Z_j$, we will say that crisis i and crisis j are the same. We take a full Bayesian approach to the problem of identifying a new crisis (in real time), given the data $\mathcal{D}$ from all previous crises and the data $\mathcal{D}_{new}$ for the new crisis. Thus, our main task consists of estimating the posterior $\pi(Z_{new} = Z_i|\mathcal{D},\mathcal{D}_{new})$ that the new crisis has the same type as each of the previously seen types in Z_i, $i = 1, \ldots, I$ and the posterior $\pi(Z_{new} \neq Z_i\ \forall i|\mathcal{D},\mathcal{D}_{new})$ of this being a new type of crisis (Z_{new} is an indicator of the type of the new crisis).

We cast this problem as an online clustering problem [20] where crises with the same label define the different clusters. Please note that this is a difficult problem from the clustering perspective, as we would like to identify the new crisis as soon as possible, which implies that we will have partial information about the new crisis (first or second state only of the Markov chain of each monitored signal). The approach we follow here will lend itself to variants of the clustering problem such as a form of semisupervised learning (some crises are definitely labeled) and others where we incorporate constraints into the clustering process (e.g., two specific crises must belong to the same cluster).

The Bayesian approach enables the computation of a full posterior on the probability of the label of the new crisis directly from the data, which in turn enables optimal decision-making. That is, given a cost for each possible action, operators can compute the expected tradeoff involved with each type of intervention [20].

In the next subsections, we describe first the clustering model and then how to compute the needed posterior.

10.4.1 Cluster Modeling

Our approach to clustering is based on a Dirichlet process mixture (DPM) model. The DPM provides natural prior specification for online clustering, allowing estimation of the number of clusters while maintaining exchangeability between observations [25]. A DPM can be obtained as the limit of a finite mixture model with Dirichlet prior distribution on the mixing proportions [26,27]. In our context, the DPM is parameterized by a scalar α controlling the expected number of types occurring in a set of crises and by a prior distribution $G_0(\{\gamma^{(j)},T^{(j)}\}_j)$ for the set of all parameters associated with each crisis type k.

We take G_0 to be the product over j of independent Dirichlet distributions for $\gamma^{(jk)}$ (with parameter vectors $a^{(j)}$), times the product over j and s of independent Dirichlet distributions for the transition matrix rows $T_{s\cdot}^{(jk)}$ (with parameter vectors $b_s^{(j)}$). The use of such a product Dirichlet prior distribution for the rows of an unconstrained transition matrix is standard practice (e.g., [28,29]).

The DPM model for the crisis types $\{Z_i\}_{i=1}^{I}$ and crisis parameters $\gamma^{(jk)}$, $T^{(jk)}$ can be described as follows, in the case where the causes (labels) of the crises are all unknown. The first crisis has type 1 ($Z_i = 1$) by definition. For subsequent crises,

$$\Pr(Z_i = z|Z_1,\ldots,Z_{i-1}) = \frac{\#\{i' < i : Z_{i'} = z\}}{i-1+\alpha} \qquad \text{for } z \in \{Z_{i'}\}_{i'<i}$$

$$\Pr(Z_i \neq Z_{i'}\ \forall i' < i \mid Z_1,\ldots,Z_{i-1}) = \frac{\alpha}{i-1+\alpha}. \tag{10.2}$$

This is called the "Chinese restaurant process"; each observation i is conceptually a guest who, upon entering a restaurant, either sits at a table that is already occupied, with probability proportional to the number of guests at that table or sits at an empty table.

Conditional on the cluster assignments Z_i, the parameters of each cluster k are independently distributed according to G_0. Thus, the DPM model can be written

$$\pi\left(\{Z_i\}_{i=1}^{I}\right) = \pi(Z_1)\prod_{i=2}^{I}\pi\left(Z_i|\{Z_{i'}\}_{i'<i}\right)$$

$$= \prod_{i=1}^{I}\left[\frac{\alpha}{\alpha+i-1}\mathbf{1}\left(Z_i = m_{i-1}+1\right) + \frac{1}{\alpha+i-1}\sum_{i'<i}\mathbf{1}\left(Z_i = Z_{i'}\right)\right] \tag{10.3}$$

$$\pi\left(\{\gamma^{(jk)}, T^{(jk)}\}_{j,k}|\{Z_i\}_{i=1}^{I}\right) = \prod_{k=1}^{m_I} G_0\left(\{\gamma^{(jk)}, T^{(jk)}\}_j\right), \tag{10.4}$$

where $m_i = \max\{Z_{i'} : i' \leq i\}$ for $i > 0$ and $m_0 = 0$. In this description, the Dirichlet process has been integrated out, obtaining a generalized Polya urn scheme [30].

10.4.2 Computing the Probability of the Crisis Label

From probability theory, we have that $\pi(Z_{new} = Z_i|\mathcal{D},\mathcal{D}_{new})$ is equal to

$$\sum_{\{Z_{i'}\}_{i'=1}^{I}} \pi(Z_{new} = Z_i|\{Z_{i'}\}_{i'=1}^{I}, \mathcal{D}, \mathcal{D}_{new})\pi(\{Z_{i'}\}_{i'=1}^{I}|\mathcal{D}, \mathcal{D}_{new}), \tag{10.5}$$

where $\mathcal{D}$ denotes the data from past crises, $\mathcal{D}_{new}$ is the data for the new crisis, and $\{Z_{i'}\}_{i'=1}^{I}$ denotes an instantiation of the labels of the I past crises. We also have the equation for the probability $\pi(Z_{new} \neq Z_i\ \forall i|\mathcal{D},\mathcal{D}_{new})$ that the new crisis belongs to a new cluster. The second factor in Equation 10.5 is proportional to the likelihood of the data multiplied by the prior, that is,

$$\pi(\mathcal{D}, \mathcal{D}_{new}|\{Z_{i'}\}_{i'=1}^{I})\pi(\{Z_{i'}\}_{i'=1}^{I}). \tag{10.6}$$

According to our modeling of the crises as Markov chains, we have that the likelihood of the data given a clustering assignment, $\pi(\mathcal{D}|\{Z_i\}_{i=1}^{I})$, is equal to

$$\int \pi(\mathcal{D}|\{Z_i\}_{i=1}^{I}, \{\gamma^{(jk)}, T^{(jk)}\}_{j,k})\,\pi(\{\gamma^{(jk)}, T^{(jk)}\}_{j,k}|\{Z_i\}_{i=1}^{I})\,d\gamma^{(jk)}\,dT^{(jk)}. \tag{10.7}$$

Using Dirichlet priors and by conjugacy, this integral is actually computed in closed form. The details are laborious from an algebra point of view but not complicated and are provided in [20].

For the inference process, we use a collapsed-space Markov chain method [31], which simulates a Markov chain with target distribution $\pi(\{Z_i\}_{i=1}^{I}|\mathcal{D})$ on the reduced space $\{Z_i\}_{i=1}^{I}$. This is made possible by the fact that we have chosen a conjugate (Dirichlet) prior for all the parameters of the clusters, so $\pi(\{Z_i\}_{i=1}^{I}|\mathcal{D})$ is available in closed form. The collapsed-space Markov chain method explores the space by either repeatedly proposing to change the value of Z_i for a single index i or repeatedly proposing to split a cluster into two or merge two clusters together (see [31] for details). In addition, to ensure that we visit all possible modes for the distribution, we modified the collapsed-space sampler with parallel tempering [32]. Parallel tempering augments the original Markov chain with several other Markov chains whose target distributions are flatter than the posterior distribution $\pi(\{Z_i\}_{i=1}^{I}|\mathcal{D})$; these chains are thus able to explore the space more effectively than the original chain. These extra chains then interact with the primary Markov chain by swapping states, which improves the convergence of the primary chain. This is a computation intensive process, and to complete the identification of a crisis in real time, we will push the collection of samples from the Gibbs sampler to be an offline process. This requires the following approximation of the expression in Equation 10.5, where $\pi(Z_{new} = Z_i|\mathcal{D},\mathcal{D}_{new})$

$$
\begin{aligned}
&= \sum_{\{Z_{i'}\}_{i'=1}^{I}} \pi(Z_{new} = Z_i|\{Z_{i'}\}_{i'=1}^{I},\mathcal{D},\mathcal{D}_{new})\pi(\{Z_{i'}\}_{i'=1}^{I}|\mathcal{D},\mathcal{D}_{new}) \\
&\approx \sum_{\{Z_{i'}\}_{i'=1}^{I}} \pi(Z_{new} = Z_i|\{Z_{i'}\}_{i'=1}^{I},\mathcal{D},\mathcal{D}_{new})\pi(\{Z_{i'}\}_{i'=1}^{I}|\mathcal{D}) \qquad (10.8)
\end{aligned}
$$

and the analogous approximation for $\pi(Z_{new} \neq Z_i\ \forall i|\mathcal{D},\mathcal{D}_{new})$. These assume that the data from the new crisis do not tell us very much about the past crisis types $\{Z_{i'}\}_{i'=1}^{I}$; this is very reasonable to assume, especially since we expect to identify the new crisis as soon as possible,* and it is quite accurate in practice as demonstrated in [20].

* Which entails that $\mathcal{D}_{new}$ will be very small in comparison to $\mathcal{D}$.

The conditional distribution $\pi(Z_{new}|\{Z_{i'}\}_{i'=1}^{I},\mathcal{D},\mathcal{D}_{new})$ of Z_{new} is expressed as

$$\pi(Z_{new}|\{Z_{i'}\}_{i'=1}^{I},\mathcal{D},\mathcal{D}_{new}) \propto \pi(Z_{new}|\{Z_{i'}\}_{i'=1}^{I})\pi(\mathcal{D},\mathcal{D}_{new} \mid Z_{new},\{Z_{i'}\}_{i'=1}^{I}) \quad (10.9)$$

where the likelihood $\pi(\mathcal{D},\mathcal{D}_{new}|Z_{new},\{Z_{i'}\}_{i'=1}^{I})$ is available in closed form as described above, and where (from the DPM model in Equation 10.3)

$$\pi(Z_{new}|\{Z_{i'}\}_{i'=1}^{I}) \propto \alpha 1\left(Z_{new} = m_I + 1\right) + \sum_{i'=1}^{I} 1\left(Z_{new} = Z_{i'}\right).$$

Given these facts, we propose the following two-step method:

COMPUTATIONAL PROCEDURE

1. After the end of each crisis, refit the clustering model by simulating the Markov chain described above, which yields sample vectors $\{Z_i^{(l)}\}_{i=1}^{I}$ from the posterior distribution $\pi(\{Z_i\}_{i=1}^{I}|\mathcal{D})$.

2. When a new crisis begins, use its data $\mathcal{D}_{new}$ to calculate the Monte Carlo estimates:

$$\hat{\pi}(Z_{new} = Z_i|\mathcal{D},\mathcal{D}_{new}) = \frac{1}{L}\sum_{l=1}^{L} \pi(Z_{new} = Z_i^{(l)}|\{Z_{i'}^{(l)}\}_{i'=1}^{I},\mathcal{D},\mathcal{D}_{new})$$

$$\hat{\pi}(Z_{new} \neq Z_i\ \forall i|\mathcal{D},\mathcal{D}_{new}) = \frac{1}{L}\sum_{l=1}^{L} \pi(Z_{new} \neq Z_i^{(l)}\ \forall i|\{Z_{i'}^{(l)}\}_{i'=1}^{I},\mathcal{D},\mathcal{D}_{new}).$$

For each l, the discrete distribution $\pi(Z_{new}|\{Z_{i'}^{(l)}\}_{i'=1}^{I},\mathcal{D},\mathcal{D}_{new})$ is first computed by enumerating Equation 10.9 over the possible values $k = 1, \ldots, m_I + 1$ of Z_{new} and normalizing. Then $\pi(Z_{new} = Z_i^{(l)}|\{Z_{i'}^{(l)}\}_{i'=1}^{I},\mathcal{D},\mathcal{D}_{new})$ and $\pi(Z_{new} \neq Z_i^{(l)}\ \forall i|\{Z_{i'}^{(l)}\}_{i'=1}^{I},\mathcal{D},\mathcal{D}_{new})$ are found by choosing the appropriate element of the vector $\pi(Z_{new}|\{Z_{i'}^{(l)}\}_{i'=1}^{I},\mathcal{D},\mathcal{D}_{new})$.

Part 1 is the slower part of the computation, but takes much less than the hours or days that typically pass between crises. The computation in part 2 above is O(LIJ), very manageable in real time.

10.4.3 Prior Specification

We select the prior hyperparameters α, $a^{(j)}$, and $b_s^{(j)}$ by combining information elicited from domain experts with information in the data. The former is formal Bayes, while the latter is empirical Bayes [33].

We elicit α from experts; results are remarkably insensitive to the choice of α. According to the DPM model, the probability that two randomly chosen crises have the same type is $1/(\alpha + 1)$. Our domain experts estimate this to be 10%, yielding $\alpha = 9$. This implies that the expected number of types in the data is 12.9 (for 27 crises), which the experts agree is reasonable.

To create a sensible prior for $a^{(j)}$, we first follow [34] in taking the prior mean of the vector $\gamma^{(jk)}$ to be the empirical frequencies over the entire data set, i.e., the empirical distribution $\gamma^* = (\gamma_1^*, \gamma_2^*, \gamma_3^*)$ of the first value of all signals in all crises observed so far. This implies that $a^{(j)} = c^{(j)}(\gamma_1^*, \gamma_2^*, \gamma_3^*)$ for some constant $c^{(j)}$ that controls the prior variance of $\gamma^{(jk)}$. Our choice of this constant will reflect the fact that the signals are selected to be indicative of crisis type, i.e., any signal has nontrivial probability of behaving similarly across crises of a particular type. For this reason, experts believe that there is a substantial prior probability for any j and k that one of the values $\gamma_t^{(jk)}$ is "close" to 1. This is formalized by specifying a 50% prior probability that $\max_t \gamma_t^{(jk)} > .9$. This choice uniquely determines the value of $c^* = c^{(j)}$, which can be found by an iterative numerical procedure that simulates from Dirichlet($c\gamma^*$), checks the proportion of samples for which one of the values is "close" to 1 as defined above and adjusts c.

Selection of $b_s^{(j)}$ is analogous. We use the data for all signals and all crises observed so far to find the empirical transition probabilities $T_{s\cdot}^*$ from each starting state s and set the prior mean of $T_{s\cdot}^{(jk)}$ equal to $T_{s\cdot}^*$. By taking the prior weight of evidence (the sum of $b_s^{(j)}$) to be equal for each row s of the transition matrix, we must have $T^{(jk)} = d^{(j)}T^*$ for some constant $d^{(j)}$. To choose $d^{(j)}$, we consider the limiting distribution $r^{(jk)} = (r_1^{(jk)}, r_2^{(jk)}, r_3^{(jk)})$ of a Markov chain with transition kernel $T^{(jk)}$. Since the signals tend to behave consistently across crises of a particular type, there is nontrivial probability that one of the values $r_t^{(jk)}$ is "close" to 1. This is formalized via a 50% probability that $\max_t r_t^{(jk)} > .9$. Once again, $d^* = d^{(j)}$ is found by simulation.

Sensitivity analysis experiments to characterize the effects of the choices above can be found in [20].

10.5 EXPERIMENTS AND RESULTS

In this section, we report on experiments conducted on real data from a small to midsize datacenter. We first describe the system and the data in the next subsection. Then, we describe two experiments: one in which we cluster the set of crises offline and then one in which we simulate a full operational setting where new crises are clustered in real time as they unfold.

10.5.1 System Under Study and Data

We have evaluated our approach on data from a commercial datacenter running a 24 × 7 enterprise-class application. It is one of four datacenters worldwide running this application, each containing hundreds of machines, serving several thousand enterprise customers, and processing a few billion transactions per day. Most machines execute the same application that is entirely contained within the machine. In this application, the incoming workload is processed on the machine in three stages: light processing in the front-end, core of the execution in the second stage, followed by some back-end processing. The requests are then distributed to the clients or to another datacenter for archival and further processing. We have no visibility to the clients or to machines in the other datacenters.

For each server, operators routinely measure about 100 signals and report the average over a sampling period. We had no choice on which signals were collected nor on the sampling period; similarly, we had no access to any other performance counters or to information allowing us to reconstruct the actual path of each job through the server. The signals include counts of alerts set up by the operators, queue lengths, latencies on intermediate processing steps, summaries of CPU utilization, and various application-specific signals.

The operators of the site designate three KPIs corresponding to the average processing time in the front end, the second stage, and one of the post-processing stages. Each KPI has an associated SLO threshold determined as a matter of business policy. A performance crisis is declared when a specific percentage of the machines violate any KPI SLO. This definition is set by the operators, and we did not tamper with it.

We use over 6 months of production data from 2008 for the experiments. We collected, in conjunction with the operators, information about 17 manually diagnosed crises. These problems range from configuration problems to unexpected increase in workloads to backlogs caused by a connection to another datacenter. These crises that were labeled by the operators required an exhaustive investigation to determine the underlying cause. In a database in the datacenter, attached to the labels, are logs providing details about the investigation and, most importantly, the

TABLE 10.1 List of Performance Crises Used in Our Experiments

Label	Labeled Instances	Instances Identified by the Model	Matched Instances
Overloaded front-end	2	3	2
Overloaded back-end	8	14	8
Database configuration error	1	2	1
Configuration error	1	0	0
Performance issue	1	0	0
Middle-tier issue	1	0	0
Request routing error	1	6	1
Whole datacenter turned off and on	1	0	0
Workload spike	1	2	1

Note: The second column lists the number of those that we verified root causes with the operators. The third column enumerates all the labels assigned by our model to 27 crises during the same period, whereas the fourth column shows the match to the operators' labeling. Note the wide range of problems (from human error to capacity issues) and the high degree of accuracy in the labeling by the model.

set of remedial actions taken and their effectiveness. As a reminder, the value of the technique we propose in this chapter is to accurately identify future crises, so that the appropriate information about remedial actions can be readily retrieved and used (or escalated in cases where new crises are identified).* In conjunction with the operators we assigned to these crises somewhat descriptive labels. These labels and the numbers of verified occurrences are depicted in the first two columns of Table 10.1. We point out that these crises manifested themselves through a combination of multiple signals taking extreme values. The labels we provide are not the actual ones attached to the operators report; for confidentiality reasons we are not able to publish these. These labels give a sense of the wide range of problems that are under study and that can be successfully identified with the techniques proposed in this chapter. In addition to these crises, we included in our experiments 10 crises that happened during the same time period. We do not have the actual ground truth labels for those. The third column in Table 10.1 presents the total number of crises and their labels identified by our model, whereas the last column contains the matches. We also had access to collected signals of 20 unlabeled crises (that occurred before the period under study). These crises were

* Note that issues such as the granularity of the diagnosis and the appropriate mapping to the set of available remedial actions are outside the scope of the technique we describe.

used to perform the selection of the relevant signals for the fingerprints as described in Section 10.3.3.

10.5.2 Experiments

We performed extensive experiments with the model and algorithms presented here in both [18,20]. Besides conducting synthetic experiments on simulated data and comparing to other model-based clustering approaches (notably [35]), we checked many aspects of the models and sensitivity to parameters. In this chapter, we report on two experiments from [20]. In the first one, described in Section 10.5.3, we establish an upper bound on what is possible in terms of accuracy by performing an offline clustering on the data described in Table 10.1. In the second experiment, described in Section 10.5.4, we simulate a fully operational setting where the identification of the crisis must be done in real time. The difference between these two settings is that in the offline case we first have full information on each of the crises (we take into account the data from the complete time series of each crisis), and second, we do not require any approximation as the real time requirement on identification is removed. This is in a sense the best we can do given the model and the data.

10.5.3 Offline Clustering

The sizes of the clusters from the posterior mode cluster assignment are shown in the fourth column of Table 10.1. This cluster assignment has 58% posterior probability and, along with the second-highest probability assignment, accounts for a total of 93.8% of the posterior probability. There is only a single difference between the two assignments, namely, a change in the labeling of one crisis, increasing the count of the biggest cluster to 15 and decreasing the count of "request routing error" to 5. We will summarize the accuracy of the posterior mode clustering assignment relative to the known causes, but this summary applies equally well to the second assignment since the crisis for which they differ has unknown cause.

The posterior mode crisis labels for the most part match the known causes, with the exception of four uncommon crisis types that are incorrectly clustered with more common types. The largest cluster obtained by our method corresponds to the cause "overloaded back-end"; all eight of the crises known to be of this type are correctly clustered together, along with six other crises (most of which have unknown cause). The "overloaded back-end" problem occurs due to poor performance of another datacenter center, one on which the servers depend. The datacenter operators do not have authority to fix the performance of that separate datacenter, explaining why this is the most common type of crisis. It is also the most important type of crisis to correctly identify, since although the problem cannot be fixed, the technicians know the best intervention for minimizing the effect of such a crisis.

The two crises of known cause "overloaded front-end" are also correctly clustered together. Similarly, the "database configuration error," "workload spike," and "request routing error" clusters are correctly identified.

Four uncommon crisis types are incorrectly clustered with more common types. For instance, the "configuration error" crisis is clustered with the "overloaded front-end" crises. This type of mistake occurs partly due to the fact that crises having different causes (e.g., closely related causes) can have the same patterns in their signals. In the most extreme case, the signals appear to be indistinguishable between the two crisis types. This result suggests to datacenter operators the need for additional signals to distinguish between these crisis types.

In the other cases of incorrectly merged crisis types, while the large majority of signals are indistinguishable between the two types, a few signals show distinct behavior. Since we have assumed the parameters of distinct crisis types to be independent a priori, the presence of distinct crisis types with similar patterns for most signals is very improbable under the prior. Such crisis types are therefore clustered together. This issue could be fixed by creating a hierarchical dependence structure between crisis types in the prior distribution. This structure would be realistic, since in fact one can define the true crisis "causes" according to a coarse division into a few main causes, or into finely subdivided causes.

Results are very insensitive to the choice of α; multiplying or dividing α by up to a factor of 20 does not change the posterior mode cluster assignment. Results are also not sensitive to the choice of the constants $a^{(j)}$ and $b_s^{(j)}$, as long as this choice is not dramatically inconsistent with the data and with the experts' belief that the signals provide substantial information about the crisis type. For instance, changing the prior median of $\max_t \gamma_t^{(jk)}$ and $\max_t r_t^{(jk)}$ from 0.90 to 0.99 or 0.85 does not change the posterior mode clustering assignment. Reducing this value to 0.80 switches the ranks of the two most probable cluster assignments, but the total posterior probability of these two cluster assignments is still high (89.9%). Reducing it further to 0.70 gives the same results as 0.80 except that the posterior probability of these two most probable cluster assignments declines to 66.7%.

We examine sensitivity to the prior mean of $\gamma^{(jk)}$ and $T_{s\cdot}^{(jk)}$ by taking the data-based prior mean described in Section 10.4.3 and mixing with a uniform distribution. When the mixture proportions are three-fourths on the data-based mean and one-fourth on the uniform distribution, the posterior mode clustering assignment is unchanged. When we reduce the weight on the data-based mean to half, the ranks of the two most probable cluster assignments are switched, but they still account for 90.3% of the posterior probability. Reducing it further to one-fourth, the first cluster and the next to last merged. Thus, the prior has to change dramatically before results change in a meaningful way.

10.5.4 Operational Setting

As described above, in the operational setting, we simulate the conditions that operators face where they need to identify a new crisis as it unfolds. We also assume a worst-case scenario where we do not have the true labels of previously seen crises. We evaluate the accuracy of online clustering relative to the offline clustering assignment, that is, treating the posterior mode cluster assignment from the offline clustering as the gold standard.

We obtain a full-data misclassification rate of 7.4%, a three-epoch misclassification* rate of 14.8%, and an average time to correct identification of 1.81 epochs. This means that, on average, the crises are identified correctly even before the technical start of the crisis. Two-thirds of the crises are identified correctly in the first epoch.

To check whether the good identification performance of our method is specific to the particular ordering of the crises, we also permute the crises and evaluate performance. Taking the average over five random permutations of the crises, we obtain a full-data misclassification rate of 5.9% (SE = 3.4%), a three-epoch misclassification rate of 11.8% (SE = 3.2%), and an average time to correct identification of 1.56 (SE = 0.07). These results are even better than for the original ordering.

10.6 DISCUSSION

The tendency to merge very small clusters (those having one instance) with larger clusters in our model points to the exploration of more complex modeling. The Dirichlet process has a single parameter α "controlling" the addition of new clusters. A more flexible approach would be to

* The term three-epoch means that we have used three state transitions for the new instance of the crisis.

use the Pitman-Yor process [36], which is a generalization of the Dirichlet process that has two parameters for the cluster assignment model and a richer ability to represent the distribution of cluster sizes. We expect that the methods given in this chapter would generalize naturally to this process, and leave this as future work.

In [18], we managed to distinguish between some of the lesser common crises using a more engineered approach, which required, among other things the exact labeling (diagnosis and root cause analysis) by operators of past crises. Notice that in our experiments in this chapter we did not have that requirement. Thus, even the online clustering proceeds without any information regarding the true labels of the past crises. However, the approach presented here extends naturally to this situation,* and to the case with partial labels, or even more complex domains where constraints between the clustering of crises is specified. We describe how to modify the approach when the causes of some of the crises are known—the partially labeled case—from this description, it should be clear how to incorporate other scenarios. The partial labeling information can be captured by indicator functions $\mathbf{1}(Z_i = Z_{i'})$ for pairs of crises i, i' that are known to have the same type (denoted by $i \backsim i'$) and $\mathbf{1}(Z_i \neq Z_{i'})$ for pairs of crises i, i', which are known to have different types (denoted by $i \nsim i'$). In this case, the prior $\pi(\{Z_i\}_{i=1}^{I})$ is proportional to the expression (10.3) multiplied by the restriction.

$$\prod_{i \sim i'} \mathbf{1}(Z_i = Z_{i'}) \prod_{i \nsim i'} \mathbf{1}(Z_i \neq Z_{i'}) \tag{10.10}$$

while the prior $\pi(\{\gamma^{(jk)}, T^{(jk)}\}_{j,k} | \{Z_i\}_{i=1}^{I})$ is unchanged. Our computational methods extend trivially to accommodate the partial labels, by simply disallowing configurations that are incompatible with those labels. We also leave the characterization of the effect of this information on the accuracy of the identification for future work.

Finally, in this chapter, we have explored the clustering by marginalizing over the parameters. However, as shown explicitly in [20], we can also compute posterior probabilities on the parameters of the models, which can be used to debug the models of each crisis and increase the operators understanding of the effects of the repair action on the evolution of the monitored signals. This, in turn, can lead to predictions regarding the time that it will take to stabilize the datacenter.

* Where operators manage to diagnose the root cause of each crisis after stabilizing the system.

10.7 CONCLUSIONS

In this chapter, we have presented an application of data mining and statistical machine learning techniques to automatically recognize performance problems in the context of the datacenter containing a very large number of machines. This is but one component in the suite of tools necessary for very large systems management, which should include enablers for software update, capacity planning, inventory management, forecasting, and diagnosis. Extensions of the approach we proposed to forecasting could involve simply finding fingerprints for the epochs preceding crises and clustering/matching those. Part of these enhancement are in our future plans. Similarly, it is the case that the fingerprints, by providing data on which signals where hot, normal, or cold, provide information pertinent for the diagnosis of a crisis. It is conceivable, therefore, that a diagnosis solution could be built around this representation.

As mentioned in the introduction to this chapter, there is a growing literature of approaches based on machine learning and statistical pattern recognition to the health management of computer systems. The contributions that our approach brings to this line of work are threefold: First, the fingerprint representation as a scalable representation of system state, which is efficient even when used in very large systems containing hundreds of thousands of computers. Second, the suitable approximations that lead to the computation of a full Bayesian posterior on the clustering. This enables optimal decision making on the selection of the appropriate actions for repair or crisis containment. Finally, the validation of the approach on production data coming from a midsize datacenter running a 24 × 7 enterprise-class application.

ACKNOWLEDGMENTS

We thank the book editors and the reviewers for their comments, which helped to improve the presentation of the material in this chapter. We also acknowledge A. Fox and H. Andersen, our coauthors in [18], for their work in the creation of the fingerprint representation.

REFERENCES

[1] D. Patterson, A. Brown, P. Broadwell, G. Candea, M. Chen, J. Cutler, P. Enriquez, A. Fox, E. Kiciman, M. Merzbacher, D. Oppenheimer, N. Sastry, W. Tetzlaff, J. Traupamn, N. Treuhaft. *Recovery oriented computing (ROC): Motivation, definition, techniques, and case studies.* Technical report, UC Berkeley, 2002.

[2] P. Bodík, A. Fox, M.I. Jordan, D. Patterson, A. Banerjee, R. Jagannathan, T. Su, S. Tenginakai, B. Turner, J. Ingalls. Advanced tools for operators at Amazon.com. In *Hot Topics in Autonomic Computing (HotAC)*, 2006.

[3] K. Glerum, K. Kinshumann, S. Greenberg, G. Aul, V. Orgovan, G. Nichols, D. Grant, G. Loihle, G. Hunt. Debugging in the (very) large: Ten years of implementation and experience. In *Proceedings of the 22nd ACM Symposium on Operating Systems Principles (SOSP 2009)*, Big Sky, Montana, 2009.

[4] J.A. Redstone, M.M. Swift, B.N. Bershad. Using computers to diagnose computer problems. In *Workshop on Hot Topics in Operating Systems (HotOS-IX)*, Elmau, Germany, 2003.

[5] I. Cohen, M. Goldszmidt, T. Kelly, J. Symons, J.S. Chase. Correlating instrumentation data to system states: A building block for automated diagnosis and control. In *Proc. 6th USENIX Symposium on Operating Systems Design and Implementation (OSDI 2004)*, San Francisco, CA, 2004.

[6] I. Cohen, S. Zhang, M. Goldszmidt, J. Symons, T. Kelly, A. Fox. Capturing, indexing, clustering, and retrieving system history. In *Proceedings of the 20th ACM Symposium on Operating Systems Principles (SOSP 2005)*, Brighton, UK, 2005.

[7] S. Zhang, I. Cohen, M. Goldszmidt, J. Symons, A. Fox. Ensembles of models for automated diagnosis of system performance problems. In *Intl. Conf. on Dependable Systems and Networks (DSN 2005)*, Yokohama, Japan, 2005.

[8] P. Bodik, M. Goldszmidt, A. Fox. Hilighter: Automatically building robust signatures of performance behavior for small- and large-scale systems. In *Proceedings of the Workshop on Tackling Computer Systems Problems with Machine Learning Techniques (SysML 2008)*, USENIX Association, 2008.

[9] C. Yuan, N.L.J.-R. Wen, J. Li, Z. Zhang, Y.-M. Wang, W.-Y. Ma. Automated known problem diagnosis with event traces. In *Proceedings of the European Conference on Computer Systems (EuroSys 2006)*, Leuven, Belgium, 2006.

[10] S. Pertet, R. Gandhi, P. Narasimhan. Fingerpointing correlated failures in replicated systems. In *SYSML'07: Proceedings of the 2nd USENIX workshop on Tackling computer systems problems with machine learning techniques*, pp. 1–6, USENIX Association, Berkeley, CA, USA, 2007.

[11] S. Duan, S. Babu. Guided problem diagnosis through active learning. In *Proceedings of the IEEE Computer Society International Conference on Autonomic Computing (ICAC 2008)*, Washington, DC, 2008.

[12] P. Reynolds, J.L. Wiener, J.C. Mogul, M.A. Shah, C. Killian, A. Vahdat. Experiences with Pip: finding unexpected behavior in distributed systems. In *Proceedings of the 20th ACM Symposium on Operating Systems Principles (SOSP 2005)*, Brighton, UK, 2005.

[13] M.Y. Chen, E. Kıcıman, A. Accardi, E.A. Brewer, D. Patterson, A. Fox. Path-based failure and evolution management. In *Proc. 1st USENIX/ACM Symposium on Networked Systems Design and Implementation (NSDI 2004)*, San Francisco, CA, 2004.

[14] P. Barham, A. Donnelly, R. Isaacs, R. Mortier. Using magpie for request extraction and workload modeling. In *Proc. 6th USENIX Symposium on Operating Systems Design and Implementation (OSDI 2004)*, San Francisco, CA, 2004.

[15] S.A. Yemini, S. Kliger, E. Mozes, Y. Yemini, D. Ohsie. High speed and robust event correlation. *Communications Magazine, IEEE*, 34(5):82–90, 1996.

[16] M. Goldszmidt, I. Cohen, S. Zhang, A. Fox. Three research challenges at the intersection of machine learning, statistical inference, and systems. In *Proc. Tenth Workshop on Hot Topics in Operating Systems (HotOS-X)*, Santa Fe, NM, 2005.

[17] B. Cook, S. Babu, G. Candea, S. Duan. Toward Self-Healing Multitier Services. In *Proceedings of the International Workshop on Self-Managing Database Systems (SMDB)*, 2007.

[18] P. Bodik, M. Goldszmidt, A. Fox, D. Woodard, H. Andersen. Fingerprinting the datacenter: Automated classification of performance crises. In *Proceedings of the European Conference on Computer Systems (EuroSys 2010)*, Paris, France, 2010.

[19] S. Guha, A. McGregor. Stream order and order statistics: Quantile estimation in random-order streams. *SIAM Journal on Computing*, 38(5):2044–2059, 2009.

[20] D. Woodard, M. Goldszmidt. Online model-based clustering for online crisis identification in distributed computing. *Journal of the American Statistical Association*, 106:49–60, 2011.

[21] P. Domingos, M. Pazzani. On the optimality of the simple Bayesian classifier under zero-one loss. *Machine Learning*, 29:103–130, 1997.

[22] N. Friedman, D. Geiger, M. Goldszmidt. Bayesian network classifiers. *Machine Learning*, 29:131–163, 1997.

[23] D.J. Hand, K. Yu. Idiot's Bayes: not so stupid after all? *International Statistical Review*, 69:385–398, 2001.

[24] K. Koh, S.-J. Kim, S. Boyd. An interior-point method for large-scale L1-regularized logistic regression. *Journal of Machine Learning Research*, 8:1519–1555, 2007.

[25] M.D. Escobar, M. West. Bayesian density estimation and inference using mixtures. *Journal of the American Statistical Association*, 90:577–588. 1995.

[26] R.M. Neal. Markov chain sampling methods for Dirichlet process mixture models. *Journal of Computational and Graphical Statistics*, 9:249–265, 2000.

[27] C.E. Rasmussen. The infinite Gaussian mixture model. In S.A. Solla, T.K. Leen, K.R. Muller, editors, *Advances in Neural Information Processing Systems*, pp. 554–560, 2000.

[28] B.P. Carlin, A.E. Gelfand, A.F.M. Smith. Hierarchical Bayesian analysis of changepoint problems. *Journal of the Royal Statistical Society, Series C*, 41:389–405, 1992.

[29] P. Diaconis, S.W.W. Rolles. Bayesian analysis for reversible Markov chains. *Annals of Statistics*, 34:1270–1292, 2006.

[30] D. Blackwell, J.B. MacQueen. Ferguson distributions via Polya schemes. *Annals of Statistics*, 1:353–355, 1973.

[31] S. Jain, R.M. Neal. A split-merge Markov chain Monte Carlo procedure for the Dirichlet process mixture model. *Journal of Computational and Graphical Statistics*, 13:158–182, 2004.

[32] C.J. Geyer. Markov chain Monte Carlo maximum likelihood. In E. Keramidas, editor, *Computing Science and Statistics, vol. 23. In Proceedings of the 23rd*

Symposium on the Interface, pp. 156–163, Interface Foundation of North America Fairfax Station, VA, 1991.

[33] B.P. Carlin, T.A. Louis. *Bayesian Methods for Data Analysis*. Chapman and Hall, Boca Raton, FL, 3rd edition, 2009.

[34] J. Zhang, Z. Ghahramani, Y. Yang. A probabilistic model for online document clustering with application to novelty detection. In Y. Weiss, editor, *Advances in Neural Information Processing Systems*, 2004.

[35] J.D. Banfield, A.E. Raftery. Model-based Gaussian and non-Gaussian clustering. *Biometrics*, 49:803–821, 1993.

[36] J. Pitman, M. Yor. The two-parameter Poisson-Dirichlet distribution derived from a stable subordinator. *Annals of Probability*, 25:855–900, 1997.

CHAPTER 11

A Combined Model-Based and Data-Driven Prognostic Approach for Aircraft System Life Management

Marcos Orchard

George Vachtsevanos

Kai Goebel

CONTENTS

11.1 INTRODUCTION

Failure prognosis—a natural extension to the fault detection and isolation (FDI) problem—has become a key issue in a world where the economic impact of system reliability and cost-effective operation of critical assets is steadily increasing. Failure prognostic algorithms aim to characterize the evolution of incipient fault conditions in complex dynamic processes, thus allowing to estimate the remaining useful life (RUL) of subsystems and components. Several examples can be used here to illustrate the range of possible applications for these algorithms: electromechanical systems, continuous-time manufacturing processes, structural damage analysis, and even fault tolerant software architectures. Most of them have in common the fact that they are highly complex, nonlinear, and affected by large-grain uncertainty.

In the particular case of critical helicopter components, failure prognosis has been addressed over the past years via a variety of model-based and data-driven approaches [1–7]. Although most of these research efforts have focused on the analysis of vibration data and the derivation of condition indicators (CIs) [6], thus far, no major advances have been done on a systematic methodology capable of integrating sensing elements, data analysis, CI selection and extraction, fault diagnosis, and prognostic algorithms within a single health monitoring platform that may be ported to existing hardware/software on-board an aircraft.

Indeed, the implementation of real-time prognosis schemes on this type of equipment can be particularly challenging, since the accuracy of RUL estimates directly depends on the quality of long-term predictions—which need to be computed in an online fashion. Moreover, complex processes usually depend on a series of inputs variables, and therefore, the evolution of damaged/degraded systems is also affected by changes in loading conditions and external uncertainty sources. For this reason, comprehensive prognostic frameworks should entail information about environmental stresses (wind speed, temperature, humidity) and process control variables (load, torque, velocity). This knowledge provides critical information about how the fault condition will develop in time, leading to smarter decisions in the sense of simultaneously mitigating the fault growth while also meeting critical performance requirements of the system.

The majority of the current reliability approaches use large amounts of historical data to fit statistical models that help to describe the failure probability distribution for a large group of similar components, spending significant computational resources and offering little room for real-time adjustments whenever the system behaves differently from what it was expected. In this sense, learning paradigms, and other data-driven techniques, offer an invaluable opportunity for the improvement of prognostic algorithms based on either first principles or statistical knowledge of the system. The incorporation of real-time information from input/output/feature measurements, with the purpose of uncertainty representation and management, directly benefits the implementation of automated contingency management systems (ACMs), as well as other automated corrective schemes. In this sense, this chapter explores the concept and introduces a combined model-based/data-driven approach to failure prognosis that relies both on degradation models of the failing component (namely, fault models) and sequential Monte Carlo (SMC) methods for state estimation (also known as particle filtering, or PF). This approach allows taking advantage of real-time measurements, as well as updating model and stress parameters to project the system evolution into the future.

PF algorithms use the concept of sequential importance sampling (in a Bayesian framework) to approximate the state probability density function (PDF) through a set of possible realizations (particles) associated to discrete probability masses (or weights), thus combining model-based a priori information with acquired observations. The data-driven component of the architecture aims to account for stochastic and time-varying load profiles—principal stress factors—through appropriate uncertainty measures and linear interpolation techniques. These measures quantify the effect of input uncertainty on the prognostic results and are also the basis for feedback correction loops to extend the RUL of faulty nonlinear systems.

The proposed PF-based methodology addresses issues of uncertainty, system nonlinearity, and non-Gaussian noise. Performance metrics are used to assist the design, and eventual implementation, of diagnostic/prognostic algorithms and to measure the effect of input uncertainty. This unified method aims to improve significantly accuracy and precision of estimates, when compared with other baseline approaches. It also provides a path toward a rigorous approach to prognostics-enhanced reconfigurable or fault-tolerant control while extending the RUL of the platform to assure mission completion without compromising the safety of the vehicle. The proposed approach is reduced to practice on a critical aircraft component to demonstrate its efficacy.

11.2 AN INTEGRATED FAULT DIAGNOSIS AND FAILURE PROGNOSIS ARCHITECTURE

We introduce in this chapter an integrated failure prognosis architecture that is applicable to a variety of aircraft systems and industrial processes [8,9]. More specifically, we are targeting a specific rotorcraft system as a prototypical testbed for proof of concept. In this architecture, online modules perform raw data preprocessing, feature—or CIs—extraction, fault diagnosis, and failure prognosis on the basis of available *ground truth* fault data, noise models, experimental data, system models, and other offline tools to tune and adapt online parameters and estimate suitable mappings.

Figure 11.1 depicts the overall architecture and distinguishes between on-board and off-board modules for an eventual on-platform implementation. This architecture suggests a hybrid and systematic approach to sensing, data processing, fault feature extraction, fault diagnosis, and failure prognosis, which may lead to an online system (hardware/software) configuration. The enabling technologies include

1. Physics-based modeling of critical components/systems that will facilitate a better understanding of the physics of failure mechanisms, while providing simulated data and stress factors for diagnosis and prognosis.

2. Novel preprocessing routines including denoising of raw data via blind deconvolution to improve signal-to-noise ratio (SNR). A thorough approach to feature or CI selection and extraction that forms the foundation for accurate and reliable fault diagnosis and failure prognosis.

3. Incipient failure diagnosis and prognosis founded on concepts from physics-based models, measurements, and Bayesian estimation techniques. This innovative approach takes advantage of hybrid modeling/measurements and estimation methods to manage uncertainty and provide early fault detection, isolation, and prediction of the time to failure of a failing component.

We detail in the following sections the major modules of the architecture.

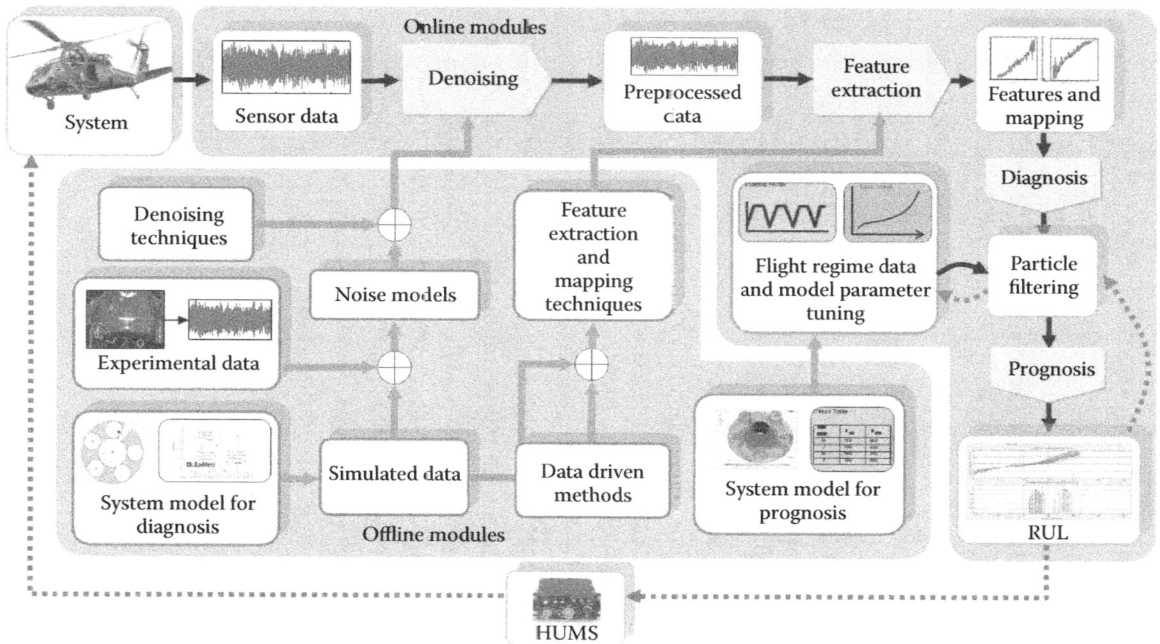

FIGURE 11.1 Overall architecture for implementation of fault diagnosis and failure prognosis algorithms.

11.2.1 Sensing and Data Processing

Modern aircraft/rotorcraft and critical industrial processes are equipped with monitoring, data acquisition, and data analysis hardware/software that are intended to assess the health of components/systems and inform the operator of impending failure conditions. Massive volumes of raw data can accumulate from a variety of distributed sensors suites, demanding proper attention to ensure that useful information, and knowledge for the application at hand, becomes available. Different applications will require diverse types of information. In particular, prognostic modules may require somewhat different information than diagnostic applications. With that in mind, data must be carefully preprocessed either in an online or offline fashion.

Knowledge discovery provides a promising technology to unearth valuable information from massive amounts of data. This capability has great value when (high-bandwidth) baseline and fault data—as well as usage and flight regime data—are acquired, on-board the aircraft, and must be processed expeditiously and accurately to support fault detection and mitigation strategies when faced with flight critical or safety critical events. The classical paradigm of progressing from *data → information → knowledge* is most relevant here and requires the development and implementation of novel techniques to trade off information content and accuracy with the computational requirements that are needed to arrive at the sought after knowledge. The value of this information must be assessed via appropriate performance metrics if useful data attributes derived through data mining tools are to describe faithfully the captured fault process. CIs or features represent certain circumstances under which an anomaly or abnormality in a system's operating state is detected; it is natural, therefore, that CI rather than raw data are used for fault diagnosis and failure prognosis.

On the one hand, fault diagnosis may be viewed as a mapping of given CI into predesigned fault classes. The same CI may be used to detect an anomaly (an unknown fault condition) when compared with healthy or baseline behaviors. On the other hand, failure prognosis involves estimating the RUL of a failing component/subsystem once a relevant fault is detected and identified. In failure prognosis problems, it is essential to track and evaluate CI exhibiting a progressive nature (with respect to a fault evaluation) on the basis of its contribution to the generation of accurate and precise RUL estimates. In fact, the success of diagnostic and prognostic algorithms depends highly on the quality of these CI. To extract CI that represent—in a compressed form—the maximum possible information content, a statistically sufficient database (both healthy and faulty data) is needed. However, data acquired on-board an aircraft are severely corrupted by noise stemming from a variety of internal and external noise sources. It is necessary, therefore, that raw data be processed first via denoising algorithms to improve the fault SNR ratio.

Typical preprocessing routines include data compression and filtering, time synchronous averaging of vibration data, and FFTs, among others. We will use a denoising methodology based on blind deconvolution that has been applied successfully for the helicopter case [10]. The process of blind deconvolution attempts to restore the unknown vibration signal by estimating an inverse filter, which is related to partially known system characteristics. This is an active field of current research in image processing [11], speech signal processing [12], but rarely applied in mechanical vibration signals. Vibration and other high-bandwidth signals are corrupted by multiple noise sources. A simplified model for such a complex signal may be defined as

$$s(t) = a(t)b(t) + n(t), \tag{11.1}$$

where $s(t)$ is the measured vibration signal, $b(t)$ is the noise-free unmodulated vibration signal, $a(t)$ is the modulating signal, and $n(t)$ is the cumulative additive noise. This model can be written in the frequency domain as

$$S(f) = A(f) * B(f) + N(f), \tag{11.2}$$

with $*$ being the convolution operation and $S(f)$, $A(f)$, $B(f)$, and $N(f)$ the Fourier transforms of $s(t)$, $a(t)$, $b(t)$, and $n(t)$, respectively. The goal is to recover $B(f)$. We propose an iterative denoising scheme that starts with $\bar{z}(t)$, an initial estimate of the inverse of the modulating signal $a(t)$, which demodulates the observed signal $s(t)$ to give a rough noise-free estimate of the vibration signal in the time domain as

$$\bar{b}(t) = s(t)\bar{z}(t). \tag{11.3}$$

Equation 11.3 can be written in the frequency domain as:

$$\bar{B}(f) = S(f) * \bar{Z}(f), \tag{11.4}$$

with $\bar{B}(f)$ and $\bar{Z}(f)$ being the Fourier transforms of $\bar{b}(t)$ and $\bar{z}(t)$, respectively. Passing $\bar{B}(f)$ through a nonlinear projection, it yields the ideal characteristics of the vibration signal $B_{nl}(f)$. Then, in the frequency domain, by minimizing the difference between $B_{nl}(f)$ and $\bar{B}(f)$:

$$min\left\|E(f)\right\| = min\left\|\bar{B}(f) - B_{nl}(f)\right\|, \tag{11.5}$$

iteratively through refining $\bar{Z}(f)$, $\bar{B}(f)$ will converge to the noise-free vibration signal. When it reaches the minimal value, $\bar{Z}(f)$ converges to $Z(f)$, and a good estimate of $B(f)$ is obtained as $B(f) = S(f) * Z(f)$.

Lastly, the estimate is transformed back into the time domain to recover the noise-free vibration signal $b(t)$. The blind deconvolution denoising scheme is illustrated in Figure 11.2 [13].

Note that the proposed scheme is implemented in the frequency domain, and the nonlinear projection, which is derived from a nonlinear dynamic model, is also given in the same frequency domain [10,13]. The blind deconvolution denoising scheme was applied to vibration data derived from a faulted planetary gear plate, a critical helicopter component. Results of 40% and 100% torque levels before and after denoising indicate a significant improvement in the SNR. Improvement in the data SNR is accompanied by similar enhancements in the CI or features. The accuracy and precision of mappings relating CI propagation to fault dimension (crack length) growth is closely related to the performance of diagnostic and prognostic algorithms. To evaluate the quality of the CI,

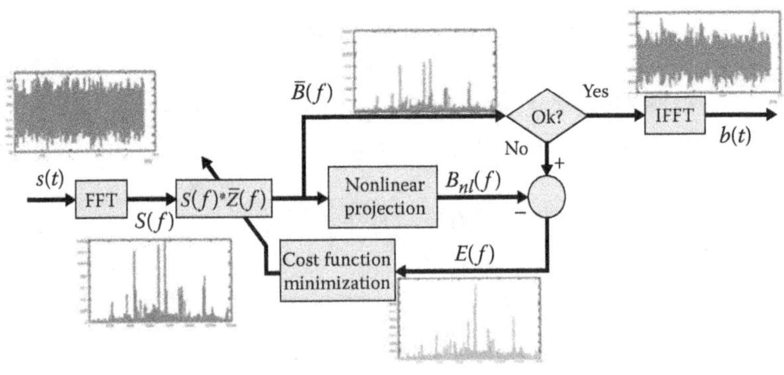

FIGURE 11.2 Blind deconvolution denoising scheme.

we define and employ two performance metrics. The first one is an accuracy measure and is defined as the correlation coefficient between the feature vector x and y as [10]:

$$CCR(x,y)=\sqrt{\frac{ss_{xy}^2}{ss_{xx}ss_{yy}}}, \tag{11.6}$$

where $ss_{xy}=\sum(x_i-\bar{x})(y_i-\bar{y})$, $ss_xx=\sum(x_i-\bar{x})^2$, and $ss_yy=\sum(y_i-\bar{y})^2$, respectively. The second is a precision measure called the percent mean deviation (PMD) and defined by

$$PMD(x,\tilde{x})=\frac{\sum_{i=1}^{n}\frac{x_i-\tilde{x}_i}{\tilde{x}_i}}{n}\times 100, \tag{11.7}$$

where n is the number of entities in the feature vector x and $\tilde{x}$ its smoothed version.

Both CCR and PDM indices have proven useful when trying to quantify the impact of denoising techniques on feature data processing. Feature data after the blind deconvolution routine enjoys the highest SNR and CCR, while simultaneously offering the lowest PMD when compared with other denoising approaches. In fact, for the case study that will be discussed in Section 11.4, the correlation coefficient and PMD for features at different operating conditions have reached performance indicators of 0.983% and 3.57.

11.2.2 Selection and Extraction of CIs

Feature or CI selection and extraction constitute the cornerstone for accurate and reliable fault diagnosis. The classical image recognition and signal processing paradigm of *data* → *information* → *knowledge* becomes most relevant and takes central stage in the fault diagnosis case, particularly since such operations must be performed online in a real-time environment.

Fault diagnosis depends mainly on extracting a set of CI from sensor data that can distinguish between fault classes of interest and detect/isolate a particular fault at its early initiation stages. These features should be fairly insensitive to noise and within fault class variations. The latter could arise because of fault location and size in the frame of a sensor. Good features must have the following attributes:

- Computationally inexpensive to measure
- Mathematically definable
- Explainable in physical terms
- Characterized by large interclass mean distance and small interclass variance
- Insensitive to extraneous variables
- Uncorrelated with other features

Past research has focused on feature extraction; whereas feature selection has relied primarily on expertise, observations, past historical evidence, and understanding of fault signature characteristics. In selecting an optimum feature set, we are seeking to address such questions as the following: Where is the information? How do fault (failure) mechanisms relate to the fundamental physics of complex dynamic systems? Fault modes may induce changes in the energy, entropy, power spectrum, and signal magnitude, among others.

Feature selection is application dependent. We are seeking those features, for a particular class of fault modes, from a large candidate set that possess properties of fault distinguishability and detectability while achieving a reliable fault classification in the minimum amount of time. Feature extraction, on the other hand, is an algorithmic process where features are extracted in a computationally efficient manner from sensor data, while preserving the maximum information content. Thus, the feature extraction process converts the fault data into an N-dimensional feature space, such that one-class of faults is clustered together and can be distinguished from other classes. However, in general, not all faults of a class need N features to form a compact cluster. It is only the faults that are in the overlapping region between two or more classes that govern the number of features required to perform classification.

We have developed a hybrid methodology for feature selection and extraction that relies on physics-based modeling of the fault modes in combination with sensor data as the latter are streaming into the processor. The physics-based models employ a finite element analysis technique jointly with a nonlinear dynamic model of the failing component's behavior to guide the selection process [14]. For example, in a typical helicopter main transmission gearbox, modeling of a crack (fault) on the planetary gear plate suggests that a good indicator may be computed from the meshing components in the frequency domain, i.e., by assessing the relative magnitudes of the dominant frequency components around a specific meshing frequency and those of the sidebands around them. Changes in the magnitudes are observed (dominants decreasing and sidebands increasing) as the crack grows. Figure 11.3 depicts the extraction process from accelerometer data [15,16].

Thus, an insight into the physics of failure mechanisms, in combination with preprocessed data, provides a reasonable and systematic approach to feature selection and extraction. We will build on these concepts to address fault modes in aircraft systems, i.e., our selected testbed. Feature evaluation and selection metrics include the similarity (or linear correlation) between the feature and the true fault (crack) size, based on the linear dependency between them. A feature is desirable for prognostic purposes if it shows a similar growth pattern to that of the ground truth

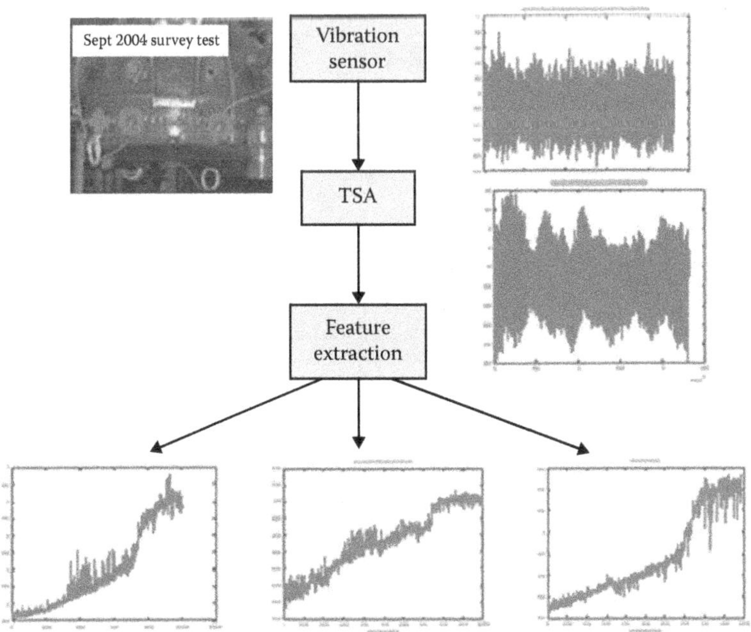

FIGURE 11.3 Feature extraction example.

data, that is, if the feature grows or decreases monotonically with respect to the severity of the fault condition. Furthermore, the better the SNR in selected features, the better the precision of prediction algorithms.

When multiple features are extracted for a particular fault mode, it might be desirable to combine or fuse uncorrelated features to enhance the fault *detectability.* One can take advantage of genetic programming algorithms to define an appropriate fitness function and using genetic operators to construct new feature populations from old ones.

11.2.3 Diagnostics and Prognostics Modules

The proposed scheme achieves robust component diagnosis by replicating multiple behaviors of a faulted system in a physics-based model. This procedure is illustrated in Figure 11.4. The simulated behaviors can be compared with particular instances of observed behaviors in an actual aircraft, while in operation, to determine the current fault or damage status of one of its subsystems, per the matching situation in the model. This is referred to as a reverse engineering approach.

We illustrate the modeling methodology in the case of a crack (fatigue damage) in the carrier plate of the planetary helicopter gearbox [14,17]. We constructed simulations using a physics-based model that replicate

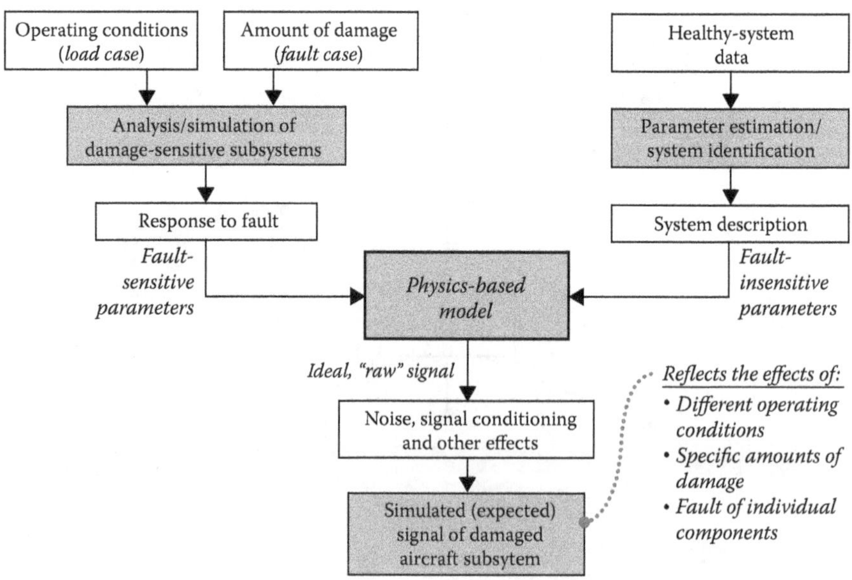

FIGURE 11.4 Suggested methodology for performing model-based simulation of faulted systems.

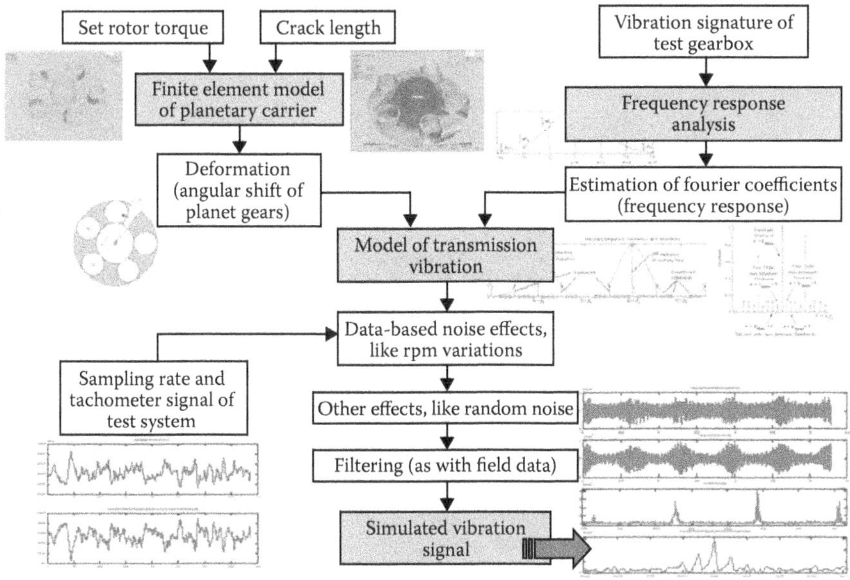

FIGURE 11.5 Application of a model-based technique for simulating the vibration of a faulted helicopter transmission.

the helicopter transmission's behavior under flight loads that then provide information on how vibration signals are expected to change when the crack is present. Actual tests are then used to observe these changes and lead to the detection of the crack. The technique of Figure 11.4 is applied to this particular example as illustrated in Figure 11.5.

Simulations can provide information about which vibration characteristics show a relationship with the crack length, which in turn makes possible to estimate the size of a crack directly from data. The procedure is illustrated in Figure 11.6.

The prognosis task attempts to estimate how quickly the damage of an aircraft subsystem will progress. We take into consideration that progression of the damage depends on how the system will be used (damage progression rates may be affected by changes in environmental conditions, load magnitude, and usage patterns), and that there is uncertainty in the estimated amount of damage when a fault is detected. We estimate the RUL—also known as the time to failure (TTF)—of the faulty component, as shown in Figure 11.7.

This novel modeling framework combines both physics-based principles and data-driven (measurement, models) techniques in a hybrid

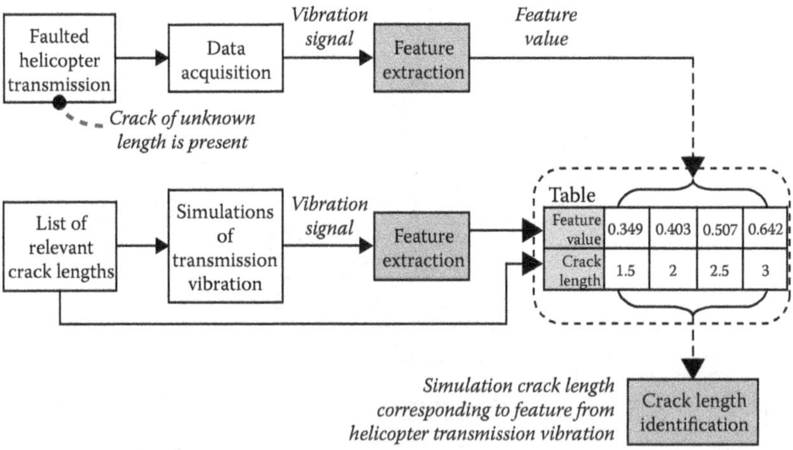

FIGURE 11.6 Application of the reverse engineering approach for performing model-based fault diagnosis in the helicopter transmission example.

setting to provide a better understanding of the physics of failure mechanisms and, on that basis, support the derivation of optimal fault features and parameters needed for fault diagnosis and prognosis.

11.3 PF ALGORITHMS IN A COMBINED MODEL-BASED/DATA-DRIVEN FRAMEWORK FOR FAILURE PROGNOSIS

The performance of a failure prognosis approach depends, to a great extent, on the ability of the dynamic model to mimic the behavior of the process under study. Linear and Gaussian dynamic models may help to

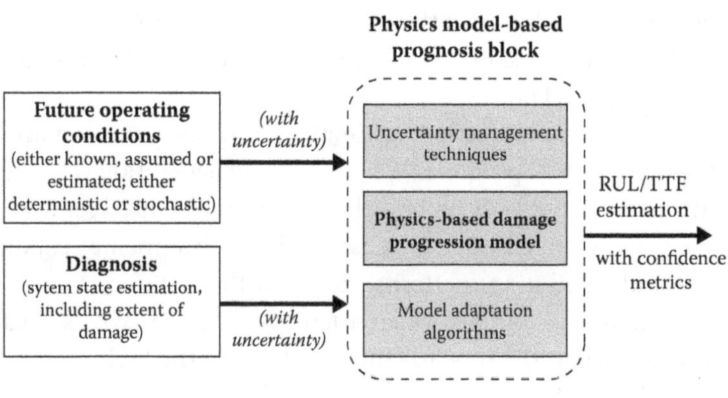

FIGURE 11.7 A general approach to realizing model-based prognostics.

describe this behavior satisfactorily when either the process complexity allows for it or when the time framework for long-term predictions is shortened. Most of the time, however, real processes require the inclusion of nonlinear dynamics or non-Gaussian stochastic components for an accurate description, especially when the time horizon required for the generation of dependable results is long enough to make evident any shortcomings introduced through linearization methodologies. For these reasons, a combined model-based/data-driven approach to prognosis not only should be capable of estimating the current condition of the system (and its model parameters), but also to adequately extrapolate the evolution of that condition in time.

11.3.1 PF Algorithms and Failure Prognosis

Nonlinear Bayesian filters, and particularly sequential Monte Carlo methods (also known as particle filters), provide a consistent theoretical framework to handle the problem of state estimation under the aforementioned conditions, i.e., to use noisy observation data to estimate at least the first two moments of a state vector governed by a dynamic nonlinear, non-Gaussian state-space model. Although we will not focus our discussion on the details associated to the implementation of PF algorithms, it suffices to say that these algorithms allow to approximate the posterior probability density $p(x_t/y_t)$ of the state vector x at time t, given a set of measurements $y_1, y_2, \ldots, y_t$ [18–20]. Note that some of the components of the state vector x may represent unknown model parameters that have to be estimated in an online fashion. On the one hand, the a priori probability density is determined by the state dynamic model that describes the discrete-time system (see Equation 11.8, where U is a vector of external inputs to the system).

$$\begin{cases} x_{t+1} = f(x_t, U_t, t) & \Leftrightarrow p(x_{t+1}/x_t, U_t) \\ y_t = h(x_t, t) & \Leftrightarrow p(y_t/x_t) \end{cases} \tag{11.8}$$

$$p(x_t/y_t) \propto p(x_t/x_{t-1}, U_{t-1}) \cdot p(y_t/x_t) \tag{11.9}$$

The posterior density $p(x_t/y_t)$, on the other hand, is approximated using a set of $N >> 1$ weighted samples $\{w_t^{(i)}, x_t^{(i)}\}_{i:1\cdots N}$, $w_t^{(i)} \geq 0$ (also referred to as particles) such that [19]:

$$p(x_t/y_t) \approx \sum_{i=1}^{N} w_t^{(i)} \cdot \delta(x_t - x_t^{(i)}). \tag{11.10}$$

As in the case of any adaptive prognosis scheme, it is assumed that there is at least one feature providing a measure that is correlated to the severity of the fault condition under analysis (fault dimension). If many features are available, they can always be combined to generate a single indicator (with the help of techniques from the computational intelligence arena, for example). Therefore, it is always possible to use state model 11.11, as a particular version of Equation 11.8, to describe the evolution in time of the fault dimension:

$$\begin{cases} x_{1,t+1} = x_{1,t} + x_{2,t} \cdot F(x_t, t, U) + \omega_{1,t} \\ x_{2,t+1} = x_{2,t} + \omega_{2,t} \\ y_t = x_{1,t} + \nu_t, \end{cases} \tag{11.11}$$

where $x_{1,t}$ is a state representing the fault dimension under analysis, $x_{2,t}$ is a state associated with an unknown model parameter, U is a vector of external inputs to the system (load/stress profiles), $F(\cdot)$ is a general time-varying nonlinear function, ω_1, ω_2, and ν are white noises (not necessarily Gaussian). The nonlinear function F may represent a model based on first principles, a neural network, or even a fuzzy system [21].

State Equation 11.11 provides the means to generate k-step ahead predictions for the expectation of the fault dimension, as well as to represent the evolution in time of the uncertainty that is associated to the state estimates. This uncertainty is characterized in the predicted conditional PDF $\hat{p}(x_{t+k}^{(i)} / x_t^{(i)})$, which describes the state distribution at the future time instant $t + k$ ($k > 1$) when the particle $x_t^{(i)}$ is used as initial condition of the system. Assuming that the current particle weights $\{w_t^{(i)}\}_{i:1\cdots N}$ are a good representation of the state PDF at time t, then it is possible to approximate the predicted state PDF at time $t + k$, by using a weighted sum of kernel functions and the law of total probabilities, as it is shown in Equation 11.12):

$$\hat{p}(x_{t+k} / x_{1:t}) \approx \sum_{i=1}^{N} w_t^{(i)} \cdot K_h(x_{t+k} - E[x_{t+k} / x_t^{(i)}]), \tag{11.12}$$

where $K_h(\cdot)$ is a scaled kernel density function, which may correspond to the process noise PDF, a Gaussian kernel, or a rescaled version of the Epanechnikov kernel [22,23].

The resulting predicted state PDF contains critical information about the evolution of the fault dimension over time. One way to represent that information is through the computation of statistics (expectations, 95% confidence intervals), the time of failure (ToF), and the remaining useful life (RUL) of the faulty system. A detailed procedure to obtain the RUL PDF from the predicted path of the state PDF is described and discussed in [23]. The general concept is as follows: The RUL PDF can be computed from the function of probability of failure at future time instants. This probability is calculated using both the long-term predictions and empirical knowledge about critical conditions for the system. This empirical knowledge is usually incorporated in the form of thresholds for main fault indicators, also referred to as the hazard zones.

In real applications, it is expected for the hazard zones to be statistically determined on the basis of historical failure data, defining a critical PDF with lower and upper bounds for the fault indicator (H_{lb} and H_{ub}, respectively). Since the hazard zone specifies the probability of failure for a fixed value of the fault indicator and the weights $\{w_t^{(i)}\}_{i:1\cdots N}$ represent the predicted probability for the set of predicted paths, then it is possible to compute the probability of failure at any future time instant (namely, the RUL PDF) by applying the law of total probabilities, as shown in Equation 11.13. Once the RUL PDF is computed, combining the weights of predicted trajectories with the hazard zone specifications, it is well known how to obtain prognosis confidence intervals, as well as the RUL expectation [23,24].

$$\hat{p}_{ToF}(t) = \sum_{i=1}^{N} w_t^{(i)} \cdot Pr(Failure / X = x_t^{(i)}, H_{ub}, H_{lb}) \tag{11.13}$$

Equations 11.11, 11.12, and 11.13 can be used to show that the a priori state PDF for future time instants, and thus the ToF PDF, directly depends on the a priori probability distribution of the load profile for future time instants. In fact, this approach has been successfully compared with other suboptimal techniques (such as those based on the extended Kalman filter) [21,23]. Most of the times, long-term predictions assume that the latter distribution is a Dirac's delta function, which basically implies a deterministic function of time for future load profiles. Although this simplification helps to speed up the prognostic procedure, it does not consider future changes in operating conditions or unexpected events that could affect the remaining useful life of the system under analysis. Monte Carlo simulation can be used to generate ToF estimates for arbitrary a priori distributions of future load conditions; however, it is not always possible to obtain these results in real time. In this sense, PF-based prognostic routines not only provide a theoretical framework where these concepts can be incorporated but also allow the use of uncertainty measures to characterize the sensitivity of the system, with respect to changes in future load distributions.

Furthermore, if a formal definition of mass probability is assigned to each possible stress condition, a ToF PDF estimate can be obtained as a weighted sum of kernels, where each kernel represents the PDF estimate of a known constant load. Indeed, if the a priori distribution of future operating conditions is given by

$$\hat{P}(U=u)=\sum_{j=1}^{N}\pi_j\cdot\delta(u-u_j), \tag{11.14}$$

where $\{u_j\}_{i:1\cdots N_u}$ is a set of deterministic functions of time, then the probability of failure at a future time t can be computed using Equation 11.15.

$$\hat{p}_{ToF}(t)=\sum_{j=1}^{N_u}\pi_j\sum_{i=1}^{N}w_t^{(i)}\cdot Pr(Failure\,/\,X=x_t^{(i)},H_{ub},H_{lb}) \tag{11.15}$$

Equations 11.8 through 11.15 represent a suitable theoretical foundation for real-time representation of uncertainty in a PF-based prognosis framework. They not only indicate how to use information from online feature measurements to update state estimates and parameters in empirical growth models but also show how to use those estimates to compute the predicted probability density distribution of the time of failure, assuming a statistical measure of uncertainty for system inputs and outputs [21,24]. The next step is to determine how to use these empirical measures to anticipate the effect that input variations have on the faulty system, particularly in terms of the resulting RUL estimates.

11.3.2 Uncertainty Measure-Based Feedback Loops for the Extension of Remaining Useful Life

The main motivation behind the definition of uncertainty measures for the outcomes of PF-based prognostic routines is to characterize in real time the effects that changes in operating conditions may cause on the system RUL. This is, however, only the first step needed to solve a much more complex issue: to establish correction loops to extend the remaining useful life of a process/subsystem, based on the current condition of the system, and a projection of its future evolution.

The aforementioned concept is illustrated in Figure 11.8, which depicts a situation where the objective is to predict the growth of given fault condition (detected at time t_{detect}) beyond the current time instant $t_{prognosis}$. In this case, the assumption of different stress profiles for the future operation of the plant may have a significant impact on the outcomes of prognostic algorithms for the system. The use of sensitivity measures to characterize the effect of input uncertainty on the uncertainty of RUL estimates provides a basis for the understanding of the modifications that the process requires to achieve the extension of its RUL. It is assumed, at this point, that the trivial solution to extend the RUL of a system (constant null load) is unfeasible, since that would indicate that the system operation is terminated (e.g., an aircraft cannot stay aloft without a nonzero level of stress being exerted on the system). Moreover, a set of feasible stress profiles and operating points is assumed to be given.

The stress sensitivity is a measure of the change in uncertainty in the RUL prediction, as a function of the uncertainty in the stress profile (input to the system). Stress sensitivity is found by adding Gaussian i.i.d. white

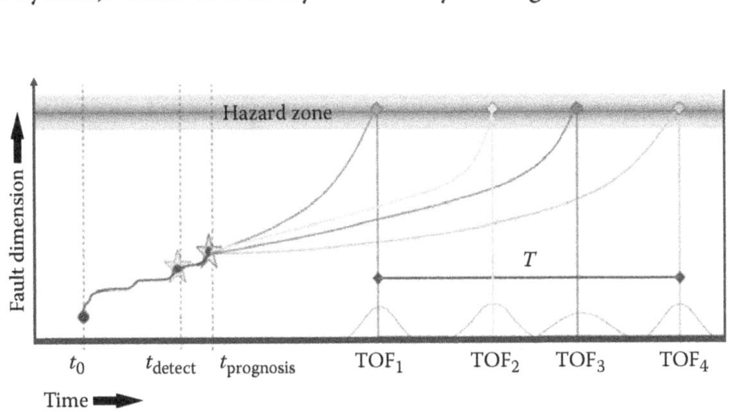

FIGURE 11.8 Predicted fault growth for different input stress levels.

noise to the median stress level U_{Base} (calculated among all feasible stress profiles) and then comparing the prognostic outcome with the RUL PDF that would be obtained when assuming the median stress level as a deterministic function of future time instants. This effect is illustrated in Figure 11.9 where the light-gray kernels show U_{Base} and the resulting RUL_{Base} PDF, whereas the gray kernels show $U_{Base+\omega}$ and the resulting $RUL_{Base+\omega}$ PDF. Since U_{Base} is a deterministic function of time, then its a priori distribution is represented as a Dirac's delta function. Stress sensitivity is measured in two ways, dispersion sensitivity (DS), defined in Equation 11.16, and confidence interval sensitivity (CIS), defined in Equation 11.17:

$$DS_\omega = \frac{stdev(RUL_{Base+\omega})}{stdev(RUL_{Base})}, \tag{11.16}$$

$$CIS_\omega = \frac{Length(CI\{RUL_{Base+\omega}\})}{Length(CI\{RUL_{Base}\})}, \tag{11.17}$$

where $RUL_{M+\omega}$ is the predicted RUL with a load factor of $U_{M+\omega}$, $U_{M+\omega}(t) = U_M(t) + \omega(t)$, and $\omega(t)$ is Gaussian white noise. Measures based on the stress sensitivity concept provide a means of determining how adjustments on the system inputs will affect the RUL prediction, without the need of running individual simulations for all possible future stress profiles.

As it has been mentioned before, the main motivation behind the definition of uncertainty measures is to provide the means to establish feedback correction loops for the extension of equipment RUL. In this sense, this section presents and analyzes a measure-based method that is proposed as

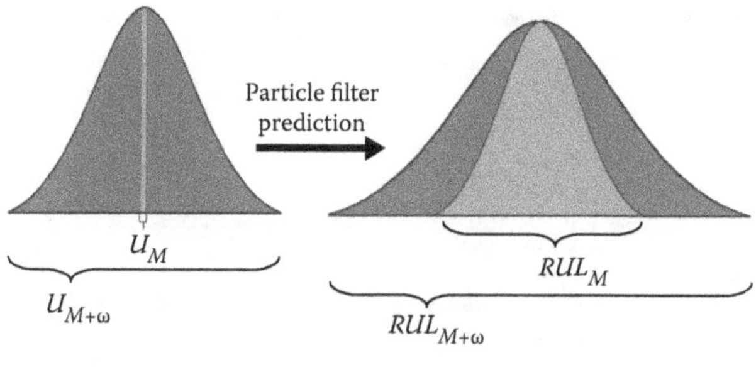

FIGURE 11.9 Illustration of PDF kernels, associated to the concept of stress sensitivity.

a general approach to implement such correction loops. The method utilizes a PF-based prognosis framework to determine the baseline PDF estimate of the remaining useful life (RUL_{Base}) and then utilizes the sensitivity measures (DS and CIS) to determine an appropriate stress level that will extend the RUL of the component to the specified desired RUL (RUL_d). Two approaches to this method are outlined below: DS- and CIS-based.

11.3.2.1 DS-Based Approach to RUL Extension

Consider a baseline RUL PDF (RUL_{Base}) that is determined through PF-based prognostic routines using the expected future stress profile U_{Base}. The proposed DS-based approach to RUL extension uses knowledge of the dispersion sensitivity measure to extend the remaining useful life from $E[RUL_{Base}]$ to RUL_d by adjusting the stress factor to a safe level (U_d). To determine U_d, the standard deviation of the RUL PDF estimate which places RUL_d in the 95th percentile of the distribution, while maintaining a mean of mean$\{RUL_{Base}\}$, must be determined. This distribution is denoted as $RUL_{Base+\bar{\omega}}$, as shown in Equation 11.18.

Using a linear fit (see Figure 11.10) to map the standard deviation of the stress to the standard deviation of the RUL PDF, it is possible to compute the standard deviation of the stress profile required to output a distribution of $RUL_{Base+\bar{\omega}}$ using Equation 11.19. The standard deviation of this stress is then utilized to determine how much the baseline stress must be reduced to attain a remaining useful life of RUL_d, as seen in Equation 11.20.

$$stdev\{RUL_{Base+\bar{\omega}}\} = \frac{RUL_D - E\{RUL_{Base}\}}{Z_{0.95}} \tag{11.18}$$

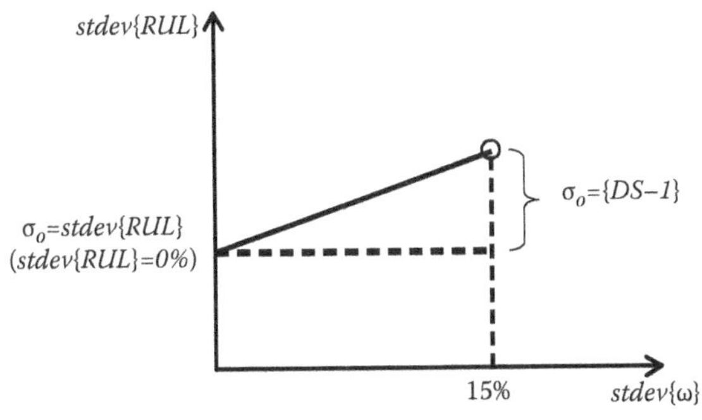

FIGURE 11.10 Linear mapping between standard deviation of stress profile and standard deviation of estimated RUL PDF.

$$stdev\{U_{Base+\bar{\omega}}\} = \left(\frac{stdev\{RUL_{Base+\bar{\omega}}\}}{stdev\{RUL_{Base}\}} - 1\right)\frac{stdev\{\omega\}}{DS-1} \tag{11.19}$$

$$U_d = U_{Base} - stdev\{U_{Base+\bar{\omega}}\} \tag{11.20}$$

11.3.2.2 CIS-Based Approach to Rule Extension

Similar to the DS-based, the CIS-based approach to RUL extension uses knowledge of the confidence interval sensitivity to extend the RUL from $E[RUL_{Base}]$ to RUL_d by adjusting the stress factor to a safe level (U_d), given a PF-based estimate for RUL_{Base} that considers a baseline stress profile U_{Base}. To determine U_d, the confidence interval length of the RUL prediction that places RUL_d at the highest end of the confidence interval of the distribution, while maintaining a mean of $mean\{RUL_{Base}\}$, must be determined. This distribution is denoted as $RUL_{Base+\bar{\omega}}$, as shown in Equation 11.21. Using a linear fit (see Figure 11.11) to map the confidence interval length of the stress to the confidence interval length of the remaining useful life, the standard deviation of the stress required to output a distribution of $RUL_{Base+\bar{\omega}}$ is determined by Equation 11.22. The standard deviation of this stress is then utilized to determine how much the baseline stress must be reduced to attain a remaining useful life of RUL_d, as seen in Equation 11.23.

$$Length(CI\{RUL_{Base+\bar{\omega}}\}) = 2(RUL_D - E\{RUL_{Base}\}) \tag{11.21}$$

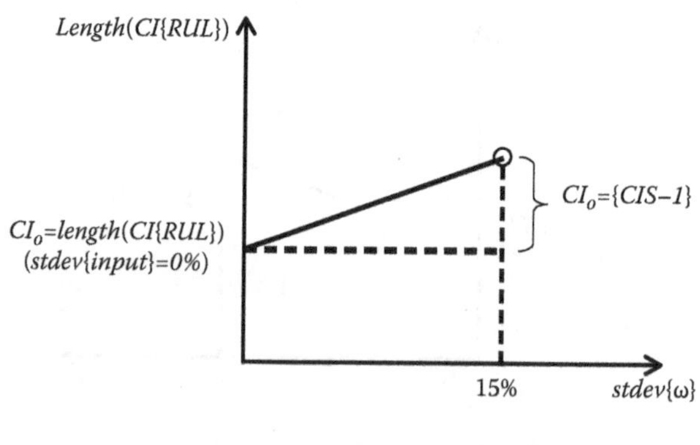

FIGURE 11.11 Linear mapping between the length of the confidence interval for stress profile and length of the confidence interval of estimated RUL PDF.

$$stdev\{U_{Base+\bar{\omega}}\} = \left(\frac{length(CI\{RUL_{Base+\bar{\omega}}\})}{length(CI\{RUL_{Base}\})} - 1 \right) \frac{stdev\{\omega\}}{CIS-1} \tag{11.22}$$

$$U_d = U_{Base} - stdev\{U_{Base+\bar{\omega}}\} \tag{11.23}$$

Both the DS- and CIS-based approaches are validated using actual fault data to demonstrate the efficacy of the proposed feedback scheme. Details of the case study that was utilized for this purpose follows next.

11.4 CASE STUDY: LOAD REDUCTION AND EFFECTS ON FATIGUE CRACK GROWTH IN AIRCRAFT COMPONENTS

An appropriate case study has been designed to test and show the potential of the proposed feedback correction strategy. This case study uses data (from a seeded fault test) that describes a propagating fatigue crack on a critical component in a rotorcraft transmission system. This particular fault mode generally leads to a critical failure condition in the aircraft, and only until very recently, there was no certain way to determine its existence save by a detailed offline inspection—a procedure that involves a significant financial cost.

In this data set, the crack has been artificially grown until it reached a total length of 1.34 inches, and after that, the gearbox was forced to operate with input changes that included variations between 20% and 120% of nominal load in a 3-min ground-air-ground (GAG) cycle (see Figure 11.12). From the material structure theory, it is well known that the crack growth evolution may be explained by using an empirical model such as the Paris' law (Equation 11.24), given the proper set of coefficients [14]:

$$\frac{dL}{dt} = C(U_t \Delta K_t)^m \tag{11.24}$$

where L is the total crack length, C and m are material-related coefficients, t is the cycle index, U_t is the parameter that includes the effect of crack closure during cycle t, and K_t is the crack tip stress variation during the cycle t (in MN/m$^{3/2}$). The implementation of a prognostic framework based on model 11.24 necessarily requires the computation of two critical time-varying parameters: K_t and U_t. Stress K_t may be estimated for a constant

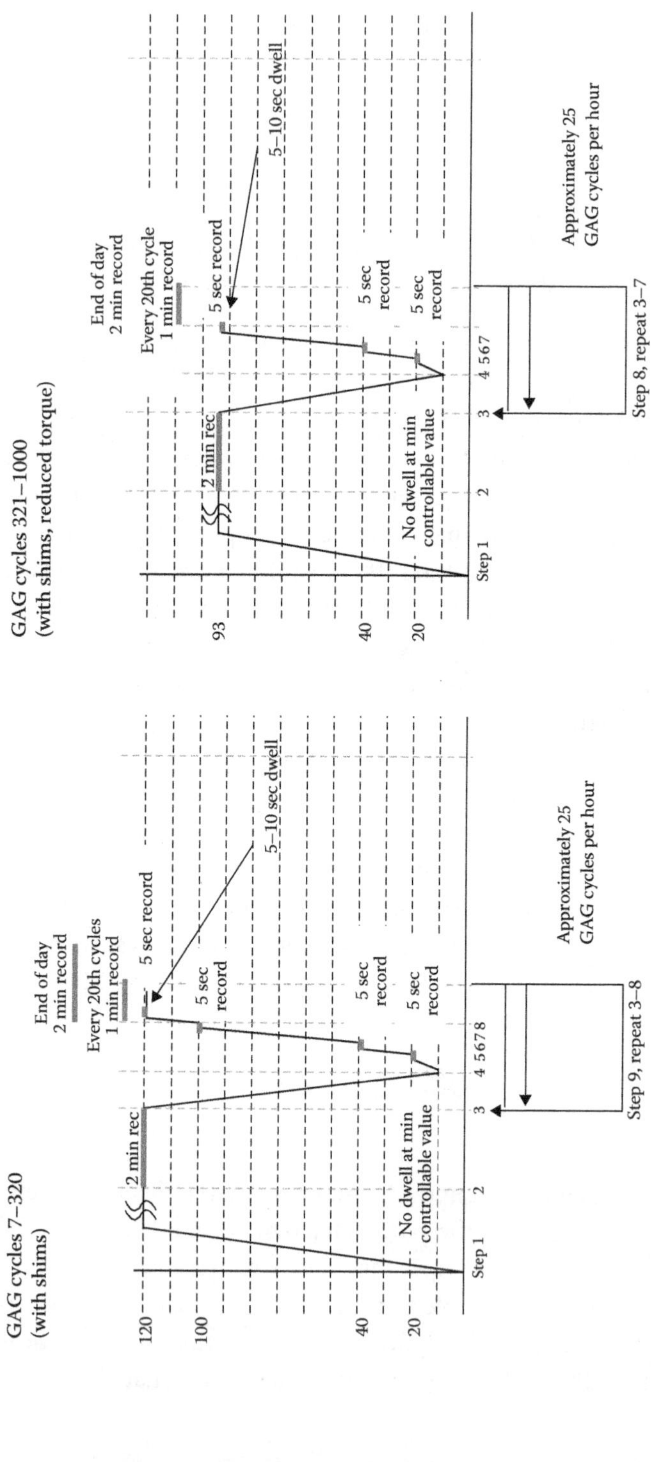

FIGURE 11.12 Loading profile (%) versus GAG cycles.

load (usually 100%) by using finite element analysis (FEA) tools such as ANSYS for different crack lengths and crack orientations. In addition, if a proportional relationship is considered between the stress on the tips of the crack and the load applied to the system, then it is possible to relate both the current crack length and load variation (per cycle) with K_t. The problem is that this modeling effort, although helpful, is insufficient to estimate the evolution of the crack length. However, the closure effect parameter U_t cannot be efficiently measured and only empirical approximations exist for certain materials. Even in the case of those materials, only upper and lower bounds may be computed, and thus it is impossible to compute expectations and/or determine statistically the validity of confidence intervals.

The inclusion of process data, measured and preprocessed in real time (see Figure 11.13), improves tremendously the prospect of what can be achieved in terms of prognostic results. The use of features based on the

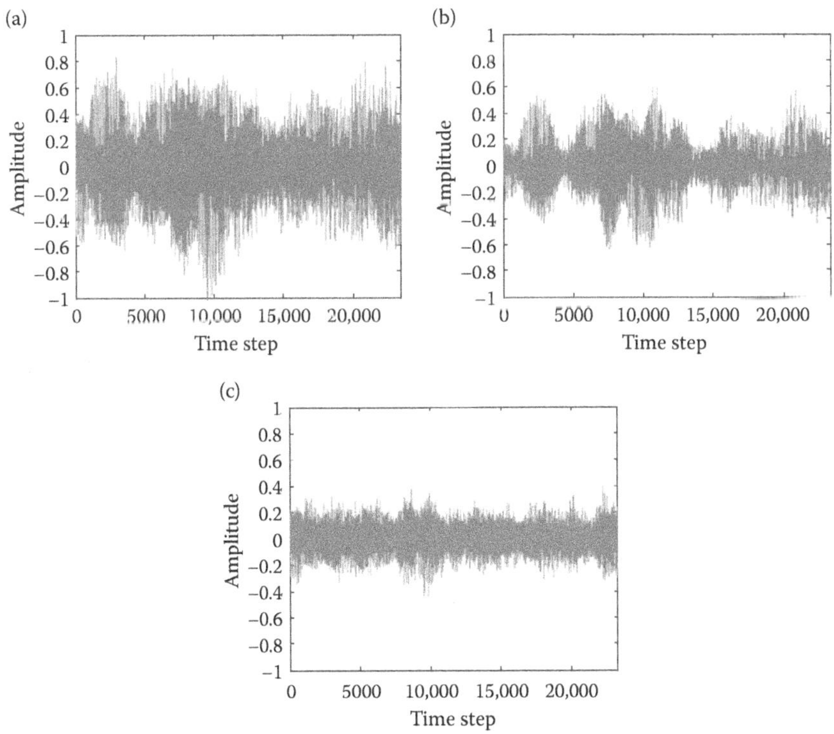

FIGURE 11.13 (a) Measured vibration signal s_t; (b) recovered noise-free vibration signal b_t; (c) noise signal n_t. Denoised vibration signal is used for purposes of feature computation.

ratio between the fundamental harmonic and the sidebands in the vibration data spectrum [14] allows the implementation of the PF-based prognosis framework that is discussed in this chapter (see Figure 11.14). In that manner, not only it is possible to estimate the expected growth of the crack but also the unknown closure parameter in crack growth model 11.24 and the RUL PDF, enabling the computation of statistics, expectations, and confidence intervals. In fact, under this premise, Equation 11.25 represents a suitable crack growth model to be used for real-time state estimation purposes [23]:

$$\begin{cases} x_{1,t+1} = x_{1,t} + Cx_{2,t} \cdot ((\Delta K_t^{inboard})^m + (\Delta K_t^{outboard})^m) + \omega_{1,t} \\ x_{2,t+1} = x_{2,t} + \omega_{2,t} \\ y_t = h(x_{1,t}) + v_t \end{cases} \tag{11.25}$$

where $x_{1,t}$ is the total crack length estimate at GAG cycle t, the state $x_{2,t}$ represents an unknown time-varying model parameter to be estimated (unitary initial condition), C and m are model constants related to material properties, and $\omega_{1,t}$, $\omega_{2,t}$, and v_t are non-Gaussian white noises. ΔK_t (inboard/outboard) is the stress variation that is effective on the tips of the crack (a function of the load profile, and the current crack length) and it can be computed through interpolation techniques. Information to be used in the interpolation procedure is obtained from offline analysis of the system, using ANSYS, for a selected subset of operating conditions [14]. Process model 11.25 is fed with denoised feature data y_t [10], which is related with the fault dimension through a bijective nonlinear mapping $h(\cdot)$ (see Figure 11.14). In this manner, this scheme allows to improve the state estimate every time a new denoised feature measurement is included, helping to ensure the enhancement of both the precision and accuracy of RUL estimated through time.

Now, consider within this framework a situation where the pilot must remain airborne for a given amount of time to reach a safe landing destination. The RUL extension methods discussed in this chapter will provide the pilot, or reconfigurable controller, with the information needed to adjust the load of the aircraft and reduce the stress on the failing component, with the purpose of extending the RUL to a desired time that ensures safe landing. Although a physics-based model for a system of these

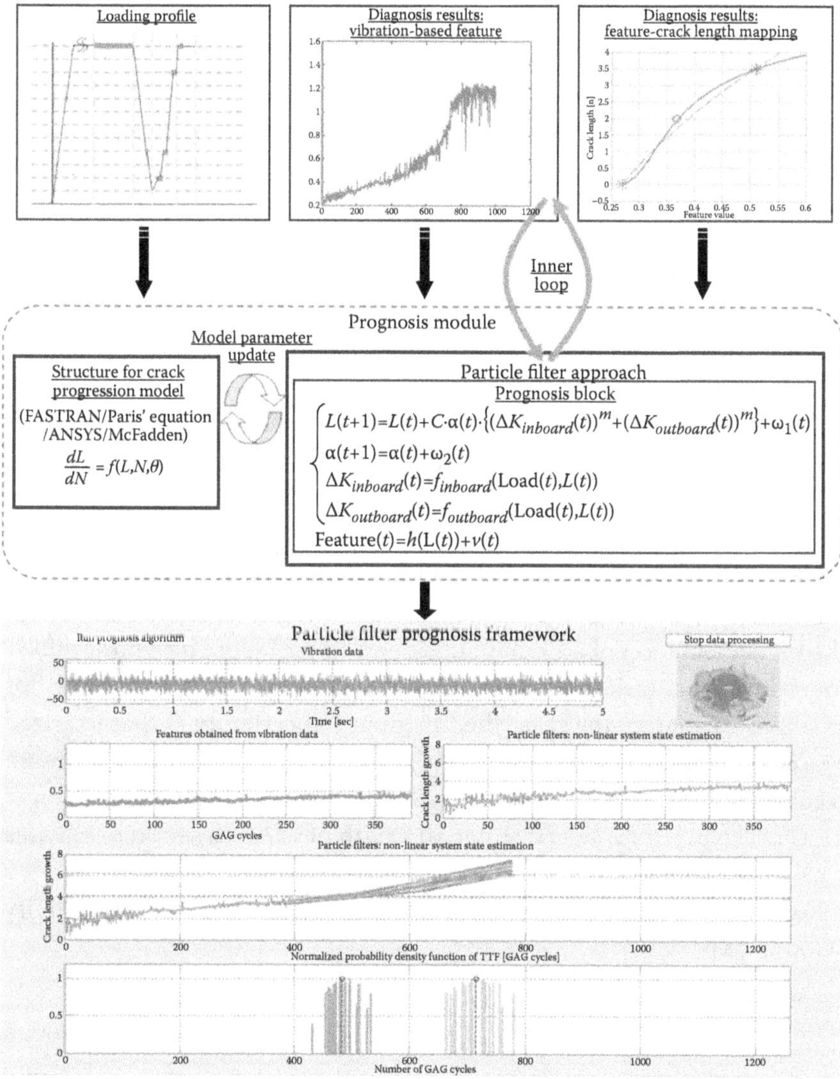

FIGURE 11.14 PF-based framework for prognosis: Crack growth in a planetary carrier plate. State estimation techniques are used to process denoised feature data, obtaining a ToF PDF estimate for selected thresholds in the feature value that define critical conditions for the system. The illustration shows the obtained results when two different thresholds are considered (gray and light-gray PDFs at the bottom).

characteristics is a complex matter, it is possible to represent the growth of the crack (fault dimension) using the simplified model 11.25, where the nonlinear mapping functions are defined on the basis of an ANSYS stress model for the inner and outer tips of the fatigue crack [14,23,24].

Under this scenario, the use of algorithms capable of estimating the RUL by only analyzing vibration-based features becomes extremely attractive and would help to dramatically decrease operational and maintenance costs, as well as the number of catastrophic events.

In the experiment, the baseline stress level was 120% of the maximum recommended torque. If this information is fed into the proposed PF-based prognosis framework, then the resulting ToF PDF (see gray PDF in Figure 11.15), computed at the 300th cycle of operation, has an expectation of 594 cycles, a standard deviation of 12.44 cycles, and a confidence interval length of 38 cycles for $\alpha = 95\%$. If we were to compute the DS and CIS measures for this system at that particular cycle of operation (300th cycle), then it is necessary to compute the statistics of the ToF PDF that results after including uncertainty in the system input. Given that the implementation of a PF-based framework for failure prognosis allows to perform this task in a simple and efficient manner, it is possible, for example, to analyze the case when the input uncertainty is characterized by zero-mean Gaussian noise (standard deviation of 15% of maximum recommended torque). The resulting ToF PDF, has a standard deviation of 41.52 cycles and a confidence interval length of 142 cycles for $\alpha = 95\%$ (see dark-gray PDF in Figure 11.15).

Considering the aforementioned information, the dispersion sensitivity is found to be

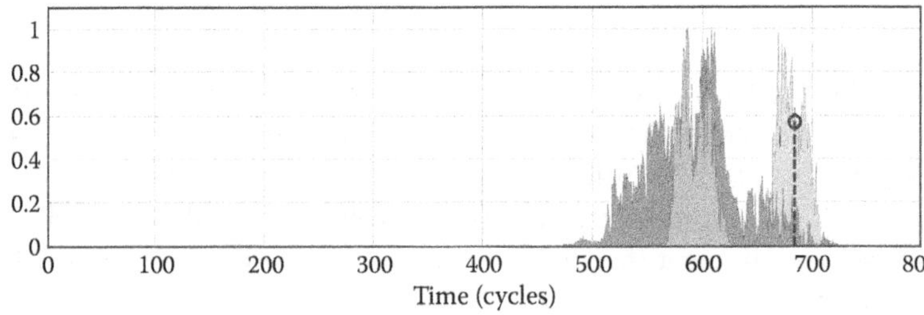

FIGURE 11.15 ToF PDF considering baseline (gray), noisy (dark-gray), and desired (light-gray) stress profiles for the problem of RUL estimation in the case study (cracked gear plate).

$$DS_{15\%} = \frac{stdev\{RUL_{Base+\omega}\}}{stdev\{RUL_{Base}\}} = \frac{41.52cycles}{12.44cycles} = 3.3362, \tag{11.26}$$

and the confidence interval sensitivity is computed as

$$CIS_{15\%} = \frac{length(CI\{RUL_{Base+\omega}\})}{length(CI\{RUL_{Base}\})} = \frac{142cycles}{38cycles} = 3.7368. \tag{11.27}$$

For this system, the desired ToF is 714 cycles (RUL of 414 cycles). If we were to use the DS-based approach to RUL extension to suggest a correction in the stress profile for the system, then the standard deviation of the noise level required to locate cycle 714 at the 95th percentile of the predicted dark-gray ToF PDF is

$$\begin{aligned} stdev\{RUL_{Base+\bar{\omega}}\} &= \frac{RUL_D - E\{RUL_{Base}\}}{Z_{0.95}} \\ &= \frac{714-594}{1.627} = 73.755. \end{aligned} \tag{11.28}$$

Inserting this value into Equation 11.19 and solving for $stdev\{U_{Base+\bar{\omega}}\}$ yields a required standard deviation of 31.64% for the input stresses. Therefore, to achieve the desired RUL of 714 cycles, the stress factor must be reduced by 31.64% from 120% to 88.36%.

Similarly, for the CIS-based approach to RUL extension, it is possible to estimate the required variation considering:

$$\begin{aligned} length(CI\{RUL_{Base+\bar{\omega}}\}) &= 2(RUL_D - E\{RUL_{Base}\}) \\ &= 2(714-594) = 240. \end{aligned} \tag{11.29}$$

Inserting this value into 11.22 and solving for $stdev\{U_{Base+\bar{\omega}}\}$ yields a required standard deviation of 29.13% for the input stress. Therefore, to achieve the desired RUL of 714 cycles, the stress factor must be reduced by 29.13% from 120% to 90.70%. Compare 88.36% and 90.70% to the actual stress factor that results in a RUL of 714, which is 93%. Clearly, both approaches for stress correction suggest a modification, for the system input, that would have translated in an appropriate extension of the remaining useful life of the system.

11.5 CONCLUSIONS

This chapter presents theoretical and practical aspects associated to the implementation of a combined model-based/data-driven approach for failure prognostics based on PF algorithms, in which the current estimate of the state PDF is used to determine the operating condition of the system and predict the progression of a fault indicator, given a dynamic state model and a set of process measurements. In this approach, the task of estimating the current value of the fault indicator, as well as other important changing parameters in the environment, involves two basic steps: the prediction step, based on the process model, and an update step, which incorporates the new measurement into the a priori state estimate.

This framework allows to estimate the probability of failure at future time instants (RUL PDF) in real time, providing information about TTF expectations, statistical confidence intervals, and long-term predictions, using for this purpose empirical knowledge about critical conditions for the system (also referred to as the hazard zones). This information is of paramount significance for the improvement of the system reliability and cost-effective operation of critical assets, as it has been shown in a case study where feedback correction strategies (based on uncertainty measures) have been implemented to lengthen the RUL of a rotorcraft transmission system with propagating fatigue cracks on a critical component. Although the feedback loop is implemented using simple linear relationships, it is helpful to provide a quick insight into the manner that the system reacts to changes on its input signals, in terms of its predicted RUL. The method is able to manage non-Gaussian PDF's because it includes concepts such as nonlinear state estimation and confidence intervals in its formulation.

Real data from a fault-seeded test showed that the proposed framework was able to anticipate modifications on the system input to lengthen its RUL. Results of this test indicate that the method was able to successfully suggest the correction that the system required.

REFERENCES

[1] I. Dar, G. Vachtsevanos. Feature level sensor for pattern recognition using an active perception approach. In *Proceedings of IS&T/SPIE's Electronic Imaging '97: Science and Technology*, 1997.

[2] A.D. Flint. A prognostic maintenance system based on the hough transformation. *Transactions of the Institute of Measurement and Control*, 16(2):59–65, 1994.

[3] G. Hadden, P. Bergstrom, B. Bennett, G. Vachtsevanos, J. Van Dyke. Shipboard machinery diagnostics and prognostics/condition based maintenance: A progress report. In *Proceedings of the Maintenance and Reliability Conference, MARCON 99*, pp. 73.01–73.16, 1999.

[4] G. Hadden, G. Vachtsevanos, B. Bennett, J. Van Dyke. Machinery diagnostics and prognostics and prognostics/condition based maintenance: A progress report, failure analysis: A foundation for diagnostics and prognostics development. In *Proceedings of the 53rd Meeting of the Society for Machinery Failure Prevention Technology*, 1999.

[5] K.A. Marko, J.V. James, T.M. Feldkamp, C.V. Puskorius, J.A. Feldkamp, D. Roller. Application of neural networks to the construction of virtual sensor and model-based diagnostics. In *Proceedings of ISATA 29th International Symposium on Automotive Technology and Automation*, pp. 133–138, 1996.

[6] T.D. Pebbles, M.A. Essawy, S. Fein-Sabatto. An intelligent methodology for remaining useful life estimation of mechanical components. In *Proceedings of the Maintenance and Reliability Conference, MARCON 99*, pp. 27.01–27.09 1999.

[7] B.J. Rosenberg. The Navy IDSS program: adaptive diagnostics and feedback analysis—precursors to a fault prognostics capability. In *Proceedings of the IEEE National Aerospace and Electronics Conference, NAECON*, vol. 3, pp. 1334–1338, 1989.

[8] M. Roemer, C. Byingston, G. Kacprzynski, G. Vachtsevanos. An overview of selected prognostic technologies with reference to an integrated PHM architecture. In *Proceedings of NASA Integrated Vehicle Health Management Workshop*, 2005.

[9] A. Saxena, B. Wu, G. Vachtsevanos. Integrated diagnosis and prognosis architecture for fleet vehicles using dynamic case-based reasoning. In *Proceedings of AUTOTESTCON 2005 Conference*, 2005.

[10] B. Zhang, T. Khawaja, R. Patrick, G. Vachtsevanos, M. Orchard, A. Saxena. Application of blind deconvolution denoising in failure prognosis. *IEEE Transactions on Instrumentation and Measurement*, 58(2):303–310, 2009.

[11] D. Kundur, D. Hatzinakos. Blind image deconvolution. *IEEE Signal Processing Magazine*, 13(3):43–46, 1996.

[12] R. Peled, S. Braun, M. Zacksenhouse. A blind deconvolution separation of multiple sources with application to bearing diagnostics. *Mechanical Systems and Signal Processing*, 14(3):427–442, 2000.

[13] B. Zhang, T. Khawaja, R. Patrick, G. Vachtsevanos, M. Orchard, A. Saxena. A novel blind deconvolution de-noising scheme in failure prognosis. *Transactions of the Institute of Measurement and Control*, 32(1):3–30, 2010.

[14] R. Patrick, M. Orchard, B. Zhang, M. Koelemay, G. Kacprzynski, A. Ferri, G. Vachtsevanos. An integrated approach to helicopter planetary gear fault diagnosis and failure prognosis. In *2nd Annual Systems Readiness Technology Conference, AUTOTESTCON 2007*, 2007.

[15] M. Orchard, B. Wu, G. Vachtsevanos. A particle filter framework for failure prognosis. In *Proceedings of WTC2005 World Tribology Congress III*, 2005.

[16] B. Wu, A. Saxena, R. Patrick, G. Vachtsevanos. Vibration monitoring for fault diagnosis of helicopter planetary gears. In *Proceedings of the 16th IFAC World Congress*, 2005.

[17] J. Keller, P. Grabill. Vibration monitoring of a UH-60a main transmission planetary carrier fault. In *Proceedings of the 59th Annual Forum AmericanHelicopter Society*, 2003.

[18] A. Andrieu, C. Doucet, E. Punskaya. Sequential Monte Carlo methods for optimal filtering. In A. Doucet, N. De Frietas, N. Gordon, editors, *Sequential Monte Carlo Methods in Practice*, Springer-Verlag, New York, 2001.

[19] M.S. Arulampalam, S. Maskell, N. Gordon, T. Clapp. A tutorial on particle filters for online nonlinear/non-Gaussian Bayesian tracking. *IEEE Transactions on Signal Processing*, 50(2), 2002.

[20] A. Doucet, N. de Freitas, N. Gordon. An introduction to sequential Monte Carlo methods. In A. Doucet, N. De Frietas, N. Gordon, editors, *Sequential Monte Carlo Methods in Practice*, Springer-Verlag, New York, 2001.

[21] M. Orchard, F. Tobar, G. Vachtsevanos. Outer feedback correction loops in particle filtering-based prognostic algorithms: statistical performance comparison. *Studies in Informatics and Control*, 18(4):295–304, 2009.

[22] C. Musso, N. Oudjane, F. Le Gland. Improving regularized particle filters. In A. Doucet, N. De Frietas, N. Gordon, editors, *Sequential Monte Carlo Methods in Practice*, Springer-Verlag, New York, 2001.

[23] M. Orchard, G. Vachtsevanos. A particle filtering approach for on-line fault diagnosis and failure prognosis. *Transactions of the Institute of Measurement and Control*, 31(3–4):221–246, 2009.

[24] M. Orchard. *On-Line Fault Diagnosis and Failure Prognosis Using Particle Filters. Theoretical Framework and Case Studies.* VDM Verlag Dr. Müller Aktiengesellschaft and Co. KG, Saarbrücken, Germany, 2009.

CHAPTER 12

Hybrid Models for Engine Health Management

Allan J. Volponi

Ravi Rajamani

CONTENTS

12.1 INTRODUCTION

Engine health management (EHM) systems deal with the monitoring, detection, isolation, predictive trending, and accommodation of engine degradation, faults, and failures. The approaches in current use draw from a wide range of engine system disciplines including materials, structures, aerodynamics, and controls. Minimizing total cost of engine ownership, and increasing asset availability and readiness, while ensuring safe operation, are key drivers for the implementation of an EHM system.

Parameter estimation in the form of system modeling is critical to system diagnostics and prognostics. The current evolution in on-board EHM systems, aimed at performing real-time engine diagnostics and prognostics, has placed a greater demand on model accuracy and implementation speed. The complexities in assembling accurate physics-based models for real-time operation have placed greater focus on the use of *hybrid* engine models employing some form of empirical modeling. Such an approach to engine modeling has previously been introduced [1,2] wherein a *hybrid* engine model architecture incorporating both physics-based and empirical components is utilized. This architecture provides a means to automatically *tune* the engine model to a particular configuration as the engine evolves over the course of its life and, furthermore, aligns the model to the particular engine being monitored to ensure accurate performance tracking and mitigates the effects of engine-to-engine variation.

Combining the physics-based and empirical approaches allows one to derive the inherent benefits from both methods. For example, embedded applications tend to use low order physics-based models due to throughput constraints. It is difficult to ensure that these models maintain their accuracy over their entire operating region when compared with detailed aerothermal models used for design purposes. By contrast, empirical models based on techniques employing statistical constructs, neural networks, support vector machines, and the like, need large amounts of training data over all operating conditions to improve accuracy. On the other hand, a hybrid model, that uses a fast physics-based model at its core, on the other hand, can negotiate changing operating points rather well. It still needs to be retrained as conditions change, but with clever automation of this function, this can be handled with acceptable overhead.

Developing a better model is only the beginning of the process. Once the model is available, this then becomes the basis for developing diagnostics algorithms. The fundamental philosophy behind the model-based diagnostics technique is that deviations in system outputs from expected behavior (as estimated by the model) are indicative of system health. These algorithms are validated against real and simulated data and then coded in software, hosted either on-board or in ground-based systems. To make sure that the deployed diagnostics systems faithfully implement the requirements of the user, sound software development processes are needed. However, this has not always been practiced within the EHM discipline. The EHM discipline shares many tools and methods with the engine controls discipline, but the use of software quality methods has typically not been as advanced. The reasons for this are not hard to find. A major reason is that the use of standard software practices has been mandated by regulatory authorities, such as the Federal Aviation Administration (FAA) in the US, at least since the 1980s for on-board controls software, mainly because this software is considered safety critical. This has led to the industry adopting rigorous software development standards in their controls departments. The same is not true of the EHM area. Monitoring and diagnostics software have had little or no rigor applied to them from a software perspective, mainly because the failure of EHM software have not had any safety-related consequences. This is changing now as EHM systems are becoming more critical to the business of engine operation. Operators depend on EHM systems to monitor fleet performance and to help them make key maintenance decisions.

It is not our intent in this chapter to describe the entire software development process for EHM systems. Instead we will only give a brief overview of a crucial element of the process: verification and validation (V&V). Software development actually begins with the development of a comprehensive set of systems requirements. This set of requirements, signed off by all stakeholders, serves as a necessary guide to, and a valuable articulation of, the way the EHM system will actually be used in practice. Once the requirements are in place, the software system is architected, designed, and coded. Each step along the way has various V&V activities associated with it. Sound software development practices require V&V activities to be planned up front and followed diligently through all phases of system development. V&V activities help in surfacing issues early in the process and prevent quality glitches from escaping into the production system.

To summarize, in this chapter, we present an overview of the development and use of hybrid models in EHM systems. We do this with an emphasis on the practical aspects of the process. We show how requirements generation and V&V are crucial processes within software development that ensure that EHM algorithms deliver value to the customer.

12.2 BACKGROUND

A practical consideration for implementing a real-time on-board module performance tracking system is the development of a high fidelity engine model capable of providing a reference level from which performance changes can be tracked. Real-time engine models made their advent with the state variable model (SVM) in the mid-1980s, which provided a piecewise linear model that granted a reasonable representation of the engine during steady-state operation and mild transients. Increased processor speeds over the next decade allowed more complex models to be considered, that were combinations of linear and nonlinear physics-based components. While the latter may provide greater fidelity over transient operation and flight envelope excursions, it bears the limitation of potential model obsolescence as performance improvements in the form of hardware modifications, bleed and stator vane schedules alterations, cooling flow adjustments, and the like, are made during an engine's life cycle. Over time, these models may deviate enough from the actual engine being monitored that the module performance estimations are inaccurate and misleading. To mitigate these effects, an alternate approach to engine modeling can be taken, utilizing a *hybrid* engine model architecture that incorporates both physics-based and empirical components. This methodology provides a means to tune the engine model to a particular configuration as the engine development matures and furthermore, aligns the model to the particular engine being monitored to ensure accurate performance tracking while not compromising real-time operation. This approach effectively removes the uncertainty introduced with engine-to-engine variability since each monitored engine would, in effect, have its own custom (*hybrid*) model representation.

Although there were variants in the individual approaches, the first attempts to incorporate on-board models utilized a simple real-time engine model, typically an SVM, coupled with a Kalman filter (KF) observer that together provided an adaptable engine model capable of tracking the monitored engine during steady-state and transient operation. The KF observer would act upon the residuals formed by the output of the SVM and the actual observed measurements to provide a set of tuners that would adapt the SVM to match the actual observations (hence driving the residuals to zero, on the average). The tuners consist of a set of engine module performance parameters such as efficiencies and flow parameters that would allow the engine states and output parameters to be adjusted in order to provide a more faithful match to the actual engine. A typical architecture for such a model is depicted in Figure 12.1, and the tuners are depicted as the "performance deltas" (or performance Δ).

The benefits derived from systems such as this are twofold: (1) engine parameter synthesis (both measured and nonmeasured) from the engine

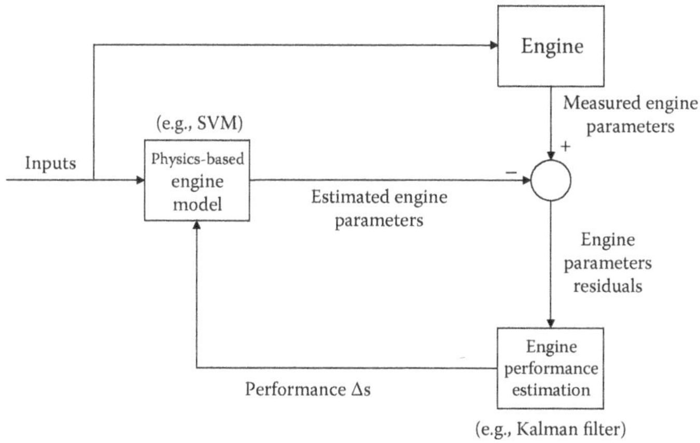

FIGURE 12.1 Typical on-board engine model architecture.

model and (2) engine module performance tracking, as provided by the tuners. The former is useful for fault detection and accommodation (FDA) in the full authority digital engine control (FADEC) and in control law implementation, while the latter has the potential to provide valuable performance deterioration information to support engine maintenance operations and logistics as well as providing a means to detect performance anomalies on-board to enhance flight safety. (FADEC is a digital engine control module on a modern aircraft engine that has "full control authority," which means that it does not need a backup hydromechanical control unit as was required on older engines.) The word *potential* in the preceding sentence is intentional. A necessary condition to achieve the performance tracking capability is that the (physics-based) engine model provides a faithful representation of the actual monitored engine, i.e., drive the measured parameters residues to zero. When the engine model does not provide an accurate representation of the monitored engine then the residual difference will be interpreted (for the most part erroneously) as changes in module performance. Hence, the tuners, which are supposed to represent physical attributes such as component efficiencies and flow parameters, instead become mathematical artifacts that help to close the gap between the model and the engine. Figure 12.2 depicts such a situation. The tuners are module efficiencies from an on-board system configured as illustrated in Figure 12.1. The data represents a flight segment of a nominal high bypass commercial turbofan engine. The climb starts around 900 sec and ends around 1150 sec in the plot. The segment consists of a level flight

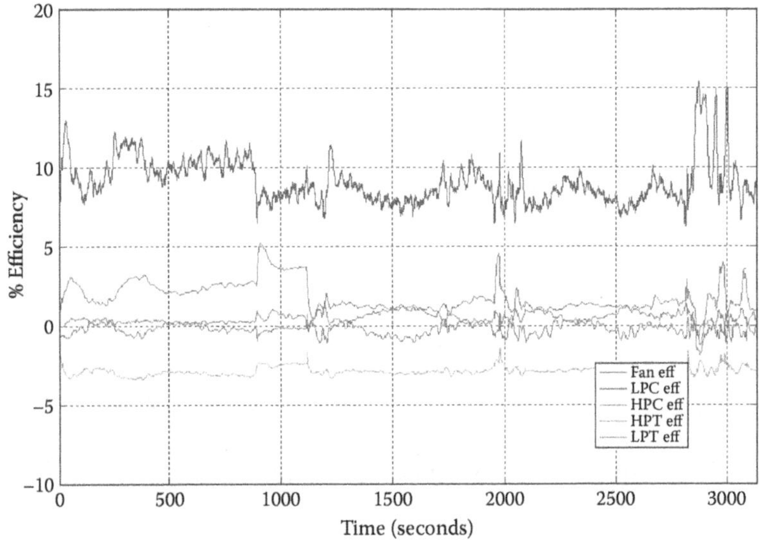

FIGURE 12.2 Example of module performance corruption due to engine model mismatch.

at 36,000-foot altitude (ALT) followed by a gradual climb to 38,000 feet and level flight thereafter. The ordinate axis represents % delta efficiency. The large magnitude and erratic behavior of these deltas are a result of an engine model to engine mismatch and as a result these deviations from zero have little to no value for true performance tracking purposes.

To help mitigate these effects, a *hybrid* engine model approach [2] has been suggested wherein the differences (or mismatch) between the physics-based model and the monitored engine are empirically modeled. One can think of this as a model initialization process where the resultant model is initialized to represent the particular engine being monitored. This will mitigate the model error effects due to engine-to-engine variation as well as deficiencies in the physics model due to its simplicity (for real-time operation) as well as model maintenance. One method to accomplish this is to construct an empirically derived *difference* model between the physics model and the particular engine being monitored. The *differences* (or residuals) in the engine gas path measured parameters can be captured in the form of regression models, autoregressive moving average (ARMA) models, artificial neural network (ANN) models, etc. When the empirical model is complete, the hybrid structure would take the general form shown in Figure 12.3. The combination of the empirical element and the

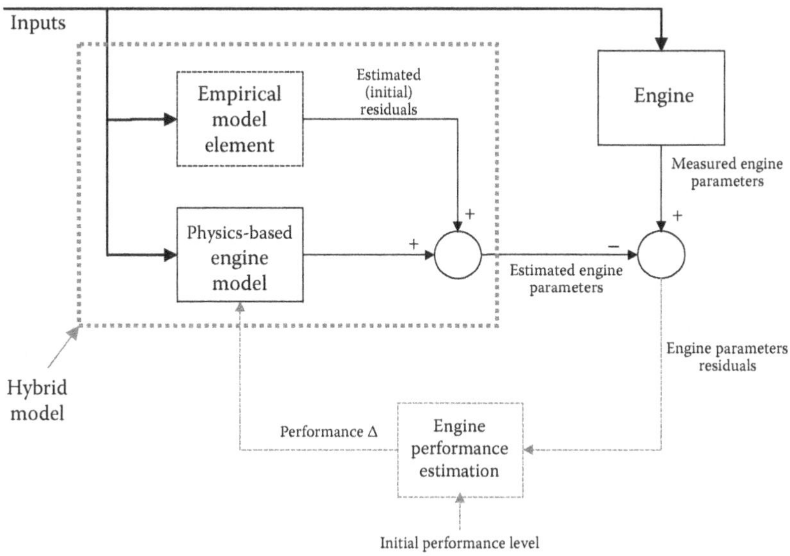

FIGURE 12.3 On-board hybrid engine model architecture.

physics-based engine model provides a more faithful representation for the particular engine being monitored. This provides more meaningful residual information from which an engine performance change assessment can be performed since potential (physics-based) model inaccuracies and shortcomings have been effectively removed by virtue of the empirical element.

The empirical model element, depicted in Figure 12.3, is developed from residuals derived from measured engine outputs and the (physics-based) model predicted outputs. Combining this empirical model with the physics-based model forms a unique hybrid model of the engine. As already mentioned, the empirical element can take many forms, including regression models, ANN models, etc.; however, to be effective, the implementation should be performed on-board in real time during actual engine operation and flight. This is a crucial enabling implementation requirement in that it forgoes the need to have an extensive data infrastructure in place to handle data transfers (for ground processing) across a fleet of engines not to mention the logistics of data maintenance and latency in incorporating the final model on an engine-to-engine basis. The primary challenge in developing a hybrid model building strategy, therefore becomes one of developing a methodology and process for capturing the physics model to engine differences within the constraints of limited memory and CPU throughput found in present day FADECs and/or diagnostic monitoring hardware boxes. In the next section, we will consider a particular empirical model building process that achieves the desired goal and conforms to these constraints.

On-board aerothermal models are an important application of hybrid models, but this is not the only application of hybrid models. They have been used successfully in other areas as well to bring real value in terms of balancing development and testing time against accuracy and ease of deployment. For example, enterprise level models, that allow us to model entire fleets—especially in the aftermarket area—have benefited immensely by combining discrete event models (governed by event probability distributions) with continuous time models (governed by ordinary differential equation) [3]. A couple of other examples can be found in [4]. However, we will concentrate on engine models and diagnostics in this chapter.

12.3 HYBRID MODEL PROCESS

A practical consideration for implementing a hybrid engine model involves the application of some form of sequential model building for the construction and specification of the empirical elements. This arises for the reason that sufficient engine data required to model the entire flight regime for a given engine/aircraft application is never available from one flight alone and may takes days or weeks to assemble. It is also desirable to encapsulate the entire model building process on-board so as to avoid the complexity (and expense) of developing and implementing a data infrastructure to enable off-board calculations. This poses an interesting challenge for the developer. The need (or desire) to establish a process that is hosted entirely on-board requires, among other things, that sufficient memory be set aside to store all the requisite engine flight data needed to develop the empirical element of the *hybrid* model. This is because sequential model building requires that *all* current and past data be available to construct an *accurate* representation at any given point in time. This situation can be accommodated through the use of submodels and data compression as described below.

One consideration that must be addressed is that measurement residuals are likely to vary with flight condition, typically Mach number (Mn) and altitude (ALT) for the same engine power condition. (Mn is calculated by the aircraft avionics system by dividing the relative forward speed of the aircraft by the local speed of sound, and ALT is estimated using ambient pressure and temperature readings.) As such, it would be reasonable to partition the flight envelope to allow individual empirical representations to be derived in lieu of using one empirical model for the entire flight regime. While it would be very convenient to have a single empirical model to represent the entire flight envelope, we have found that partitioning the flight envelope into a number of mutually exclusive regions and having different models is much more practical. This would support an incremental approach to empirical modeling in that it would not be expected that the engine would experience the entire flight regime in a single flight. In fact, it may be unlikely (especially for commercial air vehicles) that certain boundary areas of the envelope are ever traversed. A reasonable approach would be to partition the flight envelope into subregions as a function of pertinent independent flight parameters. As an illustration, one might choose ALT and Mn as the defining parameters. This would effectively capture inlet temperature and pressure variations, and Reynold's effects. Figure 12.4 below illustrates such a partition. (Several flight paths are depicted on this graph as dashed lines.)

Individual points represent where measurement data is available and residuals are computed. Over time the partition grid would become more complete and the individual (regional) models could be built. Each region

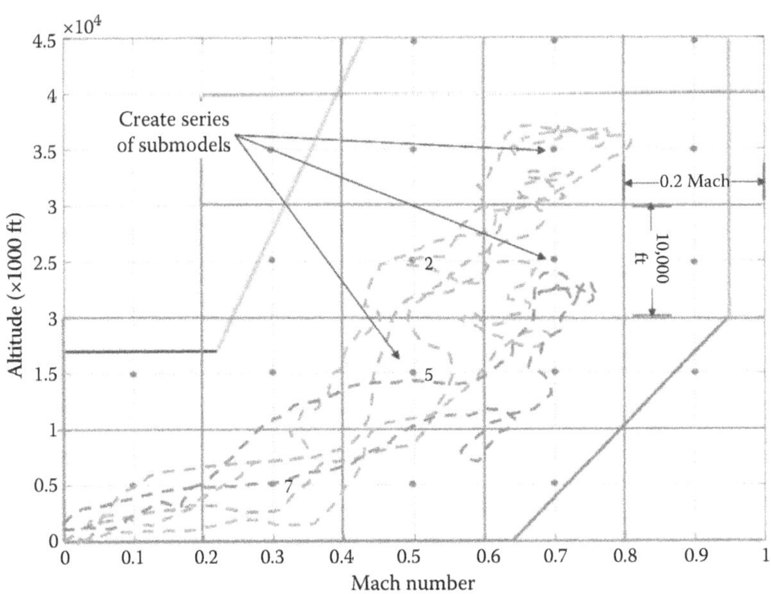

FIGURE 12.4 Sample flight regime partition.

could be represented by an individual empirical element that might take the form of, but is not limited to, a multilayer perceptron (MLP) ANN for each residual measurement under consideration. The inputs to the ANN are the same as the inputs to the engine model (ambient conditions, fuel flow, actuator positions, etc.) and the outputs are the parameter residuals. The evaluation of a partition model would require some form of continuous interpolation between models of adjacent regions so that the empirical estimates could be continuously generated as the engine traversed several flight regions. The completed empirical model (for this illustration) would be the concatenation of the individual subregion models with an appropriate regime recognition logic controlling the submodel evaluation and interpolation where required.

The partitioning of the flight envelope contributes to the concept of sequential modeling in that it allows the construction of a series of submodels to represent the overall model space. Since the grid is predefined (in order to limit the number of such submodels), it is conceivable (even likely) that insufficient data within a given grid element would be collected during a single flight to properly model that subspace. It should be clear that, no matter what particular modeling methodology is utilized, the entire set of data would have to be maintained for the proper modeling of the given subregion. Extending this across the entire partitioned flight envelope would entail the storage of an enormous amount of data that would be impractical and prohibitive.

One solution to this problem is to *compress* the data in the form of statistical and correlative information at the conclusion of each (MLP ANN) training session. The idea is to compress the data, store the compressed form of the data and establish the empirical model (i.e., train the MLP ANN) on the basis of the compressed information. It is essential that the compression process takes place in real-time, during flight whereas the empirical model build (i.e., the training of a new MLP ANN) can be accomplished after the flight has ended, still on-board, but offline) since the training process will require the concatenation of all compressed data points (within each respective subregion) to accomplish the training and would require more time than what might be available during real-time operation. The compression strategy must allow sufficient reduction in the stored data to allow training to be accomplished during the time the aircraft is on the ground (i.e., after the aircraft lands (weight-on-wheels [WoW] signal is triggered) and before the system is powered down), as well as providing a precise enough representation of the original data to permit an accurate difference model (MLP ANN) to be constructed.

One method is to establish a two-stage process. The first stage is performed in real time during flight operation. This stage performs the flight data reduction necessary to encapsulate a reduced order representation of the residual differences that are to be modeled in the second stage. The data reduction can be accomplished in real time in a sequential clustering process that leverages flight continuity to achieve drastic reductions in time and memory over conventional (generic) clustering algorithms, to create a compressed representation in the form of Gaussian mixture models (GMM) [2]. The second stage is performed on-board but offline after the stage 1 processing has been completed. This will typically be performed at the end of the flight and could be triggered by a WoW indicator. This second stage will use the cumulative collection of compact data determined during stage 1 processing from the current and previous flights to determine the residual empirical model (EM). This latter model becomes the empirical element in the hybrid model. Figure 12.5 illustrates the two-stage modeling approach.

The first stage of the process is performed in real time. It is the process of forming the GMMs. In essence, this can be thought of as a data compression phase, wherein clusters of both input and residual parameter data are formed in real time during flight and temporarily stored for subsequent processing in the second (offline, but on-board) stage of the model building procedure. In addition to forming the (compressed) data clusters, the first stage of this process also accomplishes other operations not shown in the overview. These include operations such as (a) regime recognition, i.e., knowing what part of the flight envelope the current data resides (which is important if flight envelope partitioning is used to utilize a collection of submodels in lieu of a large single model and (b) whether the current data point should be part of a new cluster or is indeed already covered by an existing cluster from a previous flight (in which case the current data point is already modeled and the configuration in Figure 12.3 can be executed using existing MLPs to track performance changes). Figure 12.5 depicts only the overview process of GMM creation (real-time) and the sequential MLP training (offline, but on-board) and does not indicate the intricacies involved in steps (a) and (b) as well as the general control of the hybrid model process. A discussion of these issues is beyond the intended scope of this overview and will not be elaborated upon. The interested reader may consult [2,5].

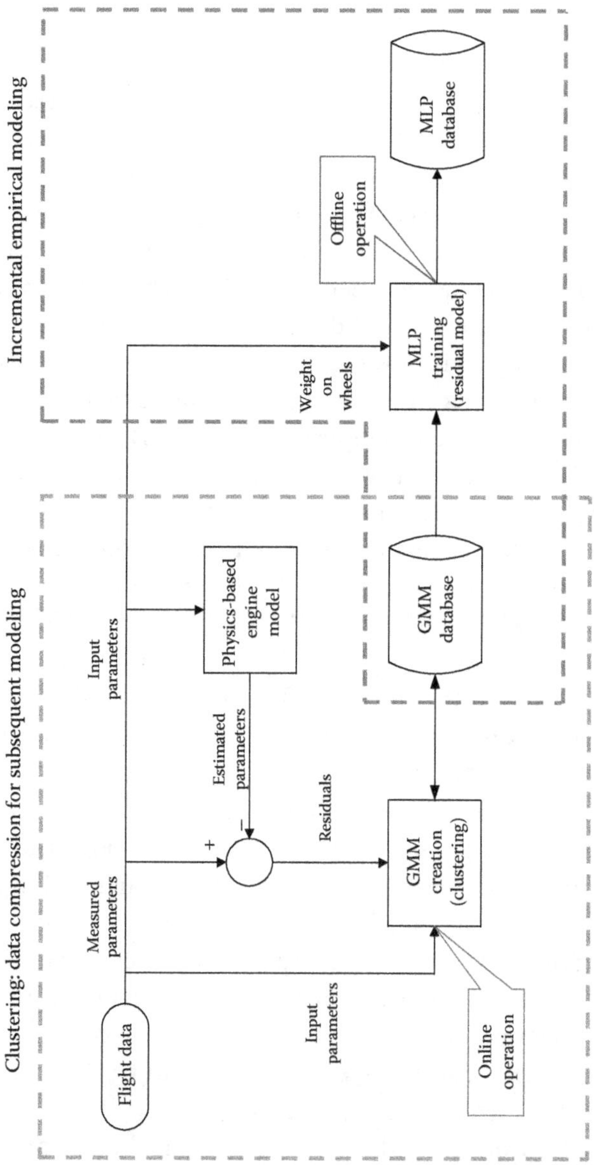

FIGURE 12.5 Two-stage empirical modeling approach overview.

What is significant is the autonomous nature of this process. What needs to be modeled and when the modeling process ceases (i.e., no more iterations are allowed) are driven by the data itself and are not predetermined. This allows the process to be deployed with a large variety of aircraft/aero-engine combinations. The process also makes no assumptions relative to the accuracy of the underlying physics-based model. It should be understood, however, that the empirical model is intended as a fine-tuning element in the hybrid modeling strategy and is intended to augment the physics-based model, not to replace it. A reasonably accurate physics model should always be the primary modeling element. A physics model is based on basic physical relationships and hence is applicable to design points as well as off-design points; it links back to the detailed aerothermal model that forms the basis for engine and controller designs, and it is validated against real data from test engines. The empirical fine-tuning compensates for (physics-based) model deficiencies, engine-to-engine variation, and supports model maintenance over the engine's life cycle.

12.4 EXAMPLE

To illustrate the concepts described above, we process a segment of a full flight of aero-engine data. The engine is a twin spool turbo fan, and the segment of the flight spans takeoff through a climb to approximately 19,400 feet. The flight segment duration is approximately 350 seconds (5.8 minutes). Plots of ALT and Mn appear in Figure 12.6.

Viewed in terms of traversing a path in a flight envelope this 5.8-minute flight segment trajectory can be plotted in terms of ALT (feet) versus Mn as in Figure 12.7.

In this particular hybrid model configuration, the flight envelope was set from 0 to 50,000 feet ALT and 0 to 1.0 Mn. This envelope does not represent any particular aircraft or engine flight envelope but is comprehensive enough to accommodate most commercial aero-engines during routine flight. This envelope was then partitioned into 21 equally sized subregimes for which hybrid submodels can be generated as described in the preceding section and numbered 1 through 21. Note that there is nothing special about this partitioning. The region can be partitioned in any way.

During this particular flight, only three subregimes were visited (no. 2, 3, and 7). Thus, the data represented will be used to determine three submodels (MLP NNs) for subregimes 2, 3, and 7. Engine and flight parameter data was collected at 20 Hz, which yielded 70,100 data points, each

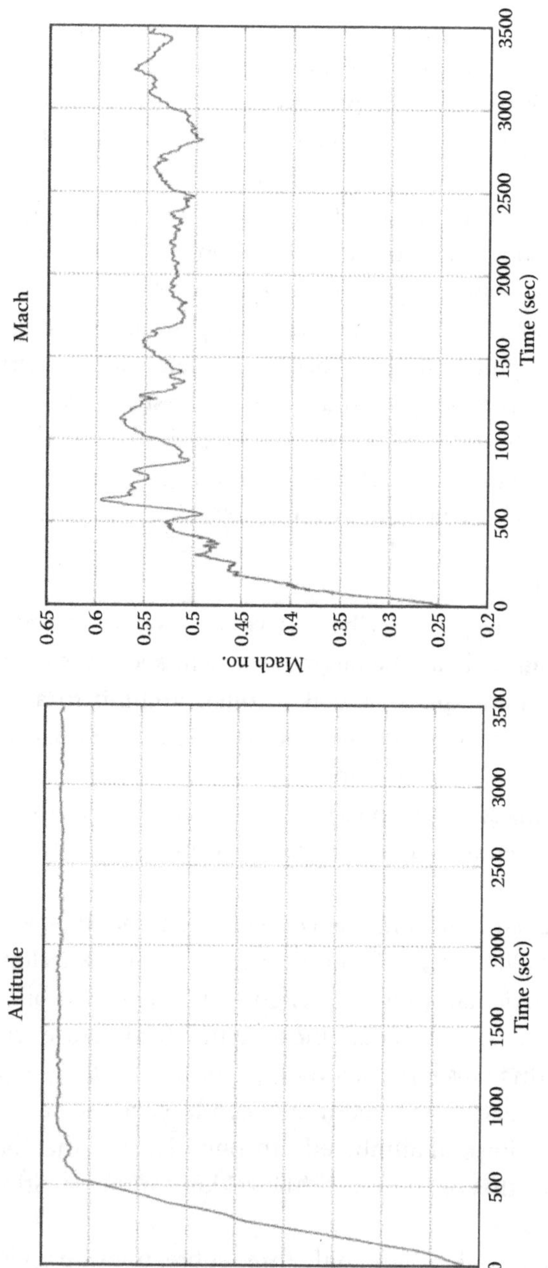

FIGURE 12.6 Altitude and Mach number.

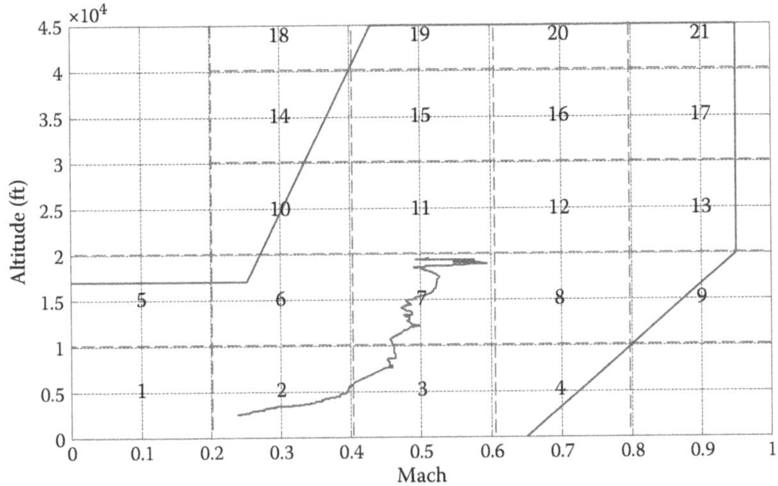

FIGURE 12.7 Flight segment trajectory (ALT vs Mn).

consisting of engine spool speeds, intercomponent temperatures and pressures, fuel flow, bleed, and variable geometry actuation feedback as well as flight and environmental conditions. In each of the three active subregimes, compression was performed in real time, which produced a total of 2586 GMMs. The breakdown according to subregime is given in the following Table 12.1.

This data was subsequently stored and used to generate the MLP NN submodels for the three subregimes traversed. It should be noted that dur ing this flight segment, the engine underwent both transient and (somewhat) steady-state operation. This can be appreciated by viewing the Fan speed parameter (N1) illustrated in Figure 12.8, which shows operation between 60% and 93% RPM.

The benefit derived from developing a hybrid model is dramatically illustrated in the plots in Figure 12.9. The top graph is a plot of various engine module performance parameter changes estimated from the gas path measurement residuals (speeds, temperatures, and pressures) calculated by comparing the actual engine flight data to the measurement

TABLE 12.1 Data Compression: Number of GMMs per Subregime

Subregime	Number of GMMs
2	121
3	163
7	2302

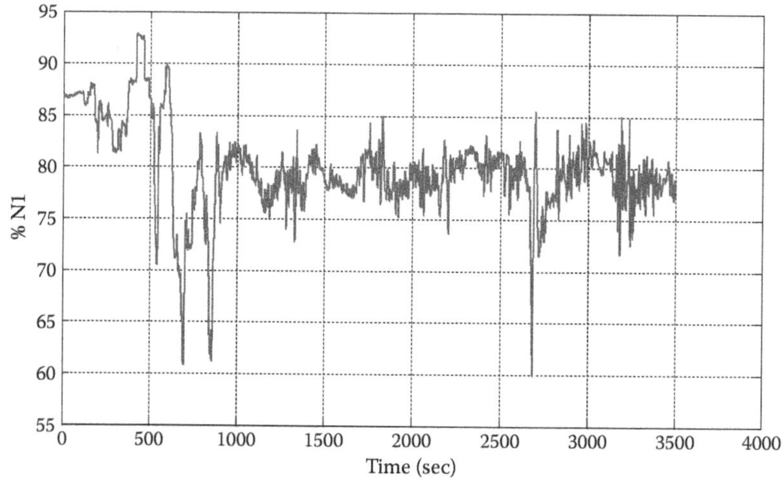

FIGURE 12.8 Fan speed (N1) illustrating transient and steady-state operation.

estimates coming solely from the on-board physics-based engine model. The ordinate in this plot is percent-of-point change, and it is evident that the information contained therein is relatively useless for diagnostic and engine heath tracking. Wide variations between ±10% for this benign flight segment are unrealistic and are the result of an engine/model mismatch, engine sensor biases, and model deficiencies. The bottom plot illustrates the same information when the MLP NN is activated. The measurement residuals are now computed relative to the hybrid model.

The stability of the performance changes in this latter plot will allow true performance changes to be tracked and anomalous behavior to be detected far more easily. To illustrate this, a simulated HPC fault was superimposed on the engine data stream at approximately 2000 seconds. We have good models for the effect various faults have on measurement parameters. Therefore, this HPC fault was simulated by modifying the measurement vector appropriately. The plot in Figure 12.10 begins with the physics model, then at approximately 1100 seconds, the MLP ANN (*hybrid* model) is activated, fault is introduced at 2000 seconds (quite visible), and the MLP ANN is deactivated at approximately 2800 seconds.

It is clear from this illustration that the *hybrid* model mitigates the effects of the engine/physics model mismatch, which in turn enables the detection of faults (anomalies) when they occur. Without the use of this type of enhancement (or data-driven model alignment), an occurring

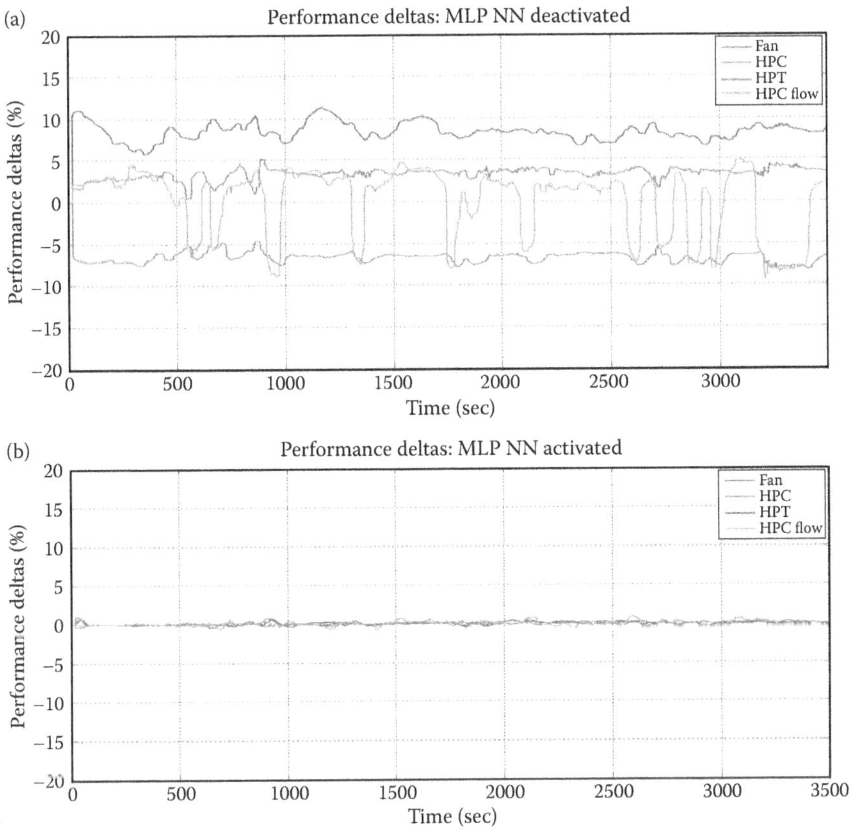

FIGURE 12.9 Module performance Δ estimates using physics model (a) and hybrid model (b).

fault hides the mismatch-driven performance variation evident in the last 700 seconds of the example depicted in Figure 12.10.

12.5 VERIFICATION AND VALIDATION

The software implementation of these algorithms is a crucial part of the process. The ultimate test of the system is how it performs in the field with real data and with all the potential failures that can occur. EHM systems are fairly new and only recently have people begun to see them as important enough to worry about the software lifecycle as it applies to them. In the past, these systems have been developed with a fairly *laissez faire* approach. Since they were not essential for the safety of the aircraft, and since they typically were not certified by regulatory agencies, a rigorous software development process was not required. The main motivation for

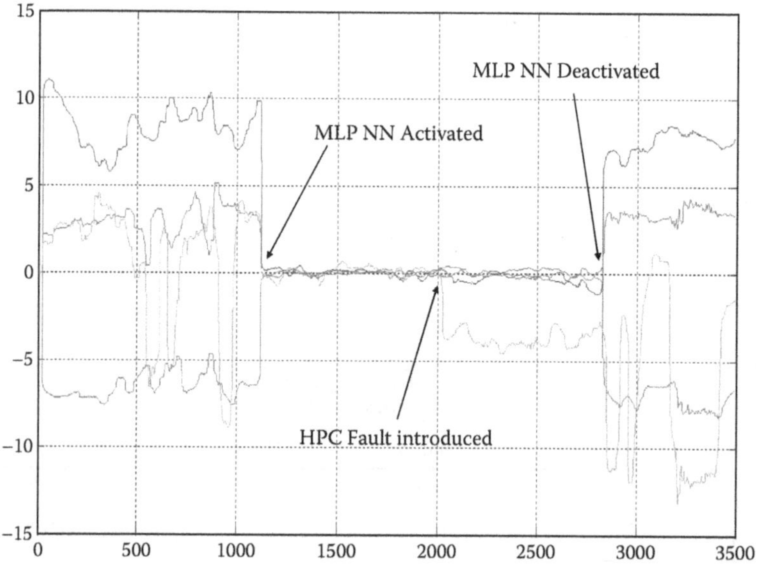

FIGURE 12.10 Hybrid model effect on performance tracking.

deploying these systems is that they provide an economic benefit to the operator. Current EHM systems give the operator (or the service provider) the ability to constantly monitor the equipment, keep on top of issues, and consequently anticipate maintenance actions, so as to lower costs over the lifetime of the equipment.

Operational interruptions due to service issues can lead to liability being assessed against the service provider. This is another reason EHM systems are becoming essential elements in the operations of the maintenance, repair, and overhaul (MRO) sector. Regulatory authorities are beginning to take notice of this shift in focus as well, as are various standards setting agencies. For example, the SAE has recently announced that a forthcoming aerospace recommended practice (ARP 5987) will address this issue [6].

A rigorous software development plan (SDP) is essential to ensuring that the software is developed in a rigorous and repeatable manner. The SDP consists of five major phases: Requirement generation, design, coding, verification and validation, and sustainment. This is a relatively generic breakdown of the process. Much of the guidance for developing software to aerospace safety standards is derived from the RTCA document, DO-178B: Software Considerations in Airborne Systems and Equipment Certification [7]. This is the document used by applicants to demonstrate to the FAA, or other regulatory agencies elsewhere in the world, that their airborne software has been developed to a high enough quality level that the risk of failure has been sufficiently mitigated.

Requirements for software systems are derived from those for the overall propulsion system. For the most part, software system requirements can be traced back to propulsion system requirements. Software requirements can be further broken down into high level requirements and detailed (low-level) requirements. High level requirements describe functional aspects of the system; i.e., they describe "what" is being designed. Low-level requirements describe the actual design of the software; i.e., "how" the software is designed. Traceability ensures that all the requirements have been implemented as software functions (forward traceability) and, conversely, that all implemented functions are linked to a requirement (backward traceability). For an interesting discussion on this topic the reader may look at a short position paper issued by the Federal Aviation Administration [8].

Low-level requirements are supposed to be presented to the system designers in sufficient detail that the job downstream of preliminary design can be accomplished efficiently and without ambiguity. In reality, requirements tend to be in flux even during the critical design or later phases, with designers and coders working in an iterative manner to get the job done.

Verification and validation tasks are key elements of the software development process. These are not very well understood even by the practitioners, and are often used interchangeably. Validation is the process by which we demonstrate that the system is designed correctly to satisfy the user's needs, i.e., that the software system is solving the right problem. Verification, on the other hand, is the process by which we demonstrate that the software performs as intended by the requirements. In other words, *validation* asks the question: *Are we building the right system*? And *verification* asks the question: *Are we building it right*? A good reference for this topic is [9]. In addition to demonstrating that all functional requirements are satisfied, the process of verification also demonstrates that the software will *only* perform as designed and that it will not inadvertently enter into a state that could lead to failure. This is typically done via a process called structural coverage. A valuable reference is [7], which provides guidance for certifying flight-critical software.

Good software development practice links requirements generation directly with the V&V process. In other words, requirements generation is not considered complete unless a means of testing the requirements is simultaneously established. This ideally would be in the form of test descriptions and scripts, but for EHM systems might also include intended behavior during typical flight conditions. The reason for this is that it is often not possible to precisely simulate the effects of a faulty system to be able to test the EHM in a realistic manner. In addition to functionally verifying the software, extensive testing allows us to be confident that no unintended failures will occur during system operation. Because this chapter is not intended to be a treatise on software development, we will leave it at that, and discuss instead how this applies to the implementation of a hybrid diagnostic system.

12.6 TRANSIENT DIAGNOSTICS AND ANOMALY DETECTION

As discussed in the previous section, a hybrid model estimates a set of critical parameters representing a new engine, correcting for engine-to-engine variations. In this section we will discuss the application of the hybrid model to transient diagnostics. And in particular, we will discuss diagnostics related to performance anomalies. The current focus of this work is on improving the detection of performance anomalies to help engine maintenance, and for tracking long term deterioration. The preferred implementation of this logic is on board so that flight safety can be enhanced en-route by picking up faults during flight. But if a sufficient number of engine parameters are recorded and downloaded, then the same functions can be performed on the ground as well. There are several advantages to doing so, the primary being that we can obviate the need to update any on-board software if changes need to be made to either the model or the diagnostics algorithms. It is not easy to manage on-board software for an entire fleet of mobile assets because even today the preferred means of updating software is through manual uploads. Add to this the problem of managing different code versions on different platforms, and it is clear that it is much easier to manage software that is ground-based.

Anomaly detection is the act of monitoring the engine parameters to determine, with a sufficient degree of confidence, if they have deviated enough from their normal ranges to be declared anomalous. Typically, this is done by finding the difference between the estimated parameters and the actual measured outputs, and comparing this difference against thresholds. Alternatively, the system can monitor absolute parameters that are normally supposed to be zero and declare an anomalous condition when they cross a threshold. But this has to be done carefully, because these "tuning" parameters will change as the engine deteriorates.

Once the system detects an anomalous condition, an attempt is made to identify the fault. In general, this is not an easy task. It helps if the signature (i.e., the vector deviations from the norm of key parameters) associated with the fault is unique, but this is very rarely the case. Normally, the diagnostic algorithm will localize the fault to an ambiguity group of faults. It is typically the job of the engineer to figure out, from a further analysis of the data, past experience with known faults, or, ultimately, by conducting a troubleshooting procedure, what specifically has triggered the fault indication (or alert).

Developing effective fault signatures is not easy. Without seeded fault tests that capture the actual effect on measured parameters, we have to depend on simulation of faults to generate these signatures. Seeded fault tests can be expensive and time consuming, and only in rare cases will engine programs authorize these tests. It is more likely that these are done on a development program or as part of a specific research program, but those are quite rare. For most faults, the best way of obtaining this information is through simulation. With a sufficiently detailed aerothermal model, various system faults can be simulated and their effects predicted.

For most programs, these modeled fault signatures are used in the fault identification system. If faults occur during the testing phase of the program, and if sufficient data is captured, then these modeled signatures are compared against the real fault signatures to confirm or adjust the stored signature. The same philosophy is used during the operational phase of the program. As engines are monitored during revenue flight, no opportunity is lost in capturing fault signatures. In addition to helping us refine our algorithms, these examples from the field are used in another important function. They are the candidates from which we draw on to develop our "golden test cases" (GTC). These GTCs are sets of real measurements that are paired with a known fault to help in the validation.

12.7 V&V OF DIAGNOSTIC ALGORITHMS

All V&V activities start with the requirements. An example of a system-level fault detection and isolation (FDI) requirement for a diagnostics system is the following: The system shall detect a 1% HPC fault with a 95%

probability of detection (P_d) and a 5% probability of false alarm (P_f). (The numbers are not necessarily those used in practice; they are used here for illustrative purposes only.)

The first step in the process is to design an algorithm that will accomplish this. The design engineer typically uses a high-level design package such as MATLAB® to design the algorithm, prototype it, and test it to ensure functionality. Once the algorithm is designed and validated, the designer will author a set of high- and low-level software requirements that will allow the software to be architected and coded so that it can implement the system-level requirement given above in a target software environment. This can be hosted on board flight hardware, or it could be off-board on standard ground-based servers.

For the sake of this example, let us assume that the design engineer follows the model-based diagnostics philosophy as outlined in the previous section. Figure 12.11 gives the block diagram of the system that generates the parameter residues. At each instance in time, i, the residue vector p_i is simply the vector of differences between the measured and the estimated outputs.

The next operation is to low-pass filter r_i to get rid of some of the noise. This gives us the filtered residue vector p_i.

$$p_i = \alpha p_{i-1} + (1 - \alpha) r_i$$

Let P be the set of signatures (in the parameter space) of m faults.

$$P = \{P_1, P_2, \ldots P_k, \ldots P_m\}$$

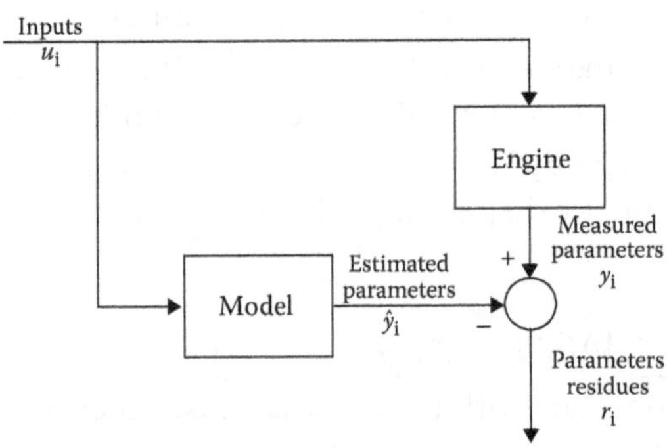

FIGURE 12.11 Residue generation for diagnostics.

where the kth signature (i.e., P_k) is associated with the 1% HPC fault. The FDI algorithm simply calculates the distance (in the least square sense) between the filtered residue vector and each signature vector. If

$$||p_i - P_k|| < \gamma_k$$

where γ_k is the fault threshold, then the algorithm declares that it is a 1% HPC fault. The thresholds may be different for different faults. This formulation of the algorithm has two design parameters associated with it: α and γ. The job of the designer is to choose these so that the requirements (95% P_d and 5% P_f) are satisfied. This would be done in a high-level modeling and simulation program such as MATLAB.

12.8 VALIDATION

As part of the design process, the designer will validate the algorithm against real and simulated data. GTCs are used extensively at this stage of development to validate the algorithm. We have built up a rich set of GTCs over the years to help us with the validation process. The advantage of using GTCs is that these are actual engine parameters recorded when the engine was known to be in a faulted state, i.e., the GTCs give us real data to verify our diagnostic algorithms. Because our experience to date has been primarily in the steady-state diagnostics area, the GTCs currently being used on a regular basis are based on steady-state data. In our application, steady state refers to stable cruise flight. The recording systems on board are programmed to recognize the cruise regime and record a set of measured parameters that are averaged over a given time interval. We emphasize that the validation process for steady-state and transient diagnostic algorithms is identical. As an example, in Figure 12.12, we show two years' worth of data for a single parameter from a single engine. This happens to be the difference between a measured and an estimated temperature but could be any other monitored parameter. The positive trend is normal for this parameter as the engine ages. Notice a downward shift in the data in the middle of the interval (around month 13). This is caused by a documented maintenance action, and hence forms the basis for a validation test. The trend-shift algorithms are required to pick up this shift during the V&V process.

Although the shift in the delta parameter in Figure 12.12 seems obvious to the naked eye, it is not easy to develop a computer algorithm to detect this consistently. In our testing we use a number of different GTCs and we average the statistics to calculate the efficacy of the algorithm.

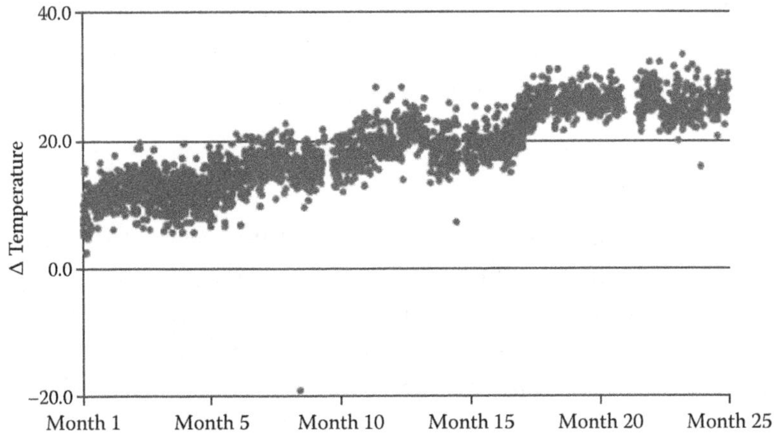

FIGURE 12.12 Temperature deviation from norm.

An example of an easier test case is shown in Figure 12.13. A persistent shift in delta speed is clearly evident. The shift far exceeds the normal variation in the parameter and most algorithms pick this up very effectively.

In cases where field data does not exist, we have developed fault simulation models that provide the designer with measurements augmented with realistic noise signatures. These simulations are based on detailed aerothermal models, and are consistent with past field experience. Hence they give us a convenient means of testing our algorithms.

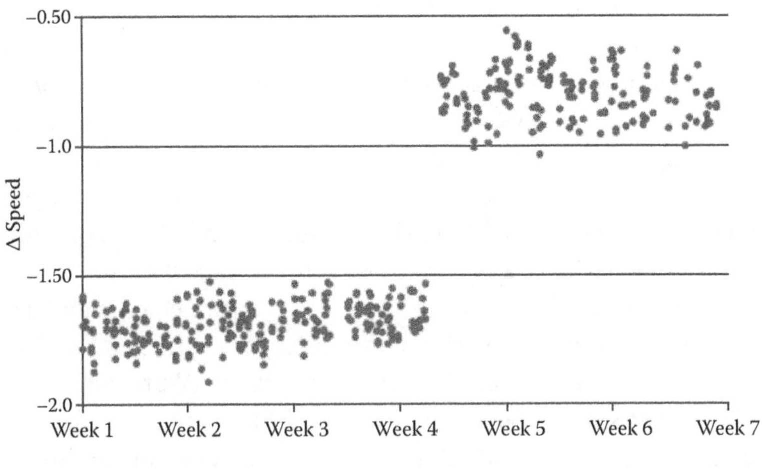

FIGURE 12.13 Persistent fault.

12.9 SOFTWARE SPECIFICATIONS AND DESIGN DESCRIPTIONS

Once the designer is satisfied with the algorithm, the next step is to develop the software requirements so that a software designer can convert the algorithm to code in the real production diagnostics system. It is not our intent in this document to explain the entire software development process. The reference documents [7,9] should be consulted for this.

Two sets of requirements are developed: high level and low level. A software requirements specification (SRS) is used to convey the high-level requirements to the software team. This is used by the software designer to develop a low-level set of requirements that are written into a set of software design descriptions (SDD).

For the algorithm given above, the SRS may consist of the following set of requirements.

A.1 The fault detection system shall use the difference between the measured and the estimated sensor signals to accomplish the FDI function.

A.2 The residue vector shall be low-pass filtered at each point in time.

A.3 The system shall store a set of vectors that represent signatures of all faults.

A.4 The filtered residue vector shall be compared to each signature vector and if it is close enough to one of them, the system shall flag that fault at the output.

A.5 And so forth.

As the reader will recognize, this is a high level set of requirements that tells the software designer the intent of the system and the major functions. This will help the software designer develop an appropriate architecture for the software system, allocate different functions to different software modules, and get the software interface control document (ICD) started. Once the system is built to satisfy the high-level requirements, the software engineer will develop the detailed design (i.e., the software design description, or SDD), which will allow the algorithm to be coded. To give an example of what the SDD might look like, take the input handling process described above. A set of requirements for this could be

B.1 At each major time-frame, the system shall create a parameter residual vector.

B.2 At each major time-frame, the system shall calculate an exponentially averaged version of the residual vector.

B.3 And so forth.

As should be clear, the SDD consists of detailed instructions that the software engineer can use to develop the code to implement the intent of the diagnostics design engineer. As another example, the 1% HPC fault isolation process described above might be specified as

C.1 An absolute distance, $d(i,j)$, shall be calculated at each time-frame, i, between $p(i)$ and each vector in the set of signatures $P(j)$, where j ranges from 1 to m.

C.2 The absolute difference shall be compared with the threshold, gamma(k).

C.3 If the difference is less than gamma(k) the system shall toggle the 1% HPC fault flag from 0 to 1.

C.4 And so forth.

Of course, this is only one example of how this might be done. In another implementation the validated high-level code (in MATLAB) might be compiled as object code and handed over to the software engineer along with specifications on how the code is to be integrated into the final system. This might come with different ICD requirements as well.

12.10 VERIFICATION

Each high- and low-level requirement needs to be verified by the software engineer. This is done typically through tests that are designed to exercise the software and compare the outputs against expected results. However, because tests can never be exhaustive, i.e., prove the absence of errors, DO-178B [7] advises a combination of tests, analysis, and reviews to verify the software.

Because it is very expensive to do these manually, a number of automated test packages have been developed that allow some of the verification to be done automatically. (However, we should point out that the level of safety mitigation required by the system will determine how extensively such automated tools may be used for verification of safety critical code [7].)

In the example above, C.3 might be tested by comparing a range of numbers around the value of gamma(k) and verifying that the fault flag is set correctly as expected. Similarly, B.2 might be tested by verifying that the weighted sum function returns the correct value for different sets of numbers.

The test for C.3 described above is an example of structural coverage, wherein all the possible inputs to the conditional are tested to ensure that the system does the right thing. In general, providing documented evidence that the software performs without flaws when all the possible conditions in the system are exercised is called multiple condition/decision coverage MC/DC [10], and is a key ingredient of verifying the most critical software (i.e., level A).

We have given a very brief exposition of what V&V of an EHM system would consist of. As mentioned above, this has not been of great importance to the EHM community in the past especially when compared to the controls community. In fact, verification is one of the most important elements of developing software for safety-critical applications, and it is expected that this kind of formal verification will become more important in the future for all EHM software.

12.11 SUMMARY

EHM systems are increasingly becoming a common feature of aerospace systems, especially aircraft engines. Many of the diagnostics algorithms that are key features of these EHM systems are model-based in nature. In this chapter, we demonstrate how we can design a hybrid model that is a combination of a physics-based model and empirical model for a real-time engine application. We also discuss briefly the V&V aspects of EHM software systems and point out that as EHM systems become more ubiquitous and begin to implement critical functions, the requirements to use rigorous V&V techniques will become an essential part of the process of deploying an EHM system.

ACKNOWLEDGMENT

We would like to thank Doug Sansom, Dean Kavalkovich, and Randy Skelding for interesting discussions that led to many improvements to the document.

REFERENCES

[1] C.M. Bishop. *Neural Networks for Pattern Recognition*. Oxford University Press, 1995.

[2] A.J. Volponi. Use of Hybrid Engine Modeling for On-Board Module Performance Tracking, ASME Paper GT2005-68169, IGTI Turbo Expo 2005, Reno, NV, June 2005.

[3] Y. Lin, A. Khibnik, K.-Y. Jeong, S. Soorapanth, W. Winston, J. Wang, R. Rajamani, P. Luh, D. Yu. Condition-based maintenance leads to value-chain benefits. In *International Workshop on IT-Enabled Manufacturing, Logistics, and Supply Chain Management*, Bangalore, India, December 2003.

[4] A.J. Volponi, R. Rajamani. Enhancing diagnostic system performance using hybrid modeling techniques, tutorial session: Advanced estimation for aircraft engines: Applications to engine controls, fault detection, and Health Management. *American Controls Conference*, NY, NY, July 2007.

[5] T. Brotherton, A. Volponi, R. Luppold, D.L. Simon. eSTORM: Enhanced self tuning on-board real-time engine model. In *Proceedings of the 2003 IEEE Aerospace Conference*, paper 1023. Big Sky, MT, March 2003.

[6] SAE E-32 Aerospace Propulsion Systems Health Management, Aerospace Recommended Practice, "Guidelines for Engine Health Management System Software Assurance Levels," ARP 5987, under review, Nov 2011.

[7] DO-178B, prepared by RTCA SC-167/EUROCAE WG-12, Software Considerations in Airborne Systems and Equipment Certification, Washington, DC, December 1. 1992.

[8] FAA Certification Authorities Software Team (CAST), Merging High-Level and Low-Level Requirements, Position Paper, CAST-15, February 2003.

[9] IEEE Computer Society, IEEE Std 1012-2004, IEEE Standard for Software Verificaiton and Validation, June 2005.

[10] A. Dupuy, N. Leveson. An empirical evaluation of the MC/DC Coverage Criterion on the HETE-2 satellite software. In *DASC (Digital Aviation Systems Conference)*, Phildelphia, PA, October 2000.

[11] E-32 Aerospace Propulsion Systems Health Management, ARP5987: Guidelines for Software Certification of Engine Health Management Systems, estimated publication date, March 2011.

CHAPTER 13

Extracting Critical Information from Free Text Data for Systems Health Management

Anne Kao

Stephen Poteet

David Augustine

CONTENTS

13.1 INTRODUCTION

The growth of transportation marks a major advance for the 20th century that has fundamentally changed the distance between people in the world and makes it an essential aspect of modern life. To maximize the safety as well as the economy of a vehicle's performance, whether it is an airplane, automobile, or spacecraft, it is crucial to take an integrated view of all vehicle-related data. These include on-board sensor data, scheduled and unscheduled maintenance records and reports, reports related to operations of the vehicle, and other environmental data (such as weather condition, how congested a highway or an airport is). To achieve success in vehicle health management and to keep operating conditions economical, all of these various factors have to be examined together. The volume of data is vast and the data sources are diverse. On the other hand, the data often have missing and incorrect values. Data mining techniques should be able to offer us a plethora of useful approaches to analyzing this data. However, a lot of technical domain knowledge (e.g., engineering design and manufacturing) is required when applying data mining techniques to this problem. The complexity in the data goes far beyond the cliché market analysis of products and customer demographic data.

In this chapter, we will focus on airplane health management. While examples used are drawn from aviation, readers will find that most discussions carry strong commonalities to other types of vehicle health management. Furthermore, while the number of data sources supporting a complete airplane health management system go well beyond the short list mentioned above and is easily in the hundreds, we will focus our discussion on two types of data sources in particular, namely, aviation safety reports and airplane maintenance log book data. These data sources contain both structured and unstructured data. However, the wealth of information is in the unstructured free text data. Our discussion will further focus on analysis of the free text data, and how that can be mined and used in conjunction with structured data.

13.2 PROBLEM DESCRIPTION

Airplane maintenance data and aviation safety data represent two very different types of information and technical challenges. The volume of maintenance data is huge. For one major airplane model (e.g., 747 or 777) at one major airline, there can be over 100,000 maintenance records. The maintenance data can come from an airline, or from the Boeing-organized consortium ARMS ISDP (Airplane Reliability and Maintainability System In-Service Data Program), which contains maintenance data from more than 20 major airlines. The data contains unscheduled maintenance data and, sometimes, scheduled maintenance data. For the current discussion, we will focus on log book data, which is the record of line maintenance (i.e., maintenance at the airport gate). The free text fields include (1) the complaint text, which records problems reported by the pilot, crew, or mechanics, and (2) the resolution text, which records the actions taken to fix each problem. The guideline is that each record should record one single problem. However, this is not always true, partly due to the complexity of the problem. The data records always contain the date (but not the time of day), the location (airport), and the airplane models (737) and submodels (737–800), as well as Air Transport Association (ATA) numbers. ATA is an international coding standard used to denote systems and subsystems, e.g., door, fuel, landing gear, air conditioning. The coding is hierarchical. Information can be coded with two digits (also known as ATA systems), three digits (ATA subsystems), or four digits (ATA sub-subsystems). Each record is typically coded with ATA systems but does not usually get more specific. The reports are primarily driven by sensor indicators that alert important system failures.

Aviation safety data are collected through open-source news reports, internal airline reports, airline reports of operating issues to the manufacturer, and self-reporting by individuals to Airplane Safety Reporting System (ASRS) managed by NASA. It is important to distinguish these data from accident investigation data. Through major efforts by airplane manufacturers and operators, airplanes are extremely safe and there have been very few accidents. In the rare cases an accident does occur, there would be an in-depth forensic analysis and report on what happened.

However, the goal of health management is not to focus on these few accidents but to analyze near-accidents and other unexpected aviation events to identify potential risk areas to further reduce the chance of accidents. Unfortunately, this task is impeded by the fact that safety data is often somewhat sketchy and lacks important details. For example, ASRS reports contain a fair number of operational issues that relate to weather, airport condition, and crew and pilot activities. However, they only contain the year and month of the reports, and not the specific date of the event, and do not give the airline involved, let alone the flight. This is by design to make the information anonymous and protect the reporter's identity, but it makes it hard to match the report with more specific airport or weather conditions obtained at the time of the incident, other than what the report explicitly states. The safety data is a lot less in volume than the maintenance data; however, it is much harder to categorize all of the records into a fixed set of categories that would be both comprehensive enough to cover all cases and detailed enough to distinguish specific issues.

To better understand human factors issues and operational issues, it is crucial to study the safety data. However, to ascertain system or part reliability, or spot potential metal fatigue and long-term wear and tear, it is important to mine the maintenance data. Other databases could also provide insight into vehicle health problems. For example, data regarding migration patterns of certain birds can help determine when and where the chance for a bird strike [1] may increase. As another example, data regarding weather or the traffic and congestion patterns at different airports could help identify the causes of certain incidents.

Both maintenance and safety data suffer from missing values, as well as duplications and pseudoduplications. A maintenance record can report on multiple actions performed on a single part or even actions performed on multiple parts (even though the guideline is one part per record). A maintenance problem can be deferred if not flight critical, resulting in multiple records for the same problem so that a simple search would overreport a problem. These records typically share the same COMPLAINT ID, but different actions may be tried and reported on in different records after the initial deferral. Worse, a maintenance problem can take more than one try to get fixed, even though it is closed out after the first try under the mistaken belief that it is resolved. In this case, there is nothing that ties the records together; they are treated as unrelated problems by the system. Similarly, since safety reports are largely self-reports, the same problem can be reported multiple times by different people, often with very diverse points of view. The lack of key discriminating information makes it hard to identify duplicates with complete confidence. Missing values are a common problem for both maintenance data and safety data. Different airlines have different practices on how to fill in various data fields. For example, how many times a nonflight critical issue is identified but deferred is not always filled out. Since safety data is self-reporting data, the reporting person may not always fill in information such as the flight phase when the event occurred (take off, cruising, or landing).

It is important not to lose sight of the goal of mining these data; it has to be able to lead to actionable results. Examples of actionable results include issuing changes in a specific maintenance procedure or issuing a warning of a specific problem. It is not sufficient to simply utilize the ATA system codes in the maintenance data to get an accurate picture of what equipment is involved. There are easily hundreds of parts in each of the ATA systems. Part numbers or equipment numbers* are not always present and cannot be used to identify the equipment. The safety data contains information that is even more open-ended, compared with the maintenance data, which are driven by equipment and sensor output. While there are attempts to categorize typical safety-related events, there is no clear way to capture or even know a priori what categories can best code the information, or even all of the questions a safety engineer may want to ask. For the maintenance data, quite often, users need to identify the parts referred to, and the total list of parts is known; however, for the safety data, users often need to do ad hoc searches to identify a potential safety-related feature, as well as run standing queries on the data to uncover trends or give alerts regarding certain situations. Thus, text mining techniques that can code ATAs automatically, or identify certain groups of terms that seem to co-occur, may be interesting and useful in helping an analyst gain insight into the data, but they do not by themselves lead to actionable results for lack of essential detail.

13.3 RELATED WORK

While search (or information retrieval) is a very old field, as is text classification, the problem at hand represents a very different set of challenges. A

* Part numbers and equipment numbers are two very different types of information: part numbers are assigned by suppliers, whereas equipment numbers are Boeing-assigned and used in various Boeing maintenance documents.

typical Google type search is either exploratory in nature, e.g., to discover out what data mining means, or is a way to find an exact match, e.g., locate Dr. Jiawei Han's paper titled "…" for download. A user is rarely interested in counting how many instances match the query in either of these cases, nor would the user be interested in all instances of the correct match. This represents a crucially important difference from a search to identify the information necessary to support vehicle health. Importantly, users in this domain expect a high recall, i.e., for the system to identify as many as possible of the correct matches, and not just a high precision, i.e., for the system to return as few mismatches as possible. The exact count is crucial because it shows the rate of a certain type of problem or part failure, which is a crucial piece of information used to quantify the risk. Typical search methods do not aim to solve this type of problem.

NASA-funded work has made major improvements in helping the user narrow down the search space and improve efficiency for search of predefined categories [2–4]. However, the work does not focus on the extraction of information at the level of detail we need to, to help engineers generate actionable intelligence. The work also does not sufficiently address engineers' needs to search on arbitrary new topics that have not been previously identified or defined.

Another important distinction is in the requirement for the user in this domain to utilize her engineering and aviation expertise. Different communities may use different terms to describe the same equipment. Also, investigation of a sensor or light indicator issue would require the user to know what concepts can be used to describe the condition. The problem is further compounded by the "noise" in the free text data. There are a fair number of nonstandard spellings in the safety data because the writers of the report are not professional writers and may not take an interest in writing the best way they can. Maintenance data is even noisier because the writers of the report are mechanics and their job is primarily to fix the airplane in a timely manner and only secondarily to write a report of what they did; they often contain more nonstandard spellings than standard spellings.

Regular expressions [5] can be used as a solution for this problem. However, they suffer from several deficiencies. First, the users are engineers and not linguists. While they have a great deal of essential domain knowledge, identifying the many ways of misspelling a term is not their forte, nor is trying to represent that in a regular expression. Note that many queries are not standing queries but ad hoc queries. It is often not practical to have an information technology (IT) team pre-engineer the optimal regular expressions. Second, regular expressions do not have context-sensitive power [6]. The abbreviation "comp" means "computer" in one context but "compressor" or "compartment" in other contexts. Regular expressions, as a general case, cannot express context. Finally, regular expression matches, while providing a kind of fuzzy match, cannot provide a score on how well it matches the target; the target either matches or it does not. So, with all of the power of regular expressions, it is still not always possible to provide a regular expression that matches exactly the patterns you want, without producing any false-positives or false-negatives.

13.4 PARTNAME MATCHING BY ANALYSIS OF TEXT CHARACTERISTICS

In this section, we will illustrate how to use a combination of knowledge-based natural language processing and various string matching algorithms to solve a high-impact part name analysis problem.*

13.4.1 The Part Name Reference Problem

The log data provides documentation on what problems have occurred and what maintenance actions have been performed. Reliability studies of parts typically are interested in all of the maintenance actions performed. The result helps determine recommended maintenance schedules for airlines and how a specific fleet or a specific airplane deviates from its class. Alternatively, a real-time health management application may be only interested in fix effectiveness and thus only focus on the actions that successfully fix each of the problems reported. In addition, it can be used to study the reliability of parts, proactively identify required upcoming maintenance, and support supply management to avoid overstocking or understocking issues. For this type of application, only replacements would be of interest, and resetting, cleaning, or adjusting actions would not be. In a complex business such as the airline industry, it can also help operation management; for example, by keeping the airline from sending an airplane to a station that is not equipped to perform any likely upcoming maintenance.

As noted above, a major part of the problem is that there are numerous spelling and ad hoc abbreviation problems in the data. Here is an example

* This is R&D work by BR&T and is not intended to be a reflection of Boeing's current practice in an operating environment.

of a part name term, “computer,” and some of the variations on how it is expressed in the log data:

computer, comptr, compter, computor, compuer, computo

Furthermore, one variant, “comp,” is ambiguous and, in addition to “computer,” can mean

compressor, compartment, compensator

Compounding the problem, a part name typically consists of multiple words, each exhibiting many variants, and the words in the part name from a given list may not all occur in a maintenance record or may occur in a different order. For example, the part name

Overhead Panel Bus Controller (L), M23112 (P11)

may occur in a maintenance record as

COMPLAINT: REF ADD 913 STS MSG **LEFT O/H PNL BUS CONTROLLER** INTERMITTENT TAGS OFF K02648Y

RESOLUTION: FIM ACTIONED AS PER MSG 23-48802 **OPBC** REPLACED IAW MM 23-93-01 GRND CHKS AND TESTS C/OUT SATIS TAGS ON B25092G

The rendering of the part name in the COMPLAINT text leaves off the equipment number (M23112) and the panel (P11) the part is located on (it being redundant with “left” in this case), realizes “L” as “left” and removes the parentheses and relocates it to the beginning of the part name. In the RESOLUTION text, the whole thing is reduced to an acronym, “OPBC,” and any indication of location (“left” or the panel) is left off (since one of these has been mentioned in the COMPLAINT text). Note that it is the less informative of these, the acronym, that is adjacent to the maintenance action performed (“replaced”). All of this makes trying to find a good match even harder.

Clearly, searching for an exact match of the string representing the part name is not going to help. While using a synonym list could help with acronyms like “O/H” and “OPBC,” it would not be a very good way of handling all the misspellings and ad hoc abbreviations illustrated by the “computer” example above. The number of parts is very large (in the thousands) and the number of spelling variations for the words in part names is also very large (often 10–20 and sometimes more than 40). This would make constructing an exhaustive list highly labor intensive, if possible at all. Furthermore, it is also hard to foresee what types of spelling variations might occur, given that one major type of error is to fuse two words together (leaving out the space in between). Traditional natural language processing approaches [7,8], which would try to parse the sentences of the log record, would fare even worse, since, in addition to requiring an entry in the lexicon for each misspelling, they would have to record grammatical information as well.

Another approach that two of the authors tried is to write regular expressions to capture the variations in spelling of the different words in a part name [5]. For example, the following regular expression captures all of the above variants of "computer," including the full word and the ambiguous abbreviation "comp," while excluding "compr" (an abbreviation of "compressor") and "compt" (an abbreviation of "compartment").

```
comp(u?t[eo]?r|ut?[eo]?r?)?
```

Regular expressions allow one to capture the generalizations that the variants of "computer" all begin with "comp" and either contain both a "t" and an "r" or have a "u" immediately following the "comp," while leaving other aspects, like the vowels, either optional or variable. Regular expressions, unfortunately, are unable to handle transposition errors in a very succinct way (one basically has to encode every possible transposition of adjacent letters). Even more importantly, manually writing these patterns for every single part name is enormously time-consuming and requires a fair amount of facility writing these patterns as well as familiarity with the data. Every time a new airplane model or series or even a customer variation of an existing series needs to be handled, a new set of part names is introduced and needs to have patterns written for them. These limitations mean that, despite its initial promise, this approach is not practical for dealing with this problem.

13.4.2 Our Solution

Our solution is to combine knowledge-based natural language processing techniques and various string matching techniques. We call our approach Partname Matching by Analysis of Text Characteristics, or P-MATCH.

First, we will briefly summarize our approach. We begin with a given list of the part names that we are interested in, the "target part name list." This is typically a subset of all the parts on an airplane, for example, the parts that can be repaired or replaced during a brief stopover between flights, i.e., the "line replaceable units." We take advantage of the structure of part names, so the first step is to parse the part names on our list according to this structure. Next, for each maintenance record, we use a list of the maintenance actions we are interested in along with the results of the part name parsing to identify a candidate part name string, a sequence of words that likely refers to a part that we are interested in. Then we use one or more fuzzy string matching algorithm to compare this candidate part name string with the part names on the target part name list. We then use the results to select the most likely part name from the target list.

Inspired by linguistic analysis of noun phrases as basically consisting of a head noun and modifiers, we parse part names into a head noun that tells us what general type of part it is (e.g., a switch), essential modifiers that determine the specific type of part it is (e.g., "outflow valve" in "outflow valve switch"), and peripheral modifiers that primarily indicate location (e.g., "aft" in "aft outflow valve switch"). The head is typically the last word in the part name, but sometimes the last word is so general that it is likely that log data may leave it out completely and the second to the last word should be treated as the head, for example, "Control Display Unit," will often be described simply as a "control display" or some variant thereof. In addition, certain types of modifiers may occur after the head in the target list of part names, although they are typically easy to recognize by their form (e.g., alphanumeric) or the fact that they occur in parentheses or after a comma. For example,

Control Display Unit (Center), N34303 (P11)

"N34303" is the equipment number of this part, rather than a part of the part name per se. "Center" is the location and a peripheral modifier. "P11" is the panel it is located on and is typically not included in the log data or, if it is, not necessarily adjacent to the part name proper.

Having parsed the target part names into heads and modifiers, the system then examines the log data. There are a number of ways in which we can start, depending on the task. One task we might be asked to perform with P-MATCH would be to find parts that had been removed, replaced, or repaired to help analyze fix effectiveness. For this, we might begin by searching for variants of the verbs representing these particular maintenance actions. By starting our search with the verb, we were able to make the search more efficient. We not only narrow the number of records that have to be examined in detail, but we also anchor our search in the text for the string representing the part name. Note that there can be quite a few variations of these verbs. Some of the variants of "replace" in the data are

> replaced, repl, rplaced, replacd, replced, repaced, replaed, replacement, replac, replaced, rplcmnt, replace, repla, replae

However, it is worth noting that the total number of verbs involved is an order of magnitude smaller than the total number of part names involved. We find this list of variants by searching a list of words extracted from a large set of the data using a fuzzy string matching algorithm.

Often, a maintenance record will contain multiple verbs (e.g., the engineer checked one part, adjusted another part, and finally replaced yet another part). When dealing with fix effectiveness, we are typically interested in the last verb in the maintenance message based on the fact that, in process descriptions, actions are typically described in the same temporal order as they occur in the real world (authors of maintenance reports never write "before we replace X, we reseated X") and we want the final action performed. An exception to this is that variants of "checked OK" frequently follow the action of replacing/removing/repairing, so it would be ignored. Another way of finding the appropriate part in a message to analyze fix effectiveness would be to organize maintenance actions into a prioritized set of categories. For example, the categories and their priority might be: removed/replaced, repaired, deferred, reset, OK, other. In this task, we find all the verbs in the maintenance message and then select the one with the highest priority as the starting point from which to search for the candidate part name. On the other hand, if we were trying to assess part reliability, we might want to find all the parts mentioned whatever the action performed was. In this case, we search for all the verbs and find associated part names near them, often yielding multiple parts per maintenance message.

The next step is to identify a candidate part name string from the maintenance record. The candidate string is a sequence of words in the maintenance text that is likely to name a part in the target part name list. In general, part names are contiguous strings of part name modifiers followed by a part name head; however, there are exceptions to this. As noted above in the case of the "left overhead panel bus controller," a fuller description of the part (e.g., the location "left") may occur earlier in the record, or even in a different field (COMPLAINT text) from the action verb that indicates which part we are actually interested in (typically located in the RESOLUTION text). These different descriptions of the part need to be synthesized. In addition, references to location ("right," "left," "forward," "aft," "Zone A," panel number) frequently occur after the head (e.g., "ECS CARD R," where "R" is "right"). Given this, we search first for a part name head in the vicinity of the verb, then search for possible modifiers of that head adjacent to it. Since the verb can be active or passive (e.g., "replaced XXX valve" or "XXX valve replaced"), both sides of the verb are searched. The head and its associated modifiers are derived from the parse of the target part name list described earlier. Because of the possibility of misspelling, the head and the modifiers are searched using a fuzzy string matching criteria. After these modifiers are collected, the rest of the message is searched for other occurrences of the head (or a variant of it, e.g., "vlv" for "valve") and, if one is found, additional modifiers adjacent to it are collected. Finally, the head and all of its modifiers are synthesized into a single part name candidate string.

Occasionally, peripheral modifiers, in addition to occurring after the head, can also occur in nonadjacent positions. For example, in one maintenance message there is a reference to the "Upper Flow Control and Shutoff Valve (L Pack), V21511 (P110)," but "LEFT PACK" occurs in the COMPLAINT text and "UPPER FLOW CONTROL VALVE" occurs in the RESOLUTION text. While it might be possible to extend the search for these peripheral modifiers to catch some of these, there is always the chance that the location does not refer to the part we are looking for but to some other part. One way to deal with this would be to ignore or downweight these location indicators during matching with part names from the target list, although we are not currently doing this.

The resulting part name candidate string is then compared with all of the part names on the target list that have the same (or equivalent) head using another fuzzy matching algorithm. We are currently using a subset of the algorithms in SecondString [9]. Similarity algorithms can be divided into several different categories: string, token, and hybrid or two-level [10]. String-based algorithms process the entire string as a sequence of characters, applying one or more approaches to measure the dissimilarity, such as Levenshtein (edit-distance), Jaro (number and order of common characters), Jaro-Winkler (Jaro with added score for common initial substrings), or Jaro-Jones (which takes into account common number-letter substitutions, such as numeral "0" and letter "O"). Token-based algorithms divide the string into tokens using one or more "space" or punctuation characters, then compare tokens ignoring their order. Examples include Jaccard (number of tokens common to both strings divided by the number of tokens in either string), TF-IDF (cosine similarity of strings represented as vectors of Inverse "Document" Frequency weighted token frequencies), and the Jensen-Shannon distance [11] (based on the Kullback-Leibler divergence of probability distributions of tokens [12]). Mixed or two-level algorithms tokenize the string, and then apply string similarity algorithms to the individual tokens. Examples include Level2Jaro (the average similarity of matching tokens using the Jaro algorithm) and Soft TFIDF (Jaro-Winkler applied to the individual tokens and TF-IDF applied to the tokens above a certain similarity level). For a more complete description of the wide variety of string similarity algorithms that are available, see [13–15]; for Jaro-Jones, see [16]. We used Jaro, Jaro-Jones, Jaccard, Jensen-Shannon, and Level2Jaro for our initial experiments.

13.4.3 Experiments

Because of the difficulty of getting aviation engineers to spend time marking up a set of maintenance records to use to assess the recall and precision of P-MATCH and its variants, we have had to limit ourselves to an initial, small-scale assessment relying on a set of records for which the authors have been able to determine part names. Please note that the authors are IT researchers, not aviation engineers.

We performed two experiments with variants on our algorithm, involving different ways of constructing the candidate string and how much reconstruction we did to the target part name string. In Experiment 1, aside from putting the head in final position, we did not try to order the modifiers in any particular way and, aside from removing punctuation and equipment numbers from the target part names, we used them just as they occurred in the target part name list. In Experiment 2, we normalized all occurrences of "right" and "left" (e.g., "r," "right hand," "r/h" were replaced by "right"), and the resulting normalized location word was placed at the beginning of the candidate string. The variants of "right" and "left" in the target were similarly normalized and put at the beginning of the target part name string.

For the test data, two of the authors manually assigned the correct part names to a sample of about 200 maintenance records. We restricted ourselves to records for which we could unambiguously determine a single part name given the text of the record. We performed two different measures of accuracy corresponding to two different ways we might want to use P-MATCH. For a purely automatic extraction of part names from the maintenance data, we score a given maintenance record as a hit for an algorithm if the algorithm ranks the correct part name for that record first among all the part names. We call this "Rank 1 Accuracy." On the other hand, if we were using P-MATCH interactively, for example, proposing the top five part names to a subject matter expert for selection, we would consider a maintenance record as a hit for a given algorithm if the correct part is ranked among the top five. We call this "Rank 5 Accuracy." For each experiment, we report on the accuracy results for both of these measures for the five SecondString algorithms we looked at.

13.4.4 Results

In Experiment 1, two of the algorithms, Jaccard and Jensen-Shannon, both token-based algorithms, scored the best, with better than 85% Rank 1 Accuracy and better than 98% Rank 5 Accuracy (see Figure 13.1). The two pure string-matching algorithms, Jaro and Jaro-Jones, performed about the same as each other and in between the token-based and hybrid algorithms, for both Rank 1 and Rank 5 Accuracy. The fact that those two

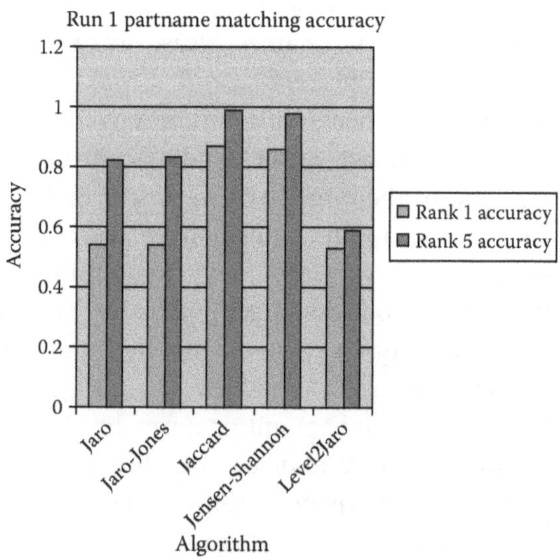

FIGURE 13.1 Results for Experiment 1.

performed the same suggests that number-letter confusions seemed to be virtually nonexistent in our test data. In fact, we expect them to be more important in comparing equipment numbers, which, when available, will make it unnecessary to match part name strings at all. The fact that they perform worse than the token-based algorithms is not too surprising: they are relatively inflexible to any changes in the order of the tokens, treating the whole part name as a single string. Token-based algorithms, on the other hand, are completely insensitive to the order of the tokens. The only hybrid algorithm we tested, Level2Jaro, did about the same as the pure string-matching algorithm in Rank 1 Accuracy and much worse than all the others in Rank 5 Accuracy. It is surprising that the pure token-based algorithms perform so much better than the hybrid algorithm. Purely token-based algorithms score nothing for misspellings or abbreviations of words; only exact matches of tokens are counted. Hybrid algorithms, on the other hand, should be able to give partial scores for misspellings and ad hoc abbreviations. While we are still investigating all the reasons for this, one possibility is that the test data we selected was not sufficiently representative and contained enough correctly spelled words to allow a token-based algorithm to get the correct match. However, this does not seem to be the case and Level2Jaro did do better than the others on a few part names where there were a significant number of abbreviations and spelling differences, e.g., “RECIRC ACN LWR FAN” for “LOWER RECIRCULATION FAN (RIGHT), (P210).” There is evidence that the hybrid algorithm is very sensitive to the number of approximate matching tokens and the degree of mismatch they show. Some examples suggest that the hybrid algorithm is producing spurious matches resulting in incorrect part names getting a higher score than the correct part name, such as “TRIM AIR RIGHT SOV” giving a higher score to “ECSM RIGHT CARD” than to “TRIM AIR PRESSURE REGULATING AND RIGHT SOV.”

In Experiment 2, with “left” and “right” normalized and moved to the front of both the candidate and target part name strings, all the algorithms showed improvements on both measures of accuracy, but to radically different degrees (see Figures 13.2 and 13.3). The two token-based algorithms, Jaccard and Jensen-Shannon, already with high scores, showed minimal improvements, while the others showed significant improvement on both measures of accuracy, ranging from 9% for the pure string matching algorithms on Rank 5 Accuracy to 51% and 54% for the hybrid algorithm on Rank 1 Accuracy and Rank 5 Accuracy, respectively. The improvement of the string matching algorithms is likely due to the movement of “right”

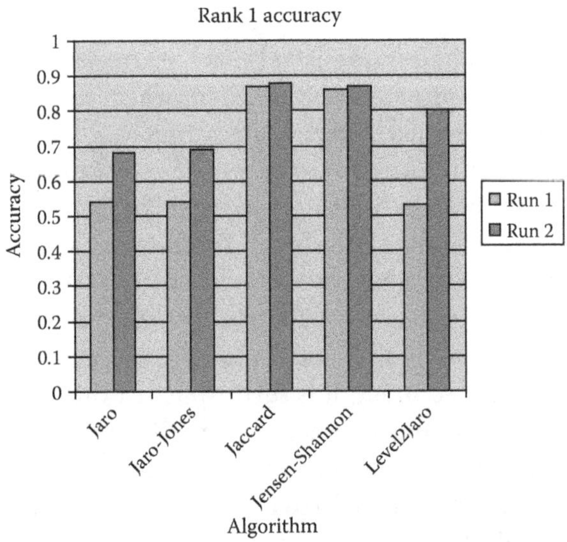

FIGURE 13.2 Comparison of Rank 1 Accuracy results for Experiments 1 and 2.

and "left" to the same position in both strings as well as the increased score from spelling out very short abbreviations like "r" and "r/h." The improvement of the hybrid algorithm is apparently all due to the spelling out of the short abbreviations, since, like the pure token-based algorithms, it is insensitive to token order.

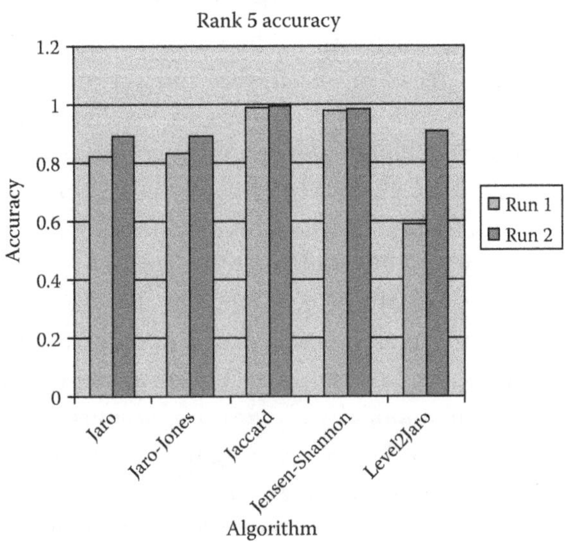

FIGURE 13.3 Comparison of Rank 5 Accuracy results for Experiments 1 and 2.

13.5 QUBIT

In this section, we will illustrate how to support user's complex search by offering a combination of text mining generated suggestions and knowledge compiled suggestions, together with information management techniques to further assist users to research high-impact problems involving aviation systems, structures, and operations.

13.5.1 The Ad Hoc Query Problem

Analysts of health management systems do not always have the luxury of a predefined list of items on which they wish to perform standing queries, such as the part-name problem described above. For example, new investigations of aviation safety-related issues usually demand users to come up with complex queries that have not been previously performed in short demand. The search may involve parts and systems as well as operational issues and environmental issues (such as volcano eruption). While users* are subject matter experts who possess a high degree of engineering knowledge, it is very difficult for them to figure out all of the spelling variations and ad hoc ways of expressing the same concept or the same part or system. The problem goes further. It is not atypical for a complex query to include more than 100 conjuncts and disjuncts in an SQL statement. Trying to rapidly refine queries of this level of complexity to get the accuracy required for the task is also a very daunting task. QUBIT (QUery BuIlder Tool) is designed to solve this problem using two integrated features, which we describe below.

13.5.2 Suggester

We utilize a number of methods to help the user come up with variations in search terms so that the users can focus on the search concept based on their engineering knowledge. It should be noted that this is not the same as providing a synonym list, which many text databases support. There are several major differences. First, a synonym list is typically obtained by having users enter the knowledge manually or by loading it from an existing knowledge base. While a knowledge base can be a source of input for a QUBIT Suggester, the goal is to use text mining techniques wherever it

* Some communities decide to have IT personnel formulate the actual queries. These IT users will communicate with subject matter experts to gain engineering knowledge. The pros and cons of this approach versus having subject matter experts perform search directly is out of the scope of this chapter. However, this alternative approach can be viewed as a direct evidence of our stated needs and challenges.

makes sense and minimize the manual work. Second, while it is possible for a user to edit the synonym list, a text database is not designed to allow the user to select or reject entries based on the goal of the search or the nature of the data being analyzed. In contrast, by design our Suggester provides multiple suggestions for the same term. The user has the option to choose what makes sense in the current search context.

Some of the major methods a Suggester may employ are discussed below. In all cases except the knowledge base, the text to be searched will be used as sample data. Depending on the size of the complete set of data and the processing speed required to make the suggesting process interactive, the whole data set or a subset of it will be used as the sample data. Typically, more than one term is used in a search. The Suggester will make suggestions on each search term. It is also possible to have a multiple word search term (e.g., "main landing gear") and the Suggester will treat this as a single term. Figures 13.4 and 13.5 show two possible incarnations of the Suggester illustrating some of the functions a Suggester might perform.

13.5.2.1 *Finding Terms with a High Co-occurrence with the Search Term*

The Suggester processes the sample data and displays a list of terms that co-occur with the search term, along with frequency counts (see the "Associated Words" column in Figure 13.4). Given a list of particular words (e.g., verbs), instances of those associated with the target term can also be shown (see the "Associated Verbs" column in Figure 13.4). Depending on the data and the application, three more parameters can be set to control this. (a) A proximity window can be set so that, for example, only words

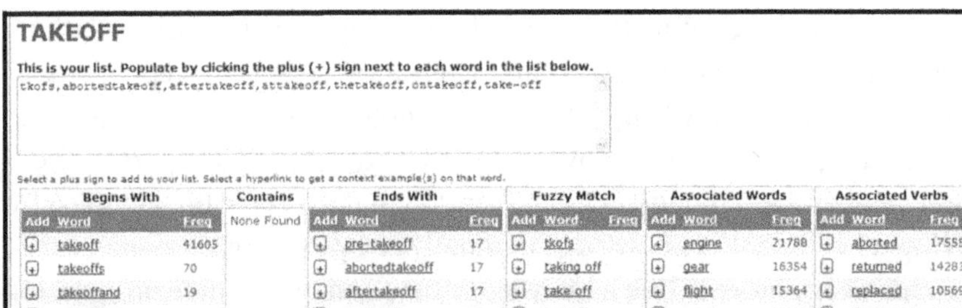

FIGURE 13.4 QUBIT Suggester, sample version 1.

that are no more than five words away from the search term are returned. This will allow the Suggester to only return co-occurring terms near the search term, and therefore more likely to be related to the search term. By setting the window to be one, only terms immediately adjacent to the search term will be returned. (b) The directionality of the window can be controlled, so that only co-occurring terms to the left or to the right of the search term are returned. Of course, this can be combined with a proximity window. Based on the syntax of English, the terms immediate to the left of a search term that is a noun will tend to be modifiers of the search term. (c) Finally, a frequency threshold can be set to limit the co-occurring terms to be displayed to the user to those with at least a certain frequency. This can be used to prevent the Suggester from returning random examples. The Suggester ranks the results by frequency and displays them to the user with a count next to each co-occurring term.

13.5.2.2 Using Fuzzy String Matching Algorithms to Generate a Potential Match

The Suggester can use string matching algorithms, such as those described in the P-MATCH section, to generate terms similar to the search term (see the "Fuzzy Match" column in Figure 13.4). Parameters such as those discussed under P-MATCH can be employed. These include the type of string matching algorithm to use, a similarity threshold, or the maximum number of results to be returned by the string matching algorithms. The Suggester can have an option to display the similarity measure or the rank of the results returned.

13.5.2.3 Using Regular Expressions to Generate a Potential Match

Some analysis tools allow users to use regular expressions to identify additional search terms. However, as described above, this is a very daunting task that is hard for subject matter experts to take on. The Suggester takes a different approach. It automatically generates certain regular expressions for each search term and finds matches in the sample data to display to the user. This is the most powerful way to deal with words fused together by error, which is very typical in noisy text data. The Suggester can identify all "words" in the sample data beginning or ending with the search term, or containing it in the middle, without any space in between (see the columns "Begins With," "Ends With," and "Contains" in Figure 13.4). Alternatively, the Suggester might just match all words in any or all of these categories, as in the "Wild Cards" column in Figure 13.5.

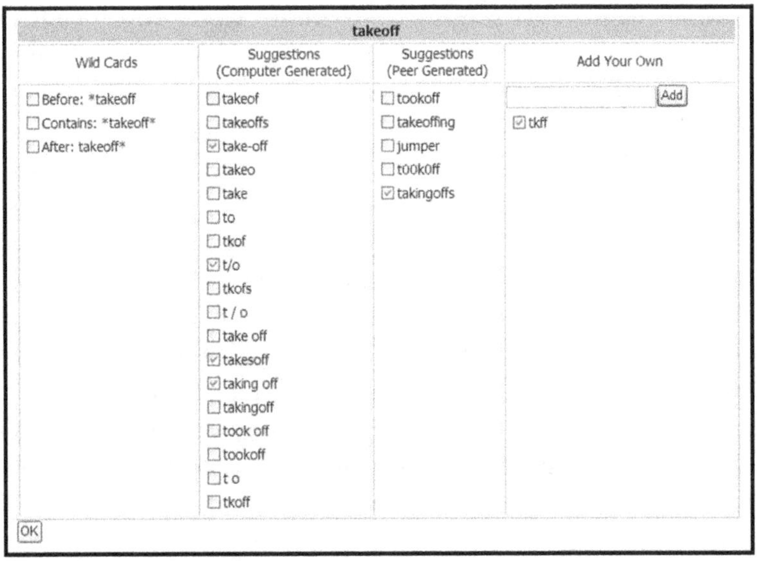

FIGURE 13.5 QUBIT Suggester, sample version 2.

Similarly, the Suggester can return examples of words with a space arbitrarily inserted into them, which frequently occurs in the data (possibly as a result of the data being moved between systems). The Suggester can also make a regular expression substituting the letter "l" with the numeral "1," if the letter "l" occurs in the search term, and similarly the letter "o" with the numeral "0." There are many other possibilities here. For example, the Suggester can replace the ending of the word with a wild card, because words are frequently abbreviated in ad hoc ways. Alternatively, regular expressions can be used to search for words that are missing one or more vowels, since that is another common means of abbreviating.

The power of the Suggester here is to allow users to benefit from the use of regular expressions, without expecting them to construct regular expressions themselves.

13.5.2.4 Using Latent Semantic Analysis or Other Methods that Generate Topic Relateness

The Suggester can use latent semantic analysis, such as TRUST [17,18] to generate terms highly correlated to the search term using the sample data. Semantically highly related terms can be identified using this method or similar text mining methods. For example, it might find words that can occasionally be used in place of each other (i.e., synonyms or near synonyms or words that are otherwise closely related semantically) such as "blower" and "fan," "circuit board" and "motherboard," or "wiring" and "harness."

13.5.2.5 Using a Knowledge Base

The Suggester can use any knowledge base that is available. This includes what a text database manager may return as a suggestion for a search term. Typically, this is a good method to handle acronyms as well as true synonyms that do not look anything like the original term, both of which are not easy to identify with the above methods. We are using synonym and acronym lists such as those produced by FAA-ASIAS as well as a Boeing internal thesaurus and acronym list.

13.5.2.6 Using Suggestions Entered by Peers

The Suggester can also allow users to use terms that have been entered by their peers using the system. There is an option for a user to enter a variant of a term that they know of or have discovered in their search of the data but that has not been suggested by the system. The system will then provide this as a peer generated suggestion to other users (see "Suggestions (Peer Generated)" in Figure 13.5). This provides a way for the system to let users leverage the expertise and experience of their peers.

Our experience shows that deciding what methods to use in the Suggester and what parameters to set is not an exact science. It depends on the data, the application, and the user's orientation. Some users requested that the Suggester return more terms that the user is likely to miss, with the understanding that a lot of the suggested terms may be incorrect, whereas other users are more bothered by a large number of incorrect suggestions and would prefer more tightly controlled suggestions. The suggestions derived from different methods can be kept and displayed separately, so that the user can make different decisions based on the method used. It is also worth noting that the Suggester provides a powerful insight into the data through analysis of the sample data, without the user doing the actual search. For example, the user does an initial search on an indicator and finds it associated with an unusual condition. Using the Suggester with the condition as a search term, the user can see what other indicators may have similar conditions without actually executing a search in the text data collection.

The user has the option to select some or all of the Suggester returned suggestions as additional search terms. The user can also save the selection as a search concept for later use. Each application can decide whether to let users share these concepts with other users or not.

13.5.3 Query Builder

Users in this domain rarely search with just one term. After using the Suggester to determine which terms best express the concepts they are interested in, the user can now concentrate on constructing the actual query based on these concepts. At this point, each concept will consist of a disjunction of terms. Depending on the application domain, there may be an advantage to further categorizing the search concepts. For example, in an application for fix effectiveness, the user often wants to organize the concepts into the following categories: part (or system), the condition of the part, the action taken in fixing the problem, and the result of the fix. In other applications, the search may be less structured and the system would offer a more flexible way to organize the concepts. Either way, the users can now put these concepts into boxes representing different conjuncts of their query and possibly one or more negative concepts, which, if found, would exclude a record from being returned. Note that if two concepts are placed in the same conjunct box (a green box in Figure 13.6), the system

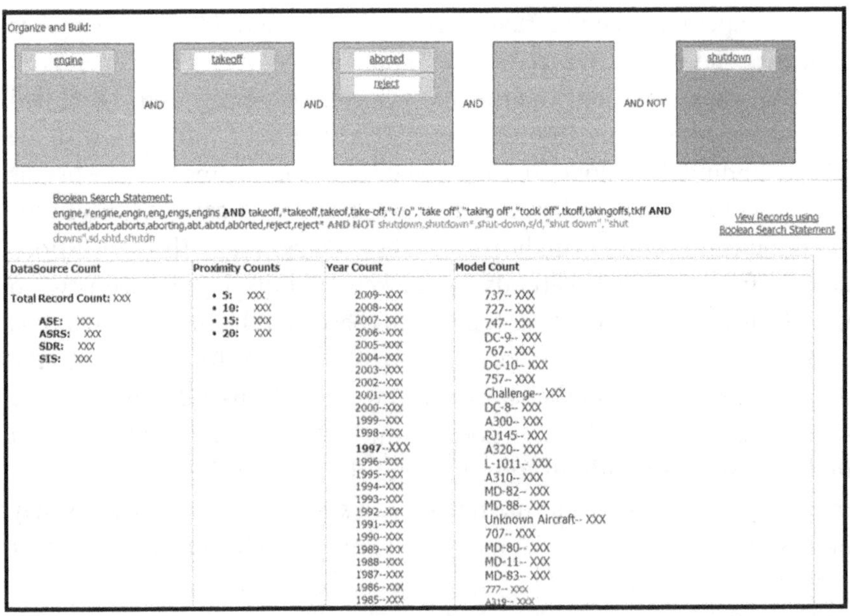

FIGURE 13.6 QUBIT Query Builder.

will treat this as a disjunct of those two concepts. For example, the query expressed in Figure 13.6 is

"engine AND takeoff AND (aborted OR rejected) AND NOT shutdown"

where each of "engine," "takeoff," etc. are concepts consisting of a disjunction of terms, typically synonyms, abbreviations, acronyms, misspellings, etc. The system will then generate a complex machine executable query based on the user's selections. The user has the option of editing the generated complex search, although this is not typically done.

The Query Builder provides a GUI to help the user formulate her query in a form that is inspired by conjunctive normal form (although not exactly the same). While conjunctive normal form is a classic concept in computer science, to the best of our knowledge, it has only been used in very few cases to support search [19] and no one has employed it in an interactive system to support complex search requiring in-depth domain knowledge. The combination of Suggester and Query Builder allows the user to focus on the concept instead of getting lost in the details of the query syntax.

13.6 FUTURE RESEARCH DIRECTIONS

There are directions we currently envision exploring for improvements in P-MATCH and other work we are doing to enhance analysis of safety or operational data.

First, we would like to look at a larger set of part names to see if the same pattern of behavior holds for them as well.

Second, we wish to continue our analysis of why certain algorithms fail on certain maintenance records or certain part names. In particular, we would like to study why the particular hybrid algorithm we looked at did so poorly. Perhaps, it was the choice of the base string-matching algorithm (Jaro) and using another (e.g., Jaro-Winkler) might have improved the performance. Or perhaps it was the way the scores are summed up, giving undue influence to spurious matches. In general, we would like to test the other string matching algorithms in SecondString.

Third, we have only looked at using a single string matching algorithm at a time for all the part names. There are several alternative approaches we would like to try. One is to see if certain algorithms work better with certain types of part names (e.g., depending on the head, length, presence of acronyms or alphanumeric patterns, etc.). A possible way of automating this selection is to use active learning, as discussed in [20]. Another approach is to use voting to combine the results of different algorithms. With voting, all algorithms could have equal influence on the results or they could be weighted differently depending on their performance on the training sample.

We could also try to take greater advantage of the parse we do of the target part names. Currently, we are just using the head and the modifiers, but we are also distinguishing peripheral modifers beyond just "right" and "left." We could use this knowledge to restrict our search for modifiers to the right of the head to just peripheral modifiers, rather than all modifiers. We could also use this knowledge to downweight the peripheral modifers so that whether something is "left," "right," "forward," or "aft" has less influence on the match than whether it is an "equipment cooling" fan or a "turbine bypass" valve.

So far, we have done little to take advantage of existing knowledge, for example, acronyms. In addition to acronyms that can be extracted from part names themselves (e.g., "Fuel Metering Unit (FMU), M76001"), we have a list of the most commonly used acronyms in the airplane maintenance domain, including some associated with part names, that have been collected from various organizations in the company. Although string matching algorithms can deal with misspellings and even ad hoc abbreviations to a certain degree, they have difficulty with acronyms. Using a precompiled list to expand acronyms in both the maintenance text and the standard part name list could facilitate the string matching algorithm. We are currently doing this only with various abbreviations of "right" and "left."

Finally, the various string algorithms we are using were developed primarily for de-duplication of public health and census data [21] and are possibly more suited for people and organization names, addresses, and phone numbers, but not for part names. We are in the process of developing of a new string matching algorithm that keys in on the structure of part names as well as the relative uniqueness of different heads and modifiers.

Going forward, there are a number of possibilities to utilize a combination of P-MATCH and QUBIT in a more tightly connected manner that is still to be explored. For example, in certain applications, QUBIT could be integrated with P-MATCH to provide a means of enlarging the standing query (e.g., find all records referring to a certain part and one or more concepts) or of adding a new part name on the fly (by finding variants of the head and modifiers, and even finding potential modifiers in the data as associated terms, i.e., treating it as a type of the more structured QUBIT query mentioned above).

13.7 CONCLUSION

In this chapter, we have discussed the need to exploit text data in addition to sensor and other numerical and categorical data to improve vehicle systems health management. We have discussed some of the problems with the data, mainly the fact that it is extremely noisy, with frequent ad hoc abbreviations and spellings. We have discussed in detail P-MATCH, an approach we have developed to find mentions of part names in this type of data. P-MATCH to illustrate how a combination of knowledge-based natural language processing techniques in conjunction with various string matching algorithms, mostly numerical, can leverage the strength of both. The following characteristics are inspired by linguistics-based NLP:

- Target part names are parsed into head nouns, core modifiers, and peripheral modifiers.
- Based on the fact that in English noun phrases nonphrasal modifiers precede the head noun, our algorithm searches to the left of the head for modifiers to include in the candidate part name, with the data-motivated exception that peripheral modifiers, primarily indicating location, can occur more "freely."
- We use the fact that in English process descriptions, actions are typically described in the same temporal order as they occur in the real world (authors of maintenance reports never write "before we replace X, we reseated X") to determine the final action taken.

P-MATCH also illustrates how to solve part name extraction, an important type of entity extraction problem with some different characteristics from typical person and place name extractions. Today's entity extraction tools with person names and place names as their primary focus cannot adequately deal with the special characteristics of specific part name matching. While P-MATCH is applied to the airline industry data today, it can benefit any industry with a large number of parts and a large amount of log data.

In addition, we have also discussed QUBIT, a method of supporting interactive ad hoc queries that exploit and rely on the user's domain knowledge. We have also briefly discussed, in the future, how the two methods may be used in conjunction with each other in a more tightly connected way.

REFERENCES

[1] A. Wade, R. Nicholson. Improving airplane safety: Tableau and bird strikes, 2010.

[2] X. Wang, H. Fang, C. Zhai. Improve retrieval accuracy for difficult queries using Negative Feedback. In *Proceedings of the 16th ACM International Conference on Information and Knowledge Management* (CIKM'07), pp. 991–994, 2007.

[3] X. Wang, C. Zhai. Mining term association patterns from search logs for effective query reformulation. In *Proceedings of the 17th ACM International Conference on Information and Knowledge Management* (CIKM'08), pp. 479–488, 2008.

[4] D. Zhang, C. Zhai, J. Han. Topic Cube: Topic modeling for OLAP on multidimensional text databases, In *Proceedings of 2009 SIAM International Conference on Data Mining* (SDM'09), pp. 1123–1134, 2009.

[5] Wikipedia Regular Expression Entry. http://en.wikipedia.org/wiki/Regular_expression.

[6] N. Chomsky. Three models for the description of language. *IRE Transactions on Information Theory* (2), pp. 113–124. 1956.

[7] J.F. Allen. *Natural Language Understanding*, 2nd ed. Benjamin Cummings, Menlo Park, CA, 1994.

[8] D. Jurafsky, J.H. Martin. *Speech and language processing: An introduction to natural language processing, computational linguistics, and speech recognition.* Pearson Education, Inc., Upper Saddle River, NJ, 2008.

[9] Secondstring Project Page. http://secondstring.sourceforge.net/. Retrieved February 20, 2010.

[10] W.W. Cohen, P. Ravikumar, S.E. Fienberg. A comparison of string distance metrics for name-matching tasks. In *Proceedings of the ACM Workshop on Data Cleaning, Record Linkage and Object Identification*, Washington, DC, 2003.

[11] Wikipedia Jensen-Shannon Entry. http://en.wikipedia.org/wiki/Jensen%E2%80%93Shannon_divergence.

[12] Wikipedia Kullback-Leibler Entry. http://en.wikipedia.org/wiki/Kullback%E2%80%93Leibler_divergence.

[13] W.W. Cohen. Record Linkage Tutorial: Distance Metrics for Text. Retrieved from http://www.cs.cmu.edu/~wcohen/Matching-2.ppt. 2006.

[14] W.E. Winkler. Overview of Record Linkage and Current Research Directions. Research Report Series, Statistics #2006-2, U.S. Census Bureau. http://www.census.gov/srd/papers/pdf/rrs2006-02.pdf, 2006.

[15] SimMetrics Project Page. http://www.dcs.shef.ac.uk/~sam/simmetrics.html. Retrieved February 20, 2010.

[16] M. Ankerst, D.H. Jones. System and method for string distance measurement for alphanumeric indicia. U.S. Patent 7,540,430, filed Sept 27, 2005, and issued June 2, 2009.

[17] S. Deerwester, S. Dumais, G. Furnas, T. Landauer, R. Harshman. Indexing by Latent Semantic Analysis. *Journal of the American Society for Information Science,* 41(6):391–440, 1990.
[18] A. Booker, M. Condliff, M. Greaves, F.B. Holt, A. Kao, D.J. Pierce, S. Poteet, Y.-J.J. Wu. Visualizing Text Data Sets. *IEEE Computing in Science & Engineering,* pp. 26–35, 1999.
[19] M. Hearst. Improving full-text precision on short queries using simple constraints. *Proceedings of the Fifth Annual Symposium on Document Analysis and Information Retrieval* (SDAIR), Las Vegas, NV, April 1996.
[20] S. Sarawagi, A. Bhamidipaty. Interactive deduplication using active learning. In *Proceedings of the eighth ACM SIGKDD international conference on knowledge discovery and data mining,* pp. 269–278. ACM, New York, 2002.
[21] W.E. Winkler. Record Linkage Software and Methods for Merging Administrative Lists. Statistical Research Report Series No. RR2001/03, U.S. Bureau of the Census, 2001.

Index

B

E

F

N

O

T

U

V

W

X